OXFORD MATHEMATICAL MONOGRAPHS

OXFORD MATHEMATICAL MONOGRAPHS

A. Belleni-Morante: *Applied semigroups and evolution equations*

I. G. Macdonald: *Symmetric functions and Hall polynomials*

J. W. P. Hirschfeld: *Projective geometries over finite fields*

N. Woodhouse: *Geometric quantization*

A. M. Arthurs: *Complementary variational principles* Second edition

P. L. Bhatnagar: *Nonlinear waves in one-dimensional dispersive systems*

N. Aronszajn, T. M. Creese, and L. J. Lipkin: *Polyharmonic functions*

J. A. Goldstein: *Semigroups of linear operators*

M. Rosenblum and J. Rovnyak: *Hardy classes and operator theory*

J. W. P. Hirschfeld: *Finite projective spaces of three dimensions*

K. Iwasawa: *Local class field theory*

A. Pressley and G. Segal: *Loop groups*

J. C. Lennox and S. E. Stonehewer: *Subnormal subgroups of groups*

D. E. Edmunds and W. D. Evans: *Spectral theory and differential operators*

Wang Jianhua: *The theory of games*

S. Omatu and J. H. Seinfeld: *Distributed parameter systems: theory and applications*

D. Holt and W. Plesken: *Perfect groups*

J. Hilgert, K. H. Hofmann, and J. D. Lawson: *Lie groups, convex cones, and semigroups*

S. Dineen: *The Schwarz lemma*

B. Dwork: *Generalized hypergeometric functions*

R. J. Baston and M. G. Eastwood: *The Penrose transform: its interaction with representation theory*

S. K. Donaldson and P. B. Kronheimer: *The geometry of four-manifolds*

T. Petrie and J. Randall: *Connections, definite forms, and four-manifolds*

R. Henstock: *The general theory of integration*

D. W. Robinson: *Elliptic operators and Lie groups*

A. G. Werschulz: *The computational complexity of differential and integral equations*

J. B. Griffiths: *Colliding plane waves in general relativity*

P. N. Hoffman and J. F. Humphreys: *Projective representations of the symmetric groups*

I. Györi and G. Ladas: *The oscillation theory of delay differential equations*

B. Amberg, S. Franciosi, and F. de Giovanni: *Products of groups*

Projective Representations of the Symmetric Groups

Q-Functions and Shifted Tableaux

P. N. HOFFMAN
University of Waterloo

and

J. F. HUMPHREYS
University of Liverpool

CLARENDON PRESS · OXFORD
1992

Oxford University Press, Walton Street, Oxford OX2 6DP
Oxford New York Toronto
Delhi Bombay Calcutta Madras Karachi
Petaling Jaya Singapore Hong Kong Tokyo
Nairobi Dar es Salaam Cape Town
Melbourne Auckland
and associated companies in
Berlin Ibadan

Oxford is a trade mark of Oxford University Press

Published in the United States
by Oxford University Press, New York

A catalogue record for this book is available from the British Library

Library of Congress Cataloging in Publication Data
(Data available on request)

ISBN 0–19–853556–2

Typeset by DCS, University of Waterloo
Printed and bound in Great Britain by
Bookcraft Ltd., Midsomer Norton, Avon

To Frances and Patricia

PREFACE

In the first decade of this century, Issai Schur initiated the study of projective representations, and developed the general theory for finite groups. In a beautiful and formidable paper, Schur (1911), he then worked out all the basic details for the symmetric and alternating groups. When our labour on this monograph began, no other complete treatment of Schur's results existed. Our main approach to the subject is quite similar to his.

The first chapter is an account of the general theory of projective representations sufficient for our purposes. The next four chapters give some results concerning the symmetric group, double covers, and representations which are crucial to the rest of the book. Chapter 6 describes, and computes the characters of, certain basic representations. This computation, due to Schur, includes one case which is equivalent to the construction of the irreducible Clifford modules. In Chapter 7 there is an account of certain symmetric functions, the Q-functions, which is just sufficient for the central results. Then, in Chapter 8, the basic representations and Q-functions are combined to prove Schur's central results, Theorems 8.6 and 8.7. It is interesting that when analysing S_n, in contrast to the case of linear representations, it is essential to develop the projective theory for A_n, the alternating group, simultaneously. The notes at the ends of these chapters clarify the extent to which we have engaged in a 'modernization' of Schur's treatment.

Chapters 7, 9 and 12, and the first part of Chapter 14, give a self-contained account of the algebraic theory of Q-functions. In Chapter 9, the classical (and one new) formulae for Q_λ are deduced from our more indirect treatment in Chapter 7. This includes a proof that Schur's Pfaffian definition agrees with the Hall–Littlewood definition. No detailed proof of this fact, essentially due to Schur, seems to exist elsewhere in the published literature. Chapter 12 includes the analogue of the Pieri formula, as well as the combinatorial formula for the coefficients of Q_λ, the skew Q-functions, and a new proof of a recent Pfaffian formula for them due to Tadeusz Jozefiak and Piotr Pragacz. The final chapter, the fourteenth, is an account of work by Bruce Sagan, John Stembridge and Dale Worley, which results in a description of how to multiply Q-functions, in the style of the Littlewood–Richardson rule. For reasons of length, certain combinatorial details in this chapter are referred to the literature. A complete account of the combinatorics associated with Q-functions would occupy an entire volume; such a volume written by an expert would be most welcome. We have given, in Chapter 13, a leisurely treatment of one basic idea, that of shifted insertion, due independently to Sagan and Worley.

For explicit computation of characters, there is a reduction rule for passing from larger to smaller n. This was discovered by Alun Morris. It is presented in Chapter 10, along with branching rules and the degree formula.

As well as the product theorem for Q-functions of Chapter 14, another significant development since we began this monograph was the discovery by Maxim Nazarov of a formula for the matrices, corresponding to group generators, for each irreducible projective representation of S_n. We give the proof of irreducibility and distinctness for these in Chapter 11, referring to the literature for the verification that they are representations.

A sizable portion of the monograph is taken up by appendices. The main chapters themselves do not depend logically on any of the appendices. All should be accessible to those with a moderate background in modern algebra. More details are left to the reader in the appendices than in the main chapters, and the appendices are less uniform in style.

Appendices 6, 8, 12, and 14 form a sequence which amounts to an independent redevelopment of Schur's theory. They depend on parts of the first four chapters, and contain alternative formulations and proofs of the main results in Chapters 4, 5, 6, 8, and 10. Appendix 6 includes a description of Clifford modules. The lengthy Appendix 8 treats an algebra which is the appropriate analogue of the well known induction algebra of (virtual) linear representations of S_n. The notes at the end of Appendix 8 correlate this to the algebra generated by the Q-functions and to products defined by other authors, in particular, Stembridge (1989) and Jozefiak (1989), two recent alternative accounts of Schur's work. It should perhaps be noted that neither of our treatments of Schur's main theorems depends on the combinatorics of shifted tableaux, on Pfaffian identities from classical algebra, or on the structure theory of semisimple superalgebras. These do occur with more or less prominence elsewhere in the book. Appendix 12 shows how to recalculate all the characters using the character-independent results of Appendix 8 and three simple Clifford module trace calculations in Appendix 6. In Appendix 14, the appropriate 'Littlewood–Richardson rule' for the algebra in Appendix 8 is stated.

In Appendix 9, we describe briefly the relationship of Q-functions to Schur and Hall–Littlewood functions. Ian Macdonald's book, Macdonald (1979) (in this series of monographs), contains a large reservoir of information on these topics. In Appendix 10, there is an outline of the combinatorial description of the blocks for the modular projective representations of S_n (an analogue of the so-called Nakayama conjecture). Some other recent results in the modular theory, due to David Benson, Alun Morris, Gerhard Michler, and Jorn Olsson are stated.

We would like to thank our colleagues mentioned by first name in this introduction for useful conversations and correspondence over the past number of years (excepting Schur, of course, who nevertheless also deserves a vote of thanks). Ian Macdonald should particularly be singled out for his encouragement and for a couple of key observations which led to some simplifications. Francis Rayner deserves our thanks for very helpful suggestions related to the various versions of Chapter 13. We must of course add the usual remark that we are claiming priority for any errors which might appear on the following pages. We thank Brenda Law and Bruce Uttley for their excellent work in converting the manuscript into a book. Research grants from NSERC (Canada) and NATO were very helpful during the course of the work. We are grateful as well to our spouses, to whom the book is dedicated, for their continuing patience and encouragement.

Waterloo, Canada P.N.H.
Liverpool, England J.F.H.
January 1992

POSTSCRIPT ON PREREQUISITES

We have stated that a moderate background in modern algebra is adequate for reading this monograph. Chapter 1 starts right in on projective representations. Normally a student would know some of the linear representation theory of finite groups before embarking on this. However, we have stated all the needed definitions and basic results on linear representations, as the need arises, with references to texts on the subject. We use only the most basic parts; in particular the theory of group algebras (twisted or straight) is not needed. Later in the book, the combinatorial material is self-contained except as noted above in reference to the final chapter, number 14. With respect to classical algebra, the book is also basically self-contained: symmetric functions are developed from scratch, and there is no dependence on the theory of Pfaffians except for one alternative proof.

CONTENTS

1

PROJECTIVE REPRESENTATIONS AND REPRESENTATION GROUPS

In this chapter, we discuss the elementary theory of projective representations of finite groups. The foundations of the theory were presented by Schur in two papers in 1904 and 1907. A subsequent paper, Schur (1911), applies this theory to the symmetric and alternating groups, and is the basis for much of the material in this book. We begin with some general definitions.

Definition. Let V be a finite-dimensional complex vector space. A (complex) projective representation of a group G on V is a function P from G into $GL(V)$, the group of automorphisms of V, such that

(i) $P(1_G)$ is the identity linear transformation of V; and

(ii) given elements x and y in G, there is a non-zero complex number $\alpha(x, y)$ such that

$$P(x)P(y) = \alpha(x, y)P(xy).$$

Taking a basis for V, we obtain a projective matrix representation $P : G \to GL(d, \mathbb{C})$, where d is the dimension of V.

Since the linear transformations $P(g)$ are invertible, it follows from (i) that, for all g in G,

$$\alpha(g, 1) = 1 = \alpha(1, g). \tag{C1}$$

Using associativity of composition and of group multiplication to evaluate $P(x)P(y)P(z)$ gives that

$$\alpha(x, yz)\alpha(y, z) = \alpha(x, y)\alpha(xy, z) \tag{C2}$$

for all x, y and z in G. Any map $\alpha : G \times G \to \mathbb{C}^{\times} := \mathbb{C} - \{0\}$ satisfying conditions (C1) and (C2) is said to be a *2-cocycle*.

Example. Let

$$G := < a, b\colon a^2 = b^2 = 1, ab = ba > ,$$

so that G is the Klein four group. Define a map $P : G \to GL(2, \mathbb{C})$ by

$$P(1) = \begin{pmatrix} 1 & 0 \\ 0 & 1 \end{pmatrix}; P(a) = \begin{pmatrix} 0 & 1 \\ -1 & 0 \end{pmatrix}; P(b) = \begin{pmatrix} i & 0 \\ 0 & -i \end{pmatrix}; P(ab) = \begin{pmatrix} 0 & -i \\ -i & 0 \end{pmatrix}.$$

It can be checked that P is a projective representation of G, and that the associated 2-cocycle α takes the values shown in the following table.

α	1	a	b	ab
1	1	1	1	1
a	1	-1	1	-1
b	1	-1	-1	1
ab	1	1	-1	-1

Definition. A linear representation of a group G on a finite dimensional vector space V is a homomorphism $R : G \to GL(V)$. Thus a linear representation is a projective representation with trivial 2-cocycle.

Before developing the theory, it may be appropriate to comment on the use of the word "projective" in this context. Given a complex vector space V, the projective space $P(V)$ associated with V is the set of equivalence classes $[v]$ of non-zero elements of V under the relation $u \sim v$ if and only if $u = \lambda v$ for some $\lambda \in \mathbb{C}^{\times}$. An element f in $GL(V)$ induces an action f^* on $P(V)$ by

$$f^*[v] = [f(v)].$$

Thus f^* may be regarded as an element of $PGL(V) = GL(V)/\mathbb{C}^{\times}I$ where I is the identity matrix. In matrix terms, suppose V has dimension d, let $Z(d)$ be the set of non-zero multiples of the identity matrix, and define $PGL(d, \mathbb{C})$ to be the quotient group $GL(d, \mathbb{C})/Z(d)$. The isomorphic groups $PGL(V)$ and $PGL(d, \mathbb{C})$ are known as projective linear groups. Given a projective representation P of degree d for the group G, the map $P' : G \to PGL(d, \mathbb{C})$ which takes an element g of G to the coset $P(g)Z(d)$ is a homomorphism. Conversely, any such homomorphism P' gives a projective representation in the original sense, by choosing a representative $P(g)$ for the coset $P'(g)$ (with the convention that $P(1) = I$). The homomorphism P' determines a 2-cocycle α by this process, but the association to P' of P and α is not unique. If we make a different choice $Q(g)$ for $P(g)$, then

$$Q(g) = \delta(g)P(g)$$

for all g in G, where $\delta(1) = 1$ and $\delta(g)$ is in $\mathbb{C}^{\times}$. If β is the 2-cocycle associated with Q, it is related to α by the rule:

$$\beta(x, y) = \delta(x)\delta(y)(\delta(xy))^{-1}\alpha(x, y)$$

for all x and y in G. Two 2-cocycles related in this way are said to be *cohomologous*. Denote the cohomology class of a 2-cocycle α by $[\alpha]$. The set of such classes forms an abelian group, the Schur multiplier, denoted $H^2(G, \mathbb{C}^\times)$ or $M(G)$, under the operation

$$[\alpha][\beta] = [\alpha\beta],$$

where

$$(\alpha\beta)(x, y) = \alpha(x, y)\beta(x, y)$$

for all x and y in G.

Example. In the previous example, let β be the 2-cocycle cohomologous to α, using the δ for which

$$\delta(1) = 1\ ,\ \delta(a) = 1\ ,\ \delta(b) = -i \text{ and } \delta(ab) = i.$$

Thus β takes the values

β	1	a	b	ab
1	1	1	1	1
a	1	-1	-1	1
b	1	1	1	1
ab	1	-1	-1	1

However, α is not cohomologous to the trivial 2-cocycle, since if it were (using some δ), and if x is in $G-\{1\}$, then $\delta(x)^2 = -1$, and so $\delta(x)$ is i or $-i$. Thus, for x and y unequal, non-identity elements of G,

$$\delta(x)\delta(y)(\delta(xy))^{-1}\alpha(x, y) \neq 1$$

since it is not real. Hence $M(G)$ contains at least two elements. In fact, $M(G)$ has order 2, but the proof of this, as well as a thorough treatment of the general theory, and most other computations of Schur multipliers, are outside the scope of this book. The interested reader should consult Huppert (1967) or Karpilovsky (1985).

The first result gives some information on the Schur multiplier.

Theorem 1.1. *For any finite group G, the Schur multiplier $M(G)$ has finite exponent dividing the order of G. Furthermore, if a cohomology class has order e, then there is a representative of that class which takes only e-th roots of unity as its values. Thus $M(G)$ is a finite group.*

Proof. For any 2-cocycle α and any elements x, y and z in G,

$$\alpha(x, y) = \alpha(x, yz)\alpha(y, z)/\alpha(xy, z) .$$

Fixing x and y and multiplying together these equations as z varies over G gives

$$\alpha(x, y)^{|G|} = \prod_{z\in G} \alpha(x, yz) \prod_{z\in G} \alpha(y, z)/ \prod_{z\in G} \alpha(xy, z).$$

Defining

$$\delta(g) = \prod_{z\in G} \alpha(g, z)$$

for any g in G, the above equation becomes

$$\alpha(x, y)^{|G|} = \delta(x)\delta(y)/\delta(xy).$$

Thus $\alpha^{|G|}$ is cohomologous to the trivial 2-cocycle and so the exponent of $M(G)$ divides $|G|$.

Now suppose that $[\alpha]$ has order e in $M(G)$, so that there is a map $\sigma : G \to \mathbb{C}^{\times}$, with $\sigma(1) = 1$, such that

$$\alpha(x, y)^{e} = \sigma(x)\sigma(y)/\sigma(xy)$$

for all x and y in G. Define a map $\delta\colon G \to \mathbb{C}^{\times}$ by taking $\delta(1) = 1$ and, for each non-identity element g of G, choosing any eth root of $\sigma(g)^{-1}$ for $\delta(g)$. It can be checked that if

$$\beta(x, y) = \delta(x)\delta(y)\delta(xy)^{-1}\alpha(x, y)$$

for all x and y in G, then β^{e} is the trivial 2-cocycle. Thus every element of $M(G)$ may be represented by a 2-cocycle chosen from among the functions from $G \times G$ to a fixed finite set (certain roots of unity), and so $M(G)$ must be a finite group.

Next we construct a particular central extension of $M(G)$ by G which contains $M(G)$ in its commutator subgroup. The motivation for this is that, as will be seen below, the study of projective representations of G can be reduced to the study of linear representations of this extension. To do this, let $[\alpha_1],\ldots, [\alpha_d]$ be a minimal generating set for $M(G)$, where $[\alpha_i]$ has order e_i and each α_i is a representative for its cohomology class chosen in accordance with Theorem 1.1 ($1 \le i \le d$). Thus, if ω_i is a primitive e_ith root of unity, then for all x and y in G

$$\alpha_i(x, y) = \omega_i^{n_i(x, y)} \quad (1 \le i \le d).$$

The conditions (C1) and (C2) on the 2-cocycles $\alpha_1,\ldots, \alpha_d$ translate into the following conditions on the integers $n_i(x, y)$, which are well defined (mod e_i):

(1) For all g in G,

$$n_i(1, g) \equiv 0 \equiv n_i(g, 1) \mod e_i.$$

(2) For all x, y and z in G,

$$n_i(x, yz) + n_i(y, z) \equiv n_i(xy, z) + n_i(x, y) \mod e_i.$$

Theorem 1.2. *Let*

$$A = \langle a_1, \ldots, a_d : a_i^{e_i} = 1 \rangle$$

be an abelian group isomorphic to $M(G)$, *and let* $\{r(g) : g \in G\}$ *be a set of symbols in bijective correspondence with the elements of* G. *Let* H *be the set of* $|A|\,|G|$ *elements of the form* $ar(g)$ *for* $a \in A$ *and* $g \in G$. *Then* H *is a group under the multiplication*

$$ar(x)br(y) = ab\Phi(x, y)r(xy)\,,$$

where

$$\Phi(x, y) = \prod_{i=1}^{d} a_i^{n_i(x, y)}\,.$$

Furthermore, A *is a central subgroup of* H *contained in the commutator subgroup of* H, *and* H/A *is isomorphic to* G.

Proof. It follows from the definition that H is closed under multiplication; condition (2) above ensures that H is associative; condition (1) shows that $r(1)$ is the identity element of H; and the inverse of $ar(g)$ is

$$a^{-1}(\Phi(g, g^{-1}))^{-1}\, r(g^{-1}).$$

Hence H is a group and A is clearly central. The map $\theta\colon G \to H/A$, defined by

$$\theta(g) = r(g)A,$$

is an isomorphism, so it only remains to show that A is contained in H', the commutator subgroup of H. To see this, note that the map $\eta\colon Hom(A, \mathbb{C}^\times) \to M(G)$ defined by $\eta(\lambda) = [\alpha]$, where

$$\alpha(x, y) = \lambda(\Phi(x, y))\,,$$

is a group homomorphism. In fact η is surjective, since any $[\alpha]$ in $M(G)$ may be written as

$$[\alpha_1]^{k_1}\,[\alpha_2]^{k_2} \ldots [\alpha_d]^{k_d}.$$

for some $k_1, \ldots, k_d$. The map $\lambda : A \to \mathbb{C}^\times$ defined on the generators $a_1, \ldots, a_d$ of A

by

$$\lambda(a_i) = \omega_i^{k_i}$$

gives a homomorphism for which $\lambda(\Phi(x, y))$ is cohomologous to α. Since its domain and codomain have equal orders, η is an isomorphism.

Now let $A_0 = A \cap H'$ and suppose, for a contradiction, that A_0 is a proper subgroup of A. Let λ be a non-trivial homomorphism from A to $\mathbb{C}^\times$ with A_0 in its kernel. Since

$$A/A_0 \cong AH'/H',$$

we may define a homomorphism λ_0 of AH', with H' in its kernel, by

$$\lambda_0(ah) = \lambda(a).$$

Let $\lambda_1 : H/H' \to \mathbb{C}^\times$ be any homomorphism which agrees with λ_0 on AH'/H'. (Such a λ_1 exists because $\mathbb{C}^\times$ is divisible and H/H' is abelian; Huppert (1967, I 13.7). Alternatively, take λ_1 to be an irreducible constituent of the character of H/H' induced from λ_0; use (C6) of Chapter 4 to see that λ_1 restricted to AH'/H' contains λ_0. We abuse notation slightly by thinking of λ_1 as a function both on H and on H/H'; similarly for λ_0.) Thus, for all x and y in G

$$\begin{aligned}
\lambda_1(r(x))\lambda_1(r(y)) &= \lambda_1(r(x)r(y)) \\
&= \lambda_1(\Phi(x, y)r(xy)) \\
&= \lambda_0(\Phi(x, y))\lambda_1(r(xy)) \\
&= \lambda(\Phi(x, y))\lambda_1(r(xy)).
\end{aligned}$$

Writing $\delta(g)$ for $\lambda_1(r(g))$, this becomes

$$\delta(x)\delta(y) = \lambda(\Phi(x, y))\delta(xy),$$

so that $\eta(\lambda)$ is the class of the trivial 2-cocycle. Since η is an isomorphism, λ is trivial. This contradiction shows that $A_0 = A$, and so A is contained in H'.

Definition. A representation group for G is any group H satisfying the following conditions:

(1) H has central subgroup A contained in H', the commutator subgroup of H.

(2) $H/A \cong G$.

(3) $A \cong M(G)$.

We have proved in 1.2 that such a group exists for any G.

Whenever a representation group H for a group G is referred to, it will be implicitly assumed that specific choices have been made for the subgroup A, and for a set $\{r(g): g \in G\}$ of coset representatives in H for H/A. The latter will be chosen with $r(1) = 1$, and, of course, such that the map $r(g)A \mapsto g$ is an isomorphism from H/A to G. We shall use the notation of Theorem 1.2 for an arbitrary representation group. The function $\Phi : G \times G \to A$ is defined by

$$r(x)r(y) = \Phi(x, y)r(xy) \ .$$

The homomorphism $\eta: \text{Hom}(A, \mathbb{C}^\times) \to M(G)$ sends λ to $[\lambda \circ \Phi]$, the class of the composite. To show that η is an isomorphism in the case of a general representation group, we again use the fact that its domain and codomain have equal cardinality. But this time the injectivity of η is verified. If λ were in the kernel of η, then, for some function δ,

$$\lambda\Phi(x, y) = \delta(x)\delta(y)/\delta(xy) \ .$$

Then the function $\mu: H \to \mathbb{C}^\times$, mapping $ar(g)$ to $\lambda(a)\delta(g)$, is easily checked to be a homomorphism which agrees with λ on A. But $A \subset H'$, so $\lambda(a) = \mu(a) = 1$ for all a, as required.

Example. Returning to the group $G = \{e, a, b, c\}$ of our previous examples, and assuming $M(G)$ has order 2, the 2-cocycle α previously defined gives rise to a group

$$H = < r(a), r(b), z : z^2 = 1 \, , r(a)^2 = z = r(b)^2 \, , r(a)r(b) = zr(b)r(a) > ,$$

so that H is the quaternion group of order 8. Taking the 2-cocycle β defined above gives another representation group for G, one which is isomorphic to the dihedral group of order 8.

We now come to the connection between representation groups and projective representations.

Theorem 1.3. *Let H be a representation group of G. Given a projective matrix representation P of G, there is a function δ: $G \to \mathbb{C}^\times$ and a linear representation R of H such that, for all g in G,*

$$P(g) = \delta(g)R(r(g)).$$

Proof. Suppose that α is the 2-cocycle associated with the projective representation P and the homomorphism λ from A to $\mathbb{C}^\times$ satisfies $\eta(\lambda) = [\alpha]$. Thus there is a function δ: $G \to \mathbb{C}^\times$ such that, for all x and y in G,

$$\lambda(\Phi(x, y)) = \delta(x)^{-1}\delta(y)^{-1}\delta(xy)\alpha(x, y).$$

Defining R by

$$R(ar(g)) = \delta(g)^{-1}\lambda(a)P(g)$$

for a in A and g in G, we see that

$$\begin{aligned} R(ar(x))R(br(y)) &= \delta(x)^{-1}\delta(y)^{-1}\lambda(a)\lambda(b)P(x)P(y) \\ &= \delta(x)^{-1}\delta(y)^{-1}\lambda(ab)\alpha(x, y)P(xy) \\ &= \delta(xy)^{-1}\lambda(ab)\lambda(\Phi(x, y))P(xy) \\ &= \delta(xy)^{-1}\lambda(ab\Phi(x, y))P(xy) \\ &= R(ab\Phi(x, y)r(xy)) \\ &= R(ar(x)br(y)), \end{aligned}$$

so that R is a homomorphism and hence is a linear representation of H.

The above calculation may be reversed to yield the following.

Theorem 1.4. *Let H be a representation group for G and let $\lambda: A \to \mathbb{C}^\times$ be any homomorphism. Suppose that R is a linear representation of H such that $R(a) = \lambda(a)I$ for each a in A. Define P by $P(g) = R(r(g))$ for all g in G. Then P is a projective representation whose associated cocycle is α, where*

$$\alpha(x, y) = \lambda(\Phi(x, y))$$

for all x and y in G.

Remark. In the example discussed several times during this chapter, the projective representation P was in fact produced, in accordance with Theorem 1.4, from the well-known 2-dimensional irreducible representation of the quaternion group of order 8.

Definition. Two projective representations P_1 and P_2 for G on vector spaces V_1 and V_2, respectively, are said to be projectively equivalent if there is an invertible linear map $T: V_1 \to V_2$, and a function $\delta: G \to \mathbb{C}^\times$ with $\delta(1) = 1$, such that, for all g in G,

$$T^{-1}P_2(g)T = \delta(g)P_1(g).$$

If δ is not needed (that is, $\delta(g) = 1$ for all g), then P_1 and P_2 are said to be linearly equivalent.

For projective representations, the notion of projective equivalence is the more natural one, when one interprets them as homomorphisms

$$P' : G \to PGL(V) \ .$$

For fixed V, it corresponds to the existence of an element $\mathcal{T}$ of $PGL(V)$ such that

$$\mathcal{T}^{-1} P'_2(g) \mathcal{T} = P'_1(g)$$

for all g in G.

The definitions apply, in particular, to linear representations, where linear equivalence is by far the more important relation.

The correspondence given in Theorems 1.3 and 1.4, between projective representations of a group G and linear representations of one of its representation groups H, can be seen quite easily to give a $1-1$ correspondence between both kinds of equivalence classes. More precisely, fix a homomorphism λ from A to $\mathbb{C}^\times$. Then there is a $1-1$ correspondence between: (1) projective equivalence classes of linear representations of H which are multiplication by $\lambda(a)$ for each a in A; and (2) projective equivalence classes of projective representations of G whose 2-cocycle is cohomologous to $\lambda \circ \Phi$. (It is evident from the definition that projectively equivalent projective representations have cohomologous cocycles.) In addition, fix a cocycle α within the class of $\lambda \circ \Phi$. Then there is also a $1-1$ correspondence between (i) linear equivalence classes of linear representations of H which map each a in A to $\lambda(a)I$, and (ii) linear equivalence classes of projective representations of G whose 2-cocycle is exactly α (linearly equivalent projective representations must have the same cocycle).

For most of this book, we shall be dealing with linear representations of fixed representation groups $\tilde{S}_n$ and $\tilde{A}_n$ of S_n and A_n, respectively. Such representations will be considered up to linear equivalence, so, by the second $1-1$ correspondence above, this will give the classification of projective representations of S_n and A_n up to linear equivalence. It will be immediate from the results, and from the earlier $1-1$ correspondence, how to obtain the appropriate statements about projective equivalence classes, if preferred.

To finish this chapter we present a simple way to produce non-trivial cocycles. This will suffice for all the material in this book on S_n and A_n. Except for the proof that $M(S_n)$ has no other non-trivial classes, the remainder of the book from Chapter 3 onwards depends only on the first two pages of this chapter, the proposition below, and the part of Chapter 2 from the definition before (2.8) to Remark (2) after (2.10).

Suppose $\theta : \tilde{G} \to G$ is an epimorphism with kernel $\{1, z\}$ which is central of order 2 and contained in the commutator subgroup of $\tilde{G}$. Choose a cross section $r : G \to \tilde{G}$; that is, a function with $\theta r(g) = g$ and $r(1) = 1$. Define

$$\alpha_r : G \times G \to \mathbb{C}^\times$$

by

$$\alpha_r(a, b) = (-1)^{n_r(a, b)}$$

where

$$r(a)r(b) = z^{n_r(a, b)} r(ab) .$$

Proposition 1.5. (*i*) *The function α_r is a cocycle for G, and $[\alpha_r]$ does not depend on the choice of r.*

(*ii*) *The class $[\alpha_r]$ is a non-trivial element of $M(G)$.*

Proof. (i) Abbreviate α_r to α and $n_r(a, b)$ to $n(a, b) \in \mathbb{Z}/2$. To prove that α is a cocycle, note that

$$z^{n(x, yz)+n(y, z)} r(xyz) = z^{n(y, z)} r(x)r(yz) = r(x)r(y)r(z) .$$

Similarly,

$$z^{n(x, y)+n(xy, z)} r(xyz) = r(x)r(y)r(z) .$$

Thus

$$n(x, yz)+n(y, z) = n(x, y)+n(xy, z) ,$$

for all x, y and z in G, and so

$$\alpha(x, yz)\alpha(y, z) = \alpha(x, y)\alpha(xy, z) .$$

Using the fact that $r(1) = 1$ yields easily that

$$\alpha(g, 1) = 1 = \alpha(1, g) .$$

Thus α is a cocycle.

If s is another cross section, then

$$r(a) = z^{m(a)} s(a)$$

for some $m(a) \in \mathbb{Z}/2$. Then

$$z^{n_r(a, b)} = r(ab)^{-1} r(a)r(b) = z^{m(ab)-m(a)-m(b)} z^{n_s(a, b)} ,$$

so

$$n_r(a, b) - n_s(a, b) = m(ab) - m(a) - m(b) .$$

Thus

$$\alpha_r(a, b)/\alpha_s(a, b) = \delta(ab)/\delta(a)\delta(b)$$

for $\delta = (-1)^m$, as required.

(ii) Again abbreviate α_r to α and n_r to n. Suppose, for a contradiction, that for some δ and all a and b,

$$\alpha(a, b) = \delta(ab)/\delta(a)\delta(b) . \tag{1}$$

For any a, we have

$$1 = r(1) = r(a^{-1}a) = z^{n(a^{-1},a)}r(a^{-1})r(a) .$$

Thus

$$r(a)^{-1} = r(a^{-1})z^{n(a^{-1},\, a)} . \tag{2}$$

Now write z as a product of commutators in $\tilde{G}$:

$$\begin{aligned} z &= r(a_1)r(b_1)r(a_1)^{-1}r(b_1)^{-1}\ldots r(a_s)^{-1}r(b_s)^{-1} \\ &= z^N r(a_1b_1a_1^{-1}\ldots b_s^{-1}) \end{aligned} \tag{3}$$

where

$$N = n(a_1, b_1) + n(a_1b_1, a_1^{-1}) + \ldots + n(a_1b_1a_1^{-1}\ldots b_sa_s^{-1}, b_s^{-1})$$
$$+ \sum_i [n(a_i^{-1}, a_i) + n(b_i^{-1}, b_i)] .$$

Applying $r\theta$ to both sides of (3) yields

$$r(a_1b_1a_1^{-1}\ldots b_s^{-1}) = r(1) = 1 ,$$

and so $N = 1$. But $(-1)^N$ is

$$\alpha(a_1, b_1)\alpha(a_1b_1, a_1^{-1})\ldots\alpha(a_1b_1a_1^{-1}\ldots a_s^{-1}, b_s^{-1})\prod_i \alpha(a_i^{-1}, a_i)\alpha(b_i^{-1}, b_i).$$

Substituting (1) for each α yields

$$\delta(a_1^{-1})^2\delta(a_1)^2\delta(b_1^{-1})^2\delta(b_1)^2\delta(a_2^{-1})^2\delta(a_2)^2\ldots = -1 ,$$

that is,

$$\alpha(a_1^{-1}, a_1)^2\alpha(b_1^{-1}, b_1)^2\ldots = -1 ,$$

a contradiction, since α takes only values ± 1.

Notes

1. Our discussion was restricted to projective representations over the complex field. It is possible to consider projective representations over any field K. Such representations determine 2-cocycles $\alpha : G \times G \to K^{\times}$. The group $H^2(G, K^{\times})$ of equivalence classes of 2-cocycles depends on the field K. The structure of this cohomology group can be described in terms of the Schur multiplier $M(G)$ in the following cases:

(1) (Schur, 1904). When K is algebraically closed of characteristic zero, $H^2(G, K^{\times})$ is isomorphic to $M(G)$.

(2) (Asano, Osima and Takahasi, 1937). When K is algebraically closed of non-zero characteristic p, $H^2(G, K^{\times})$ is isomorphic to the p-complement of $M(G)$ (the direct sum of the q-primary parts for $q \neq p$).

(3) (Asano, 1939). For the field of reals, $H^2(G, \mathbb{R}^{\times})$ is isomorphic to a direct sum of $(s+t)$ copies of a cyclic group of order 2, where s is the 2-rank of $M(G)$ and t is the 2-rank of the commutator quotient group G/G'.

The questions of relating projective representations to linear representations of a covering group, and of constructing a representation group, are more complicated in the case of a general field K. For example, $H^2(G, K^{\times})$ is sometimes infinite. See Karpilovsky (1985) for this, and for a treatment of projective representations as modules over a certain ring known as the twisted group algebra.

2. Projective representations arise naturally in the study of linear representations. Given a normal subgroup N of G, the decomposition of the restriction of an irreducible linear representation R of G is determined by theorems of A.H. Clifford (1937), as follows. First, no matter what the field K is, this restriction is a direct sum of certain irreducible representations of N with equal multiplicities. That is,

$$R \downarrow N = e(S_1 \oplus S_2 \oplus \ldots \oplus S_r),$$

where $S_1, S_2, \ldots, S_r$ are non-isomorphic irreducible representations of N, and e is a positive integer. Furthermore, there is a subgroup L of G, of index r, containing N, such that the representation R is induced from an irreducible representation of L. This enables one to deduce information about representations of G from information on proper subgroups unless $r = 1$ (so that $L = G$). When $r = 1$, if K is algebraically closed, a further result of Clifford shows that R can be written as a tensor product $P_1 \otimes P_2$, where P_1 is a projective representation of G/N with degree e, while P_2 is a projective representation of G whose restriction to N is S_1. The cocycles of P_1 and P_2 are inverses of each other.

3. Another discipline where projective representations occur is quantum theory. Roughly speaking, a quantum mechanical system has an associated complex Hilbert space. The space of states of the system is identified with the projective space of this Hilbert space; that is, the state vector is only well-defined up to multiplication by a non-zero complex number. One expects therefore that any symmetry group of the system will act on the space of state vectors to give rise to an infinite dimensional projective representation. See Mackey (1980).

2

REPRESENTATION GROUPS FOR THE SYMMETRIC GROUP

In this chapter, we shall determine the Schur multiplier of the symmetric group S_n and give presentations for its representation groups. To find an upper bound for $M(S_n)$, we shall present a portion of Schur's method to calculate $M(G)$ starting from a presentation of G. For this, two basic results in group theory, Theorems 2.1 and 2.3 below, are quoted without proof. Alternatively, there is a direct approach to this upper bound which uses the twisted group algebra and the existence of a projective representation for any given cocycle; for a treatment, see Jozefiak (1989), pp. 198–200.

The first quoted result consists of some consequences of the classification theorem for finitely generated abelian groups.

Theorem 2.1. *Let A be a finitely generated abelian group of rank r, and let B be a subgroup of A.*

(*a*) *If A is free abelian, then so is B.*

(*b*) *If B has finite index in A, then B also has rank r.*

(*c*) *If B can be generated by r elements and has finite index in A, then B is free abelian of rank r.*

For the proof, see I.13.14 of Huppert (1967).

Corollary 2.2. *Let D be a finite subgroup of an abelian group A. Suppose that A/D is free abelian of rank r. Then A is finitely generated of rank r, and D is its torsion subgroup.*

Proof. Choose $a_1,\ldots, a_r$ so that $a_1D,\ldots, a_rD$ is a basis for A/D. Then A is generated by the set $\{a_1,\ldots, a_r\} \cup D$, and has rank r. Let B be the subgroup of A generated by $a_1,\ldots, a_r$, so that B has finite index in A. By Theorem 2.1 (c), B is free abelian. An element of $B \cap D$ is a torsion element in the free abelian group B, so $B \cap D$ is trivial. Thus A is the direct product of B and D, and so D is the torsion subgroup of A.

The other result which we quote without proof is due to Schur.

Theorem 2.3. *Let G be a group in which $Z(G)$, the centre of G, has finite index. Then G' is a group of finite order.*

For the proof, see VI.2.3 of Huppert (1967).

In his 1907 paper, Schur showed that if a finite group G is presented as a quotient of a finitely generated free group F by a normal subgroup R, then $M(G)$ is isomorphic to $(F' \cap R)/[F, R]$. The full force of this result is not needed for our present purposes. It will only be necessary for us to show that $M(G)$ is an epimorphic image of $(F' \cap R)/[F, R]$.

Proposition 2.4. *Suppose that G is isomorphic to F/R, where F is a free group with normal subgroup R. Note that $[F, R]$ is a subgroup of $F' \cap R$ and is normal in F.*

(i) If G is also isomorphic to H/A for a group H with $A \subset Z(H) \cap H'$, then there is an epimorphism from $(F' \cap R)/[F, R]$ to A.

(ii) Suppose that G is finite and that F is free on r generators. Then $(F' \cap R)/[F, R]$ is the torsion subgroup of $R/[F, R]$, and the latter is a finitely generated abelian group of rank r.

Proof. (i) We shall construct an epimorphism $\lambda : F' \cap R \rightarrow A$ with $[F, R] \subset \ker \lambda$. Let $\alpha : F \rightarrow G$ and $\beta : H \rightarrow G$ be epimorphisms with kernels equal to R and A respectively. The solid arrows in the diagram below are the given inclusions and surjections. The dotted arrows will be constructed in the following arguments:

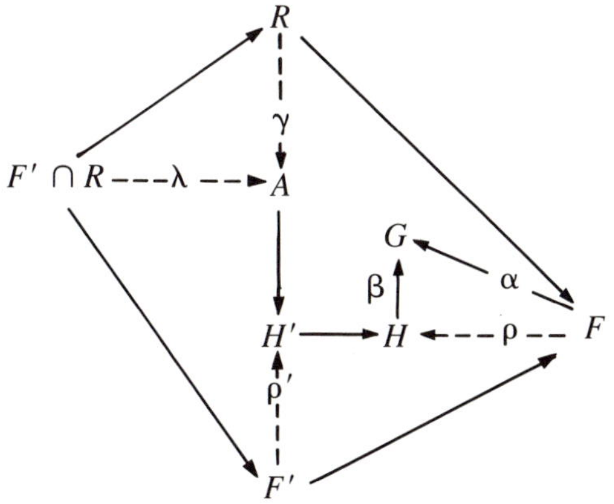

Since β is onto and F is free, we can choose some ρ with $\beta\rho = \alpha$. For $h \in H$, let $\beta(h) = \alpha(f)$. It follows easily that $h = \rho(f)a$. Thus H consists of products ba

with $b \in \text{Im}\,\rho$ and $a \in A$. The commutator $[b_1a_1, b_2a_2]$ of two such elements equals $[b_1, b_2]$, so

$$H' = (\text{Im}\,\rho)' \subset \text{Im}\,\rho.$$

Thus

$$\text{Im}\,\rho = \text{Im}\,\rho \cup H' \supset \text{Im}\,\rho \cup A,$$

which generates H. Thus $\text{Im}\,\rho = H$, i.e. ρ is surjective. If $r \in R$, then $\beta\rho(r) = \alpha(r) = 1$, so $\rho(r) \in \ker\beta = A$. Thus we may define $\gamma(r) = \rho(r)$, and then define λ to be the restriction of γ (or ρ). Now $\rho[F, R] \subset [\rho F, \rho R] \subset [H, A] = 1$, so $[F, R] \subset \ker\lambda$. It remains to prove that λ is surjective. Define $\rho'(x) = \rho(x)$ for $x \in F'$. Then ρ' is onto, since ρ is. Given $a \in A$, choose $x \in F'$ with $\rho'(x) = a$. Then $\alpha(x) = \beta\rho(x) = \beta(a) = 1$, so $x \in \ker\alpha = R$. Thus $x \in F' \cap R$ and $\lambda(x) = a$, as required.

(ii) Clearly $R/[F, R]$ is abelian. We have

$$(R/[F, R])/(F' \cap R/[F, R]) \cong R/F' \cap R \xrightarrow{\mu} F/F',$$

where μ is induced by the inclusion $R \to F$. Then μ is injective with finite cokernel isomorphic to G/G'. But F/F' is free abelian of rank r, so $R/F' \cap R$ is as well, by 2.1 (a). Now $Z(F/[F, R])$ contains $R/[F, R]$, which has finite index $|G|$ in $F/[F, R]$. By 2.3, $(F/[F, R])'$ is finite, and therefore so is its subgroup $F' \cap R/[F, R]$. Now applying 2.2 with $A = R/[F, R]$ and $D = F' \cap R/[F, R]$ yields the result.

Corollary 2.5. *Let G be a finite group isomorphic to F/R, where F is a finitely generated free group. Then $M(G)$ is an epimorphic image of $(F' \cap R)/[F, R]$, the torsion subgroup of $R/[F, R]$.*

Proof. Let H and A in Proposition 2.4 be the groups constructed in Theorem 1.2.

This result allows information to be deduced concerning the Schur multiplier of G starting from a presentation of G. Before illustrating this for S_n, we state a result which has an obvious proof.

Lemma 2.6. *Suppose x and y are elements of a group satisfying $xy = yx$ and $x^3 = y^2$. Then the group generated by x and y is a cyclic group generated by $u = x^{-1}y$ with $u^2 = x$ and $u^3 = y$.*

Theorem 2.7. *The Schur multiplier of S_n has order at most 2, and is trivial if $n \leq 3$.*

Proof. We start from the following well-known presentation for S_n, in which x_i corresponds to the transposition interchanging i and $i+1$ (see I.19.7 of Huppert, 1967). The group S_n is isomorphic to the quotient of the free group F on $x_1, ..., x_{n-1}$ factored by the normal subgroup R generated by the elements

$$x_i^2, \quad 1 \leq i \leq n-1;$$
$$(x_i x_{i+1})^3, \quad 1 \leq i \leq n-2;$$
$$x_i^{-1} x_j^{-1} x_i x_j, \quad 1 \leq i, j < n-1 \text{ and } |i-j| > 1.$$

Let $K = F/[F, R]$ and $N = R/[F, R]$, so that S_n is isomorphic to K/N. Writing k_i for the coset $x_i[F, R]$ $(1 \leq i \leq n-1)$, we see that $\{k_1, ..., k_{n-1}\}$ is a set of generators for K. The elements g_i, m_i and ℓ_{ij} defined below are in N, since they are the projections of elements in R:

$$g_i = k_i^2, \quad 1 \leq i \leq n-1;$$
$$m_i = (k_i k_{i+1})^3, \quad 1 \leq i \leq n-2;$$
$$\ell_{ij} = [k_i, k_j], \quad 1 \leq i, j \leq n-1 \text{ and } |i-j| > 1.$$

In fact, these elements generate N, since they are the projections of the given generators for R.

Since S_n is n-fold transitive, for any i and j with $|i-j| > 1$ and any $n \geq 4$, there is an element s in S_n with

$$s(1) = i, \; s(2) = i+1, \; s(3) = j \text{ and } s(4) = j+1.$$

If t is an element of K such that tN corresponds to s, it follows that there exist a_i and a_j in N such that

$$t^{-1} k_1 t = k_i a_i \text{ and } t^{-1} k_3 t = k_j a_j.$$

Since N is a subgroup of $Z(K)$,

$$\begin{aligned} \ell_{13} &= t^{-1} \ell_{13} t \\ &= t^{-1}[k_1, k_3] t \\ &= [t^{-1} k_1 t, t^{-1} k_3 t] \\ &= [k_i a_i, k_j a_j] \\ &= \ell_{ij}, \end{aligned}$$

so that all the ℓ_{ij}'s are equal to ℓ, say, with ℓ in N. Also,

$$\begin{aligned} k_i^2\ell^2 &= (k_i\ell)^2 \\ &= (k_j^{-1}k_ik_j)^2 \\ &= k_j^{-1}g_ik_j \\ &= k_i^2, \end{aligned}$$

and so $\ell^2 = 1$. There is also a relation between the g's and the k's since

$$k_ik_{i+1}k_i = m_ik_{i+1}^{-1}k_i^{-1}k_{i+1}^{-1}.$$

Squaring gives

$$g_i^2g_{i+1} = m_i^2g_{i+1}^{-2}g_i^{-1}$$

so that

$$m_i^2 = g_i^3g_{i+1}^3 \quad (1 \leq i \leq n-2).$$

It now follows by Lemma 2.6 that N may be generated by $\{d_1, d_2, ..., d_{n-1}, \ell\}$, where $d_1 = g_1$, and

$$d_{i+1} = g_i^{-1}g_{i+1}^{-1}m_i \quad (1 \leq i < n-2).$$

By Proposition 2.4, N has rank $n-1$. Since $\ell^2 = 1$, Corollary 2.2 implies that the torsion subgroup of N has order 2, and so by Corollary 2.5, the order of $M(S_n)$ divides 2.

If $n < 4$, it is easily checked, by calculations as above, that N has trivial torsion subgroup, and so $M(S_n)$ is trivial. This completes the proof.

In order to establish a lower bound for $|M(S_n)|$, we construct a group of order $2(n!)$, which will subsequently be proved to be a representation group for S_n as long as $n \geq 4$.

Definition. Let $\tilde{S}_n$ be the group with generators $z, t_1, ..., t_{n-1}$ and relations

$$z^2 = 1;$$
$$zt_j = t_j z, \quad 1 \le j \le n-1;$$
$$t_j^2 = z, \quad 1 \le j \le n-1;$$
$$(t_j t_{j+1})^3 = z, \quad 1 \le j \le n-2;$$
$$t_j t_k = z t_k t_j, \quad \text{for } |j-k| > 1 \text{ and } \quad 1 \le j, k \le n-1.$$

Evidently, the second relation follows from the third, and z need not appear as a generator.

Theorem 2.8. *The group $\tilde{S}_n$ has order $2(n!)$. The subgroup $\{1, z\}$ is central, and is contained in the commutator subgroup of $\tilde{S}_n$, provided n is at least four. The quotient group of $\tilde{S}_n$ by $\{1, z\}$ is isomorphic to S_n.*

Proof. The only fact which is not immediate is the order asserted for $\tilde{S}_n$. Since the given relations together with $z = 1$ give the presentation for S_n we have already used, this comes down to showing that the given relations do not force z to be 1. One way to do this is to write down certain matrices which satisfy the relations in a non-trivial way. We shall ignore degenerate cases and take $n > 2$. Several details left to the reader below are given in the proof of Proposition 6.1. Let

$$A = \begin{pmatrix} 0 & i \\ i & 0 \end{pmatrix}, B = \begin{pmatrix} 0 & -1 \\ 1 & 0 \end{pmatrix}, C = \begin{pmatrix} 1 & 0 \\ 0 & -1 \end{pmatrix}.$$

so that

$$A^2 = B^2 = -I_2 \text{ and } C^2 = I_2.$$

We use the tensor product (or Kronecker product) of matrices (see Serre, 1977, p. 8) to define, for each positive m, a set of $2m+1$ matrices of size 2^m. For $1 \le k \le m$, let

$$M_{2k-1} = C^{\otimes(m-k)} \otimes A \otimes I_2^{\otimes(k-1)}$$
$$M_{2k} = C^{\otimes(m-k)} \otimes B \otimes I_2^{\otimes(k-1)}.$$

Define

$$M_{2m+1} = iC^{\otimes m}.$$

Denoting the identity $2^m \times 2^m$ matrix by I, it follows that

$$M_j^2 = -I \text{ and } M_jM_k = -M_kM_j \tag{*}$$

for $1 \le j < k \le 2m+1$. Now let m be the integer part of $(n-1)/2$ and define, for $1 \le k \le n-1$,

$$T_k = (2k)^{-1/2}\left(-(k-1)^{1/2}M_{k-1} + (k+1)^{1/2}M_k\right).$$

The proof is then completed by checking that,

$$T_k^2 = -I\,,$$

$$T_jT_k = -T_kT_j \text{ for } k > j+1\,,$$

and

$$(T_kT_{k+1})^3 = -I\,.$$

These follow directly from (*) without going back to the definition of M_k. Checking the last relation can be made less tedious by observing that

$$T_{k+1}T_k = -T_kT_{k+1} + I \text{ for } 1 \le k \le n-2\,.$$

Thus the defining relations for $\tilde{S}_n$ are satisfied with $t_k = T_k$ and $z = -I$, as required.

Remark. The matrix representation of $\tilde{S}_n$ given above is known as the basic representation or spin representation. It plays a crucial role in the sequel. A more general context for this representation is the following.

Let V be a real vector space of dimension ℓ with a non-degenerate symmetric bilinear form f. The Clifford algebra $A(V,f)$ is the associative algebra over $\mathbb{R}$ with generators $1, e_1, \ldots, e_\ell$ such that

$$1e_k = e_k1 = e_k,\ \ 1 \le k \le \ell;$$

$$e_k^2 = f(e_k,e_k)1,\ \ 1 \le k \le \ell;$$

$$e_je_k + e_ke_j = 0,\ \ 1 \le k, j \le \ell\,,\ k \ne j;$$

where $\{e_1, \ldots, e_\ell\}$ is an f-orthogonal $\mathbb{R}$-basis for V. The Clifford algebra has as $\mathbb{R}$-basis all elements

$$e_1^{\delta_1}e_2^{\delta_2}\ldots e_\ell^{\delta_\ell}$$

where each δ_i is either 0 or 1. Thus $A(V,f)$ has dimension 2^ℓ. Now specialize to the case where $f(e_i, e_i) = -1$ for all i, and denote $A(V,f)$ as $A(\ell)$. Using the matrices that occur in the proof of Theorem 2.8, we see that when $m = [\ell/2]$, mapping e_i to M_i provides a matrix representation of $A(\ell)$ of degree 2^m. When ℓ is odd the matrix representation given by mapping e_i to $-M_i$ is a non-equivalent representation.

The group of units of $A(\ell)$ has a subgroup $Pin(\ell)$, with a central subgroup Z of order 2 contained in the commutator subgroup of $Pin(\ell)$, such that $Pin(\ell)/Z$ is isomorphic to the orthogonal group $O(\ell)$. The lifting to $Pin(n-1)$ of the natural embedding of S_n in $O(n-1)$ provides the spin representation.

Details about Clifford algebras and their representations may be found in Atiyah, Bott and Shapiro (1964), Porteous (1966), and in Appendix 6 of this book, where another proof of Theorem 2.8 is provided.

Theorem 2.9. *For $n \geq 4$, the Schur multiplier of S_n is cyclic of order 2, and $\tilde{S}_n$ is a representation group for S_n.*

Proof. Let A be the central subgroup $\{1, z\}$ of $\tilde{S}_n$. Since z is a commutator, and $\tilde{S}_n/A \cong S_n$, we need only prove that $M(S_n)$ has order 2, in order to see that $\tilde{S}_n$ is a representation group. By Theorem 2.7, it remains only to find a cocycle $\alpha : S_n \times S_n \to \mathbb{C}^\times$ which is not cohomologous to the trivial cocycle. But this is immediate from Proposition 1.5 since z is the commutator of t_1 and t_3.

Combining this result with Theorem 1.3 gives the following.

Corollary 2.10. *For $n \geq 4$, let α be a 2-cocycle of S_n with $[\alpha] \neq [1]$. Then any projective representation of S_n with 2-cocycle α may be lifted to a linear representation of $\tilde{S}_n$ in which z is represented as $-I$.*

Remarks. For $n \leq 3$, $\tilde{S}_n$ is not a representation group of S_n since $M(S_n)$ is trivial. In fact, $\tilde{S}_1$ is cyclic of order 2; $\tilde{S}_2$ is cyclic of order 4; and $\tilde{S}_3$ is metacyclic of order 12; that is,

$$\tilde{S}_3 = \langle s, t : s^6 = 1, s^3 = t^2, t^{-1}st = s^{-1} \rangle.$$

The Schur multiplier of A_n is quite complicated to determine, so we state the result without proof. The interested reader can find the details in Karpilovsky (1985) or Schur (1911). The result is the following.

Theorem 2.11. *For any positive integer n*

$$M(A_n) \cong \begin{cases} 1 & \text{if } n \le 3; \\ C_6 & \text{if } n = 6 \text{ or } 7; \\ C_2 & \text{for all other } n; \end{cases}$$

where C_k denotes the cyclic group of order k.

When $n \geq 4$, the cocycle α constructed in Theorem 2.9 restricts to a cocycle on A_n which is not a coboundary by Proposition 1.5, since z is the square of the commutator of t_1t_2 and t_2t_3. The projective representations of A_n which are dealt with in the remainder of this book all correspond to that cohomology class, the element in $M(A_n)$ of order 2.

Using the matrices $T_1, \ldots, T_{2m+1}$ which give the spin representation of $\tilde{S}_n$, define R_k to be iT_k (where $i^2 = -1$). Then

$$R_j^2 = I, \quad 1 \le j \le n-1;$$

$$(R_jR_{j+1})^3 = I, \quad 1 \le j \le n-2;$$

$$R_jR_k = -R_kR_j, \text{ for } |j-k| > 1 \text{ and } 1 \le j, k \le n-1.$$

It follows, as in the proof of Theorem 2.8, that S_n has a double cover $\hat{S}_n$ which is generated by a central involution z and $s_1, \ldots, s_{n-1}$, where

$$s_j^2 = 1, \quad 1 \le j \le n-1;$$

$$(s_js_{j+1})^3 = 1, \quad 1 \le j \le n-2;$$

$$s_js_k = zs_ks_j, \text{ if } |j-k| > 1 \text{ and } 1 \le j, k \le n-1.$$

Thus $\hat{S}_1$ is cyclic of order 2, $\hat{S}_2$ is the Klein four-group and $\hat{S}_3$ is dihedral of order 12. The following is again due to Schur.

Theorem 2.12. *The group $\hat{S}_n$ is a representation group for S_n if $n \geq 4$. The groups $\tilde{S}_n$ and $\hat{S}_n$ are not isomorphic except when n is 1 or 6.*

Outline of proof. Schur constructs an isomorphism θ between $\tilde{S}_6$ and $\hat{S}_6$ by defining θ on generators as follows:

$$\theta(t_1) = s_1s_3s_5; \qquad \theta(t_2) = s_3s_2s_1s_4s_3s_2s_5s_4s_3;$$

$$\theta(t_3) = s_1s_4s_3s_5s_4; \qquad \theta(t_4) = s_1s_2s_1s_3s_2s_1s_5;$$

$$\theta(t_5) = s_1s_3s_4s_3s_5s_4s_3.$$

In general, any isomorphism between $\tilde{S}_n$ and $\hat{S}_n$ would induce an automorphism on the central quotient S_n. For n different from 6, any automorphism of S_n maps the conjugacy class $\mathcal{T}$ of transpositions to itself. (For $n = 6$, there is an outer automorphism mapping $\mathcal{T}$ to the class of elements whose cycle type is three disjoint transpositions.) Since the elements of $\tilde{S}_n$ which map onto $\mathcal{T}$ have order 4, while the elements of $\hat{S}_n$ which map onto $\mathcal{T}$ have order 2, it follows that $\tilde{S}_n$ is not isomorphic to $\hat{S}_n$ in general.

Remark. Yet another result of Schur (1907) is that if G is any finite group with G/G' having invariants $n_1, \ldots, n_r$, and $M(G)$ having invariants $m_1, \ldots, m_s$, then the number of non-isomorphic representation groups of G is at most

$$\prod_{\substack{1 \le i \le r \\ 1 \le j \le s}} g.c.d.\,(n_i, m_j).$$

This shows that, except for $n = 1, 2, 3$ and 6, there are exactly two representation groups for S_n, namely $\tilde{S}_n$ and $\hat{S}_n$.

Define $\tilde{A}_n$ and $\hat{A}_n$ to be the inverse images of A_n under the projections of $\tilde{S}_n$ and $\hat{S}_n$, respectively, to S_n. These are representation groups of A_n for $n \ne 1, 2, 3, 6, 7$. Using Schur's bound above, one may deduce that $\tilde{A}_n$ and $\hat{A}_n$ are isomorphic.

Notes

1. The formula given after 2.3 for the Schur multiplier provides a proof that $H^2(G, K^\times)$ is isomorphic to $M(G)$ for any algebraically closed field K of characteristic zero.

2. Schur's upper bound for the number of isomorphism types of representation groups shows that a perfect finite group (one for which $G = G'$) has, up to isomorphism, a unique representation group. A group H is *quasisimple* if H is perfect and the quotient group of H by its centre is a simple group. A group G is *semisimple* if either G is a central product of quasisimple groups or $G = \{1\}$. The *layer* of a finite group G is the largest normal subgroup of G which is semisimple. The analysis of layers proved to be a significant constituent in the classification of finite simple groups (completed in 1980). Toward this end, the calculation of the Schur multipliers of the known simple groups was the subject of much effort in the 1960's and 70's. An indication of the complexity of these calculations can be given by considering the situation which arose with the Mathieu group M_{22} of order 443,520. Its Schur multiplier was first calculated as C_3, then corrected to C_6, and finally (correctly) shown to be C_{12} by Mazet (1979). Among the classical simple groups, one of the most complicated Schur multipliers occurs for the group $G = PSL(3,4)$, for which $M(G)$ is $C_{12} \times C_4$.

3. Our calculation of $M(S_n)$ is based on Proposition 2.4, which is a variation of a result due to Jones; see Wiegold (1982).

4. There are cohomology groups $H^n(G; A)$ for all $n \geq 0$ and all G-modules A, with extensive applications to, for example, topology and number theory. The interested reader should consult MacLane (1967), Ch. IV, and the references therein.

3

A CONSTRUCTION FOR GROUPS

In this chapter, we shall describe a binary operation within a certain class of objects, each of which is a group with extra structure. This construction can be used to determine the structure of the centralizer of an element of $\tilde{S}_n$, and so we shall be able to deduce information about the conjugacy classes of $\tilde{S}_n$. In later chapters, the construction will play a role in the definition of certain representations.

Definition. Let $\mathcal{G}$ be the class of triples (G, z, σ), where G is a finite group, z is an element of order 2 in the centre of G, and σ is a homomorphism from G to $\mathbb{Z}/2$ with $\sigma(z) = 0$. When z and σ are unambiguous, we often denote an object in $\mathcal{G}$ simply as G.

The canonical example of an object in $\mathcal{G}$ is provided by the group $\tilde{S}_n$ defined before Theorem 2.8. In this case, σ is the composite $s \circ \theta$, where $(-1)^s$ is the sign homomorphism on S_n, and θ is the map $\tilde{S}_n \to S_n$ defined by taking the generator t_i to the transposition interchanging i and $i+1$.

Proposition 3.1. *Suppose given (G_1, z_1, σ_1) and (G_2, z_2, σ_2) in $\mathcal{G}$. Let $G_1 \,\tilde{\times}\, G_2$ be the cartesian product $G_1 \times G_2$ with multiplication defined by*

$$(x_1, x_2)(y_1, y_2) = (z_1^{\sigma_2(x_2)\sigma_1(y_1)} x_1y_1, x_2y_2).$$

Then $G_1 \,\tilde{\times}\, G_2$ is a group, and

$$\{(1, 1), (1, z_2), (z_1, 1), (z_1, z_2)\}$$

is a central subgroup of $G_1 \,\tilde{\times}\, G_2$.

Proof. The identity element is $(1, 1)$. The fact that $\sigma_1(g_1) = \sigma_1(g_1^{-1})$ implies that

$$(g_1, g_2)^{-1} = (z_1^{\sigma_1(g_1)\sigma_2(g_2)} g_1^{-1}, g_2^{-1}).$$

Since σ_1 and σ_2 are homomorphisms, the associativity of the product of

(g_1, g_2), (h_1, h_2) and (k_1, k_2) follows using the identity

$$\sigma_2(g_2)\sigma_1(h_1) + \sigma_2(g_2h_2)\sigma_1(k_1) = \sigma_2(h_2)\sigma_1(k_1)+\sigma_2(g_2)\sigma_1(h_1k_1).$$

It is easily checked that the given subgroup is central.

The group $G_1 \tilde{\times} G_2$ does not have a unique natural structure as an object in $\mathcal{G}$. However, the quotient $G_1 \tilde{\curlyvee} G_2$ of $G_1 \tilde{\times} G_2$ below gives the appropriate object in $\mathcal{G}$ for our purposes.

Proposition 3.2. *Given* (G_1, z_1, σ_1) *and* (G_2, z_2, σ_2) *in* $\mathcal{G}$, *let* Z *denote the central subgroup* $\{(1, 1), (z_1, z_2)\}$ *of* $G_1 \tilde{\times} G_2$. *Define* $G_1 \tilde{\curlyvee} G_2$ *to be* $G_1 \tilde{\times} G_2 / Z$. *Then* $G_1 \tilde{\curlyvee} G_2$ *can be given the structure of an object in* $\mathcal{G}$ *by defining* z *to be* $(z_1, 1)Z$ *and letting*

$$\sigma\big((g_1, g_2)Z\big) = \sigma_1(g_1) + \sigma_2(g_2).$$

The proof of this is clear.

Theorem 3.3. *Let* G_1, G_2 *and* G_3 *be objects in* $\mathcal{G}$. *Then*:

(*a*) *the subsets* $\{(g_1, 1)Z: g_1 \in G_1\}$ *and* $\{(1, g_2)Z: g_2 \in G_2\}$ *are normal subgroups of* $G_1 \tilde{\curlyvee} G_2$, *and are isomorphic to* G_1 *and* G_2, *respectively*;

(*b*) $G_1 \tilde{\curlyvee} G_2 \cong G_2 \tilde{\curlyvee} G_1$: *and*

(*c*) $G_1 \tilde{\curlyvee} (G_2 \tilde{\curlyvee} G_3) \cong (G_1 \tilde{\curlyvee} G_2) \tilde{\curlyvee} G_3$.

Proof. It is easy to check (a). An isomorphism to establish (b) is given by

$$(g_1, g_2)Z \mapsto (z^{\sigma_1(g_1)\sigma_2(g_2)} g_2, g_1)Z,$$

and the map required for (c) is that taking $(g_1, (g_2, g_3))$ to $((g_1, g_2), g_3)$.

Later we shall use the additional facts that the above isomorphisms map z to z, and commute with σ, in the sense that the isomorphisms leave $\ker \sigma$ invariant.

Example. Let a, b and n be positive integers with $n \geq a+b$. We regard $\tilde{S}_a$ as the subgroup of $\tilde{S}_n$ which is the double cover of the symmetric group on $\{1, 2,..., a\}$, and $\tilde{S}_b$ as the double cover of the symmetric group on $\{a+1, a+2, ..., a+b\}$. Thus the generators for $\tilde{S}_b$ are now denoted $t_{a+1}, t_{a+2}, ..., t_{a+b-1}$. As subgroups of $\tilde{S}_n$, we may form $<\tilde{S}_a, \tilde{S}_b>$, the subgroup generated by $\tilde{S}_a \cup \tilde{S}_b$. This is a group with generators $z, t_1, ..., t_{a-1}, t_{a+1}, ..., t_{a+b-1}$, with z a central element of order 2, and relations

$$t_i^2 = z, \quad 1 \le i \le a+b-1, \ i \ne a;$$
$$(t_i t_{i+1})^3 = z, \quad 1 \le i \le a-2 \text{ or } a+1 \le i \le a+b-2;$$
$$t_i t_j = z t_j t_i, \quad 1 \le i, j \le a+b-1, \ |i-j| > 2, \ i \ne a \ne j.$$

On the other hand, $\tilde{S}_a$ and $\tilde{S}_b$ are both objects in $\mathcal{G}$, and so we can form $\tilde{S}_a \,\tilde{\curlyvee}\, \tilde{S}_b$. It will be seen that the map taking t_i to $(t_i, 1)Z$ for $1 \le i \le a-1$, and to $(1, t_i)Z$ for $a+1 \le i \le a+b-1$, is an isomorphism between $<\tilde{S}_a, \tilde{S}_b>$ and $\tilde{S}_a \,\tilde{\curlyvee}\, \tilde{S}_b$.

To see this, first consider the obvious method for finding a presentation for $G \,\tilde{\curlyvee}\, H$, given presentations for G and H. As generators, take all the elements $(x, 1)Z$ and $(1, y)Z$, as x varies over the generators for G, and y over the generators for H. For relations, one replaces x by $(x, 1)Z$ in each relation for G, replaces y by $(1, y)Z$ in each relation for H, and adds to these two a third set, consisting of all the relations

$$\big((1, y)Z\big)\big((x, 1)Z\big) = z^{\sigma(x)\sigma(y)}\big((x, 1)Z\big)\big((1, y)Z\big).$$

Applying this with $G = \tilde{S}_a$ and $H = \tilde{S}_b$, the generators and relations just above match up exactly under the claimed isomorphism with those given two paragraphs above. This completes the proof that $<\tilde{S}_a, \tilde{S}_b>$ and $\tilde{S}_a \,\tilde{\curlyvee}\, \tilde{S}_b$ are isomorphic.

This argument may be iterated to give the following result.

Theorem 3.4. *Let $n = n_1 + n_2 + \ldots + n_r$ be a partition of n, and let $\tilde{S}_{n_1}, \tilde{S}_{n_2}, \ldots, \tilde{S}_{n_r}$ be subgroups of $\tilde{S}_n$ whose projections into S_n act on pairwise disjoint subsets of $\{1, 2, \ldots, n\}$. Then there is an isomorphism*

$$<\tilde{S}_{n_1}, \ldots, \tilde{S}_{n_r}> \to \tilde{S}_{n_1} \,\tilde{\curlyvee} \ldots \tilde{\curlyvee}\, \tilde{S}_{n_r}.$$

Corollary 3.5. *Let u and v be elements of $\tilde{S}_n$ such that $\theta(u)$ and $\theta(v)$ move disjoint subsets of $\{1, 2, \ldots, n\}$. Then $uv = vu$ unless both $\theta(u)$ and $\theta(v)$ are odd permutations, in which case $uv = zvu$.*

Proof. Suppose $\theta(u)$ moves only the integers in a subset X of $\{1, 2, \ldots, n\}$, and $\theta(v)$ moves only the integers in Y. By hypothesis $X \cap Y$ is empty. Thus if $\tilde{S}_a$ is the double cover of the symmetric group on the integers in X, and $\tilde{S}_b$ that of the integers in Y, then

$$<\tilde{S}_a, \tilde{S}_b> \cong \tilde{S}_a \,\tilde{\curlyvee}\, \tilde{S}_b.$$

Identifying u with $(u, 1)$ and v with $(1, v)$, we see that, in $\tilde{S}_a \,\tilde{\times}\, \tilde{S}_b$,

$$(u, 1)(1, v) = (u, v) \text{ and}$$
$$(1, v)(u, 1) = (z^{\sigma(v)\sigma(u)}u, v),$$

giving the result.

This corollary will be useful in calculating the conjugacy classes in $\tilde{S}_n$ and $\tilde{A}_n$. The following is a general result on conjugacy classes in a double cover.

Theorem 3.6. *Let H be a group with central subgroup $Z = <z>$ of order 2, and let $\theta: H \to H/Z$ be the natural epimorphism. Let C be a conjugacy class in H/Z. Then its inverse image, $\theta^{-1}(C)$, is either a union of two conjugacy classes in H, or is itself a conjugacy class in H. This latter case arises precisely when there is an element g in $\theta^{-1}(C)$ such that zg is conjugate to g.*

Proof. Suppose a is an element of $\theta^{-1}(C)$, and b is conjugate to a. Since $\theta(a)$ is conjugate to $\theta(b)$, the element b is also in $\theta^{-1}(C)$. It follows that $\theta^{-1}(C)$ is a union of conjugacy classes. On the other hand, if a and b are in $\theta^{-1}(C)$, then $x^{-1}\theta(a)x = \theta(b)$ for some x. Writing x as $\theta(d)$, we see that $\theta(d^{-1}ad) = \theta(b)$. Thus $d^{-1}ad$ is either b or zb. We have shown that if a and b are elements of $\theta^{-1}(C)$, then a is conjugate to either b or zb. If there is an element b of $\theta^{-1}(C)$ which is conjugate to zb, then $\theta^{-1}(C)$ is a single conjugacy class in H. (It follows that for each element h of $\theta^{-1}(C)$, h is conjugate to zh.) However, if there is an element b in $\theta^{-1}(C)$ which is not conjugate to zb, since every element of $\theta^{-1}(C)$ is conjugate to exactly one of b or zb, $\theta^{-1}(C)$ is a union of two conjugacy classes.

Definition. In the notation of Theorem 3.6, we say that the conjugacy class C of H/Z *splits* in H if $\theta^{-1}(C)$ is a union of two conjugacy classes of H.

We shall determine which conjugacy classes of S_n split in $\tilde{S}_n$, and which conjugacy classes of A_n split in $\tilde{A}_n$. At this point, it is convenient to give the well-known results which determine the conjugacy classes in S_n and A_n. Recall that the cycle type of a permutation π in S_n is the sequence

$$(1^{a_1} 2^{a_2} \dots k^{a_k}),$$

where the decomposition of π into disjoint cycles contains a_i cycles of length i.

Theorem 3.7. *Two elements π and ρ of S_n are conjugate if and only if π and ρ have the same cycle type. Any conjugacy class C of even permutations in S_n is either a conjugacy class in A_n, or divides into two conjugacy classes of equal cardinality. This latter case occurs precisely when $n > 1$ and each element in C*

can be expressed as a product of disjoint cycles whose lengths are pairwise distinct odd integers; that is, the cycle type $1^{a_1}\, 2^{a_2} \ldots$ *of C satisfies*

$$a_k = \begin{cases} 0 & \text{if } k \text{ is even,} \\ 0 \text{ or } 1 & \text{if } k \text{ is odd.} \end{cases}$$

For the proof, see 1.2.6 and 1.2.10 of James and Kerber (1981).

The following is the result for $\tilde{S}_n$.

Theorem 3.8. *The conjugacy classes of* S_n *which split in* $\tilde{S}_n$ *are: (a) the classes of even permutations which can be written as a product of disjoint cycles with no cycles of even length; and (b) the classes of odd permutations which can be expressed as a product of disjoint cycles with no two cycles of the same length (including length 1). Expressed in cycle type notation, these conditions are:*

(a) $a_{2i} = 0$ *for all i;*

(b) $a_i \leq 1$ *for all i, and the number of even parts is odd.*

Proof. Let t be an element of $\tilde{S}_n$ and consider four cases.

(1) Suppose that $\theta(t)$ is an even permutation all of whose cycles have odd length. In this case the order, k, of $\theta(t)$ is odd. Thus $t^k = z^i$ for some i, and

$$(zt)^k = z^k t^k = z^{1+i},$$

since k is odd. It follows that zt and t have different orders and so cannot be conjugate. Applying Theorem 3.6, the class of $\theta(t)$ splits.

(2) Suppose $\theta(t)$ is an even permutation with a cycle of even length. In this case, write t as uv with $\theta(u)$ being a cycle of even length. Since $\theta(u)$ is an odd permutation, $\theta(v)$ is also odd, and so by Corollary 3.5, $uv = zvu$. Thus

$$ut = uuv = uzvu = ztu,$$

so that t is conjugate to zt. By Theorem 3.6, the class of $\theta(t)$ does not split.

(3) Next consider the case when $\theta(t)$ is an odd permutation with cycle type $\ell_1, \ell_2, \ldots, \ell_r$, where $\ell_i \neq \ell_j$ if $i \neq j$. Since $\theta(t)$ is an odd permutation, an odd number k, say, of the integers $\ell_1, \ell_2, \ldots, \ell_r$ must be even. Now write $t = u_1 \ldots u_r$, where $\theta(u_i)$ is a cycle of length ℓ_i $(1 \leq i \leq r)$. It can be easily checked that since $\ell_1 + \ell_2 + \ldots + \ell_r = n$, the centralizer in S_n of $\theta(t)$ is

$$\langle\theta(u_1)\rangle \times \langle\theta(u_2)\rangle \times \ldots \times \langle\theta(u_r)\rangle.$$

If $\theta(u_i)$ is an even permutation, Corollary 3.5 implies that u_i commutes with u_j for all j, and so u_i commutes with t. However, if $\theta(u_i)$ is odd, $u_i u_j$ is $z u_j u_i$ for $k-1$ values of j, again by Corollary 3.5. Since $k-1$ is even, u_i commutes with t in this case also. Thus the centralizer of t in $\tilde{S}_n$ is generated by $z, u_1, \ldots, u_r$ and has order $2\ell_1\ell_2\ldots\ell_r$. Since t has the same number of conjugates in $\tilde{S}_n$ as $\theta(t)$ has in S_n, the class of $\theta(t)$ splits.

(4) Finally, suppose that $\theta(t)$ is an odd permutation with two cycles ρ_1 and ρ_2, say, of length k. Suppose that ρ_1 is the cycle $(x_1\ldots x_k)$ and ρ_2 is the cycle $(y_1\ldots y_k)$. Let π be the $2k$-cycle

$$\pi = (x_1y_1x_2y_2\ldots x_ky_k)\,,$$

so that $\pi^2 = \rho_1\rho_2$. Now let u be an element of $\tilde{S}_n$ such that $\theta(u) = \pi$, and define v to be tu^{-2}. Since $\theta(v)$ and π are disjoint permutations and both are odd, Corollary 3.5 gives that $uv = zvu$. Thus

$$ut = uvu^2 = zvu^3 = ztu,$$

so that t is conjugate to zt, and the class of $\theta(t)$ does not split.

Examples. When $n = 2$, both classes of S_2 split in $\tilde{S}_2$.

When $n = 3$, the cycle types in S_3 are (1^3), $(2\ 1)$ and (3), and each of these splits in $\tilde{S}_3$.

When $n = 4$, the cycle types in S_4 are (1^4), $(2\ 1^2)$, $(2\ 2)$, $(3\ 1)$ and (4). Of these, (1^4), $(2\ 2)$ and $(3\ 1)$ correspond to even permutations, and so (1^4) and $(3\ 1)$ split in $\tilde{S}_4$, while $(2\ 1^2)$ and (4) correspond to odd permutations, and so (4) splits in $\tilde{S}_4$.

When $n = 5$, the classes in S_5 which split in $\tilde{S}_5$ are of cycle types

$$(1^5),\ (3\ 1^2),\ (5),\ (3\ 2),\ (4\ 1).$$

We next turn our attention to $\tilde{A}_n$.

Theorem 3.9. *The conjugacy classes of A_n which split in $\tilde{A}_n$ are: (a) the classes of permutations whose decompositions into disjoint cycles have no cycles of even length; and (b) the classes of permutations which can be expressed as a product of disjoint cycles with at least one cycle of even length and with no two cycles of the same length (including length 1). In cycle type notation, these conditions are:*

(a) $a_{2i} = 0$ for all i;

(b) $a_i \leq 1$ for all i, and $a_{2i} = 1$ for at least one value of i.

Proof. Let t be an element of $\tilde{A}_n$. There are three cases to consider.

(1) Suppose that all the cycles in $\theta(t)$ have odd length. Then the argument in case (1) of the proof of Theorem 3.8 shows that the class of $\theta(t)$ splits in $\tilde{A}_n$.

(2) Now suppose that $\theta(t)$ can be written as a product of disjoint cycles of distinct lengths $\ell_1,...,\ell_r$, with $\ell_1,...,\ell_k$ even integers for some $k \geq 1$, and $\ell_{k+1},...,\ell_r$ odd. Writing $t = u_1...u_r$ where $\theta(u_i)$ has length ℓ_i $(1 \leq i \leq r)$, the centralizer D of $\theta(t)$ in S_n is

$$<\theta(u_1)> \times <\theta(u_2)> \times ... \times <\theta(u_r)>.$$

Not all the elements in D are even permutations. In fact, the centralizer C of $\theta(t)$ in A_n is the subgroup consisting of elements of the form

$$\theta(u_1)^{n_1}...\theta(u_r)^{n_r}$$

where the sum $n_1+n_2+...+n_k$ is an even integer. For $1 \leq i \leq r$, we may write t as u_iv_i (where $v_i = u_i^{-1}t$) so that $\theta(u_i)$ and $\theta(v_i)$ are disjoint permutations. It follows by Corollary 3.5 that $u_iv_i = zv_iu_i$ for $1 \leq i \leq k$, and that $u_iv_i = v_iu_i$ if $k+1 \leq i \leq r$. Hence $tu_i = zu_it$ for $1 \leq i \leq k$, and t commutes with u_i for $i > k$. The centralizer of t in $\tilde{A}_n$ is therefore $\theta^{-1}(C)$, and the number of conjugates of t in $\tilde{A}_n$ is equal to the number of conjugates of $\theta(t)$ in A_n. The class of $\theta(t)$ therefore splits in this case.

(3) Finally, consider the case when $\theta(t)$ has two cycles ρ_1 and ρ_2, say, of length k, and $\theta(t)$ also has at least one cycle π, say, of even length. Write $t = uv$, where $\theta(u) = \pi$, so that $\theta(u)$ and $\theta(v)$ are odd permutations. Corollary 3.5 implies that $tu = zut$ so that t is conjugate to zt. To show that the class of $\theta(t)$ does not split, we find a further element w of $\tilde{S}_n$ which centralizes t and such that $\theta(w)$ is odd. It will then follow that t is conjugate to zt via uw, and $\theta(uw)$ is even, so uw is in $\tilde{A}_n$. To construct w, let ρ be the $2k$-cycle obtained by interlacing ρ_1 and ρ_2 (as in the definition of π in case (4) of the proof of Theorem 3.8). Thus ρ^2 is $\rho_1\rho_2$. If w is an element of $\tilde{S}_n$ with $\theta(w) = \rho$ and $x = tw^{-2}$, it follows that $\theta(x)$ is even, since $\theta(w)$ is odd. Corollary 3.5 implies that

$$tw = xw^3 = wxw^2 = wt\ ,$$

as required.

Examples. There are three conjugacy classes in A_3 corresponding to cycle types (1^3) and (3), since the class containing (3) divides into two by Theorem 3.7. Each of these classes splits in $\tilde{A}_3$.

In A_4, the class of cycle types $(3\ 1)$ divides into two and the other classes correspond to cycle type (1^4) and (2^2). Of these, (1^4) and the two of type $(3\ 1)$

split in $\tilde{A}_4$.

In A_5, the classes are of type (1^5), $(2^2\ 1)$, $(3\ 1^2)$ and (5). The class (5) divides into two. The classes of cycle type (1^5) and $(3\ 1^2)$ split in $\tilde{A}_5$, as do the two of cycle type (5).

To conclude this chapter, we count the number of conjugacy classes which split in $\tilde{S}_n$ and $\tilde{A}_n$. For any positive integer n, let a_n denote the number of partitions of n into distinct parts with an even number of even parts. Let b_n denote the number of partitions of n into distinct parts with an odd number of even parts.

Corollary 3.10. *The number of classes of S_n which split in $\tilde{S}_n$ is $a_n + 2b_n$. The number of classes of A_n which split in $\tilde{A}_n$ is $2a_n + b_n$, provided $n > 1$.*

Proof. It is clear that $a_n + b_n$ is equal to the number of partitions of n into distinct parts. It is well-known that this is equal to be number of partitions on n into odd parts. (See Andrews (1976) or Goulden and Jackson (1983), 2.5.20.) Now apply Theorem 3.8, noting that case (a) of that result gives $a_n + b_n$ classes of S_n which split in $\tilde{S}_n$, while case (b) gives b_n classes which split. Since the classes arising from these two cases are disjoint, the result for $\tilde{S}_n$ follows.

To establish the result for $\tilde{A}_n$, we let c_n be the number of partitions of n into distinct odd parts. By Theorem 3.7, c_n is the number of classes of S_n which divide into two classes of A_n of equal size. Of the $a_n + b_n$ classes of S_n counted in case (a) of Theorem 3.9, the number which remain single classes in A_n is $a_n + b_n - c_n$. The remaining c_n divide into two classes of equal size. The number of conjugacy classes arising in case (b) of Theorem 3.9 is $a_n + c_n$, so the total number of classes of A_n which split in $\tilde{A}_n$ is $2a_n + b_n$, as required.

Examples. The above numbers for small n are given as follows:

n	2	3	4	5	6	7	8
a_n	0	1	1	1	2	2	3
b_n	1	1	1	2	2	3	3
a_n+2b_n	2	3	3	5	6	8	9
$2a_n+b_n$	1	3	3	4	6	7	9

Notes

The twisted central product $G_1 \tilde{\curlyvee} G_2$ was introduced in Hoffman and Humphreys (1985). It is closely related to the calculation of $H^2(A \times B; \mathbb{Z}/2)$ in terms of the cohomology of groups A and B, where the alternative description of $H^2(A; \mathbb{Z}/2)$, as classes of double covers of A, is used. The special case of this construction when G_1 and G_2 are representation groups of symmetric groups appears implicitly in Schur (1911). The formulae for the numbers of conjugacy classes in $\tilde{S}_n$ and in $\tilde{A}_n$ are due to Schur (1911).

4

REPRESENTATIONS OF OBJECTS IN $\mathcal{G}$

In this chapter, we shall discuss the representation theory of an object (G, z, σ) in $\mathcal{G}$. To avoid unnecessary repetition, the word "representation" will be used to mean "linear representation" as defined in Chapter 1. We shall restrict attention to representations over the complex numbers. Character theory can then be used in many proofs. This is a suitable point, therefore, for a short digression recalling the main results of character theory. The following six paragraphs, labelled (Di), will each conclude with a statement of a property of characters labelled (Ci). The reader should consult Huppert (1967) for proofs of the non-obvious facts below. The references given in parentheses are to that book.

(D1) Given a representation R of a group G, the character θ of R is defined by

$$\theta(g) = \operatorname{tr} R(g)$$

for all g in G. The next result follows directly from the fact that similar matrices have the same trace.

(C1) The character θ of a representation R of G is a class function. That is, for all g and x in G,

$$\theta(gxg^{-1}) = \theta(x).$$

(D2) Two representations R_1 and R_2 on vector spaces V_1 and V_2 are (linearly) equivalent if there is a bijective linear transformation $T:V_1 \to V_2$ such that, for all g in G,

$$R_1(g) = T^{-1}R_2(g)T.$$

We then have the following result (V.5.5):

(C2) Two representations of a group G are equivalent if and only if they have the same character.

(D3) Recall that a representation R of a group G on a non-zero vector space V is irreducible if there is no subspace U of V, other than $\{0\}$ and V itself, such that U is invariant under each of the linear transformations $R(g)$ ($g \in G$). Let $Irr(G)$ denote the set of irreducible characters of G; that is, the characters of a complete set of inequivalent irreducible representations of G. The space of complex-valued class functions on G can be given an inner product by

$$<\theta, \phi>_G = \frac{1}{|G|} \sum_{g \in G} \theta(g)\overline{\phi}(g),$$

where $\overline{\phi}(g)$ is the complex conjugate of $\phi(g)$. Then (V.5.9) gives the following:

(C3) The set $Irr(G)$ is an orthonormal basis for the space of complex-valued class functions on G. In particular, $Irr(G)$ is linearly independent, and $|Irr(G)|$ is the number of conjugacy classes of G.

(D4) The character table of G is a square array, with rows labelled by irreducible characters, and columns by conjugacy classes. The entry $\theta(x)$ is in the row labelled θ and the column labelled by a representative x of a conjugacy class C. (By (C1) this entry is independent of the choice of representative x in C.) Using the inner product defined in (D3) (that is, weighting by conjugacy class sizes), the rows of the character table are orthogonal. The columns of the table satisfy the second orthogonality relations (V.5.8):

(C4) Let $x_1, ..., x_r$ be representatives for the distinct conjugacy classes in G. Then, for each i and j,

$$\sum_{\theta \in Irr(G)} \theta(x_i)\overline{\theta}(x_j) = \delta_{ij} |C_G(x_i)|,$$

where δ_{ij} is 0 or 1 according as i is different from j or not, and $C_G(x)$ is the centralizer of x in G. In particular, taking $x_i = x_j$ to be the identity element, the sum of the squares of the dimensions of the irreducible representations is $|G|$.

(D5) Let H be a subgroup of G. Let $x_1, ..., x_k$ be a (right) transversal for H in G (so that $Hx_1, ..., Hx_k$ are the distinct right cosets of H in G). Let ϕ be a class function on H, and define

$$(\phi \uparrow G)(g) = \dot{\phi}(x_1 g x_1^{-1}) + \dot{\phi}(x_2 g x_2^{-1}) + ... + \dot{\phi}(x_k g x_k^{-1}),$$

where $\dot{\phi}(y) = 0$ if y is not an element of H, and $\dot{\phi}(y) = \phi(y)$ if $y \in H$. Then $\phi \uparrow G$ is a class function on G, called the induced class function. By (V.16.3):

(C5) If ϕ is the character of a representation Q of H, then $\phi \uparrow G$ is the character of a representation, denoted $Q \uparrow G$, of G. These are called the induced character and the induced representation, respectively.

(D6) Given a representation R of G with character θ, restriction of the representation to any subgroup H of G gives a representation $R \downarrow H$ with character $\theta \downarrow H$, the restriction of the function θ to H. The result connecting restriction and induction is known as Frobenius reciprocity (V.16.5):

(C6) Let H be a subgroup of G, with θ the character of a representation of G, and ϕ the character of a representation of H. Then

$$<\theta, \phi \uparrow G>_G = <\theta \downarrow H, \phi>_H.$$

This actually holds for all class functions θ and ϕ.

Now consider the representations of an object (G, z, σ) in $\mathcal{G}$.

Definition. An irreducible representation R of G is *negative* if $R(z) = -I$.

Proposition 4.1. *Let G be an object in $\mathcal{G}$ and let $R_1, \ldots, R_t$ be a complete list of non-equivalent irreducible negative representations of G. Suppose that R_i has degree d_i. Then*

$$d_1^2 + d_2^2 + \ldots + d_t^2 = |G|/2 .$$

Proof. Let R be any irreducible representation of G on a vector space V. By Schur's lemma (V.4.3 of Huppert, 1967), the linear transformation $R(z)$ is a scalar multiple of the identity. Since $z^2 = 1$, this scalar is either 1 or -1. The irreducible representations which represent z as the identity linear transformation may be regarded as the irreducible representations of $G/<z>$. By (C4) above, the sum of the squares of the degrees of these "positive" irreducibles is

$$|G/<z>| = |G|/2 .$$

Hence, using (C4) again, the sum of the squares of the dimensions of the negative irreducible representations is

$$|G| - |G|/2 = |G|/2,$$

as required.

For the group $\tilde{S}_n$, the positive irreducible representations are essentially the irreducible representations of S_n. Corollary 2.10 shows that the irreducible negative representations of $\tilde{S}_n$ determine projective representations of S_n corresponding to a 2-cocycle which is not cohomologous to the trivial 2-cocycle. These projective representations are irreducible, where irreducibility is defined exactly as with linear representations. By the discussion after Theorem 1.4, a more precise version is the following. There is a $1-1$ correspondence between (1) linear equivalence classes of irreducible projective representations of S_n with a fixed non-trivial cocycle, and (2) linear equivalence classes of irreducible negative linear representations of $\tilde{S}_n$.

We now consider the influence of the homomorphism σ on the representation theory of (G, z, σ).

Definitions. Let R be a representation of (G, z, σ). The *associate representation* R^a is given by

$$R^a(g) = (-1)^{\sigma(g)} R(g) .$$

A representation R of (G, z, σ) is *self-associate* if R is equivalent to R^a.

Remarks. (i) If $\ker \sigma = G$, every representation is self-associate.

(ii) The character θ^a of the associate R^a agrees with the character θ of R on $\ker \sigma$, and takes the value $-\theta(g)$ for g in $G \backslash \ker \sigma$. It follows by (C2) that a representation R is self-associate if and only if θ is zero on $G \backslash \ker \sigma$.

Definitions. Let N be a normal subgroup of a group G. Given an element g of G and a representation Q of N, the *conjugate representation* Q^g is defined by

$$Q^g(x) = Q(gxg^{-1})$$

for all x in N.

A representation Q of a normal subgroup of G is *self-conjugate in* G if and only if Q^g is equivalent to Q for all g in G.

Remarks. (iii) If (G, z, σ) is an object in $\mathcal{G}$ and σ is zero, every representation of $\ker \sigma$ is, of course, self-conjugate.

(iv) Let Q be a representation of a normal subgroup N of G. If $g \in G$, and both h and x are in N, then

$$Q(hgxg^{-1}h^{-1}) = Q(h)Q(gxg^{-1})Q(h)^{-1} ,$$

and so Q^g is equivalent to Q^{hg}. In particular, Q^h is equivalent to Q for all h in N.

It follows that for (G, z, σ) in $\mathcal{G}$ and Q a representation of $\ker \sigma$, there are either one or two equivalence classes of representations conjugate to Q. For $g \in \ker \sigma$, Q^g is unique up to equivalence independently of g, and we shall denote it Q^c, and its character ϕ^c, where ϕ is the character of Q. Thus Q is self-conjugate if and only if $\phi^c = \phi$.

Theorem 4.2. *Let G be an object in $\mathcal{G}$, and suppose that G_0, the kernel of σ, is a proper subgroup of G. Let R be an irreducible representation of G. If R is self-associate, the restriction of R to G_0 has two non-equivalent irreducible constituents, and they are conjugates of each other. If R is not self-associate, its restriction to G_0 is irreducible and self-conjugate, and is equivalent to the restriction of R^a to G_0.*

Proof. Suppose first that R is self-associate and let θ be its character. It follows by remark (ii) above that θ vanishes off G_0. Using (C3), we have

$$\begin{aligned} 2 &= 2 <\theta, \theta> \\ &= \frac{2}{|G|} \sum_{g \in G} \theta(g)\bar{\theta}(g) \\ &= \frac{1}{|G_0|} \sum_{g \in G_0} \theta(g)\bar{\theta}(g) \\ &= <\theta \downarrow G_0, \theta \downarrow G_0> . \end{aligned}$$

Since $1^2 + 1^2$ is the only decomposition of 2 as a sum of squares, the restriction of θ to G_0 is a sum $\phi + \psi$, say, of two inequivalent irreducible characters of G_0. By (C6),

$$<\phi \uparrow G, \theta> = <\phi, \theta \downarrow G_0> = 1 .$$

Thus $\phi \uparrow G = \theta + \zeta$ for some character ζ, possibly zero. Similarly, $\psi \uparrow G = \theta + \eta$ for some character η. But then

$$\begin{aligned} 2\theta(1) + \zeta(1) + \eta(1) &= (\phi \uparrow G)(1) + (\psi \uparrow G)(1) \\ &= 2\phi(1) + 2\psi(1) \\ &= 2(\theta \downarrow G_0)(1) \\ &= 2\theta(1) . \end{aligned}$$

Thus $\zeta = 0 = \eta$, since the dimensions $\zeta(1)$ and $\eta(1)$ of the corresponding representations are zero. Now for any $g \in \ker \sigma$

$$(\theta \downarrow G_0)(g) = \theta(g) = (\phi \uparrow G)(g) = \phi(g) + \phi^c(g) ,$$

by the definition (D5) of induced characters. Thus $\theta \downarrow G_0 = \phi + \phi^c$, and so $\psi = \phi^c$, as required.

On the other hand, suppose that θ is the character of an irreducible representation of G which is not self-associate. Let ϕ be any irreducible constituent of $\theta \downarrow G_0$. Then for some character α, possibly zero,

$$\theta \downarrow G_0 = \phi + \alpha = \theta^a \downarrow G_0 .$$

By (C6)

$$<\phi \uparrow G, \theta> = <\phi, \theta \downarrow G_0> \geq 1 .$$

Similarly,

$$<\phi \uparrow G, \theta^a> \geq 1 .$$

Thus for some character β, we have

$$\phi \uparrow G = \theta + \theta^a + \beta ,$$

since θ and θ^a are irreducible and inequivalent. Then

$$\begin{aligned} 2\phi(1) &= (\phi \uparrow G)(1) = \theta(1) + \theta^a(1) + \beta(1) \\ &= 2\phi(1) + 2\alpha(1) + \beta(1) . \end{aligned}$$

Thus $\alpha(1) = \beta(1) = 0$, so α and β are zero, giving

$$\theta \downarrow G_0 = \phi = \theta^a \downarrow G_0 .$$

Finally, for $g \in G_0$ and $g_1 \in G \backslash G_0$,

$$\phi^c(g) = \phi^{g_1}(g) = \phi(g_1 g g_1^{-1}) = \theta(g_1 g g_1^{-1}) = \theta(g) = \phi(g) ,$$

so $\phi^c = \phi$, as required.

Corollary 4.3. *Let* (G, z, σ) *be an object with* σ *non-zero. Then*:

(*i*) *the number, c, of self-associate irreducible negative representations of G, is the same as the number of conjugate pairs of non-self-conjugate irreducible negative representations of* ker σ; *and*

(*ii*) *the number, d, of associated pairs of non-self-associate irreducible negative representations of G, equals the number of self-conjugate irreducible negative representations of* ker σ. (*By abuse of notation, we are using "representation" to mean "equivalence class of representations"*.)

Proof. We may confine our attention to negative representations since the restriction of a negative representation of G to $G_0 = \ker \sigma$ is negative, and since inducing a negative representation of G_0 gives a negative representation of G.

(i) The first part of 4.2 gives a map, from the set of self-associate irreducibles θ to the relevant set of conjugate pairs $\{\phi, \phi^c\}$, determined by $\theta \downarrow G_0 = \phi + \phi^c$. The equation $\phi \uparrow G = \theta$ in its proof shows that this map is injective. To prove surjectivity, let $\{\phi, \phi^c\}$ be a conjugate pair. We shall find an irreducible θ with $\theta \downarrow G_0 = \phi + \phi^c$; the second part of 4.2 guarantees that θ is self-associate. Let θ be any irreducible constituent of $\phi \uparrow G\ [= \phi^c \uparrow G]$. By (C6)

$$<\theta \downarrow G_0, \phi> \;=\; <\theta, \phi \uparrow G> \geq 1\,.$$

Similarly

$$<\theta \downarrow G_0, \phi^c> \geq 1\,.$$

Thus

$$\theta \downarrow G_0 = \phi + \phi^c + \alpha$$

for some character α (possibly zero). Now

$$2\phi(1) = (\phi \uparrow G)(1) \geq \theta(1) = (\theta \downarrow G_0)(1) = 2\phi(1) + \alpha(1)\,.$$

Hence $\alpha(1) = 0$, so $\alpha = 0$, as required.

(ii) Similarly, the second part of 4.2 gives an injective map from the relevant set of pairs $\{\theta, \theta^a\}$ to the set of self-conjugate negative irreducibles ϕ of G_0. Given such a ϕ, we must find an irreducible character θ for G such that $\theta \downarrow G_0 = \phi$; the first part of 4.2 guarantees that θ will not be self-associate. Now

$$\begin{aligned}
<\phi \uparrow G, \phi \uparrow G> &= |G|^{-1}\sum_{g\in G} |\phi \uparrow G)(g)|^2 \\
&= |G|^{-1}\sum_{g\in G_0} |\phi(g) + \phi^c(g)|^2 \\
&= |G|^{-1}\sum_{g\in G_0} |2\phi(g)|^2 \\
&= |G|^{-1} 4|G_0| = 2\,.
\end{aligned}$$

Thus $\phi \uparrow G = \theta + \gamma$ for a pair of distinct irreducible characters θ and γ of G. Then

$$<\theta \downarrow G_0, \phi> \;=\; <\theta, \phi \uparrow G> \;=\; 1\,.$$

Thus $\theta \downarrow G_0 = \phi + \mu$ for some character μ. Similarly, $\gamma \downarrow G_0 = \phi + \nu$ for some ν. But then

$$2\phi(1) = (\phi \uparrow G)(1) = \theta(1) + \gamma(1) = 2\phi(1) + \mu(1) + \nu(1),$$

so $\mu = \nu = 0$ and $\theta \downarrow G_0 = \phi$, as required.

Remark. The results in 4.2 and 4.3 are essentially a special case of the theory, relating representations of a group to those of a normal subgroup, in Clifford (1937).

Proposition 4.4. *With notation and hypotheses as in Corollary 4.3, the number $c+2d$ is the difference between the numbers of conjugacy classes in G and in $G/\{1, z\}$.*

Proof. This follows from (C2), since the total number of irreducibles for G is the number of conjugacy classes in G, and the number of non-negative irreducibles is the number of conjugacy classes in $G/\{1, z\}$.

Remark. By the same argument, $2c + d$ is the analogous difference for $\ker \sigma$ and $\ker \sigma/\{1, z\}$.

Example. As an example we shall calculate the character tables for $\tilde{S}_4$ and $\tilde{A}_4$. The central quotient $\tilde{S}_4/<z>$ is the symmetric group S_4. The character table of S_4 is as shown in Table 4.5. The columns are labelled by cycle types of partitions (by Theorem 3.7), and the rows by irreducible characters.

Table 4.5 The character table of the symmetric group S_4.

	(1^4)	$(2\,1^2)$	(2^2)	(31)	(4)
θ_1	1	1	1	1	1
θ_2	1	-1	1	1	-1
θ_3	2	0	2	-1	0
θ_4	3	1	-1	0	-1
θ_5	3	-1	-1	0	1

For readers unfamiliar with character theory, we outline one way to obtain this table. Note firstly that, given a quotient group G/N of a group G and a representation R' of G/N, a representation R of G may be "lifted" from R' by defining $R(g)$ to be $R'(gN)$. The character θ of R is given by $\theta(g)=\theta'(gN)$, and θ is irreducible if and only if θ' is irreducible. The group S_4 has a normal subgroup A_4, with quotient group cyclic of order 2. A group of order 2 has two irreducible representations obtained by mapping the generator to 1 or -1. Lifting these to S_4 gives the characters θ_1 and θ_2, the trivial and the sign characters.

The character table of S_3 is as shown in Table 4.6. It is obtained by writing down the lifted characters from the quotient S_3/A_3 to obtain the first two rows, and using the second orthogonality relations (C4) to obtain the third row.

Table 4.6 The character table of the symmetric group S_3.

(1^3)	$(2\ 1)$	(3)
1	1	1
1	-1	1
2	0	-1

This table can be used in two ways. Firstly, the quotient group of S_4, by the normal subgroup $N = \{1, (12)(34), (13)(24), (14)(23)\}$, is isomorphic to S_3. The character θ_3 is obtained by lifting the character of S_3 of degree 2. Secondly, characters of the subgroup S_3 may be induced to S_4. To calculate such an induced character, use N as a transversal. Using the definition of induced character in (D5), we see that inducing the trivial character of S_3 to S_4 gives a character ζ of degree 4 which is 2 on transpositions, takes the value 1 on 3-cycles, and is zero otherwise. It can be checked that the inner product of ζ with itself is 2, so that ζ is a sum of two irreducible characters. By Frobenius reciprocity, θ_1 is a constituent of ζ, so $\zeta-\theta_1$ is an irreducible character, which we call θ_4. The table may be completed using column orthogonality to determine θ_5. Alternatively, observe that for any one-dimensional representation λ of a group G, and any irreducible representation R with character θ, an irreducible representation λR may be defined by $(\lambda R)(g) = \lambda(g)R(g)$, with character $\lambda(g)\theta(g)$. Applying this with $\lambda = \theta_2$ and $\theta = \theta_4$ gives the character θ_5.

Now consider the group $\tilde{S}_4$. Using the results of Chapter 3, we see that $\tilde{S}_4$ has eight conjugacy classes. The two conjugacy classes of S_4 which do not split in $\tilde{S}_4$ are denoted $(2\ 1^2)$ and (2^2). The same notation is used for their inverse images in $\tilde{S}_4$. Primes are used to distinguish between the two classes in a pair of $\tilde{S}_4$-classes which arise by splitting from a single S_4-class. The first five rows of the character table are stolen from the table for S_4. The remaining three rows correspond to negative representations. By Theorem 3.6, the negative characters vanish on classes which do not split. The values of a negative character on a pair of classes arising from a splitting are negatives of each other.

Table 4.7 The character table of $\tilde{S}_4$.

	$(1^4)'$	$(1^4)''$	$(2\,1^2)$	(2^2)	$(31)'$	$(31)''$	$(4)'$	$(4)''$
ζ_1	1	1	1	1	1	1	1	1
ζ_2	1	1	-1	1	1	1	-1	-1
ζ_3	2	2	0	2	-1	-1	0	0
ζ_4	3	3	1	-1	0	0	-1	-1
ζ_5	3	3	-1	-1	0	0	1	1
ζ_6	2	-2	0	0	1	-1	$\sqrt{2}$	$-\sqrt{2}$
ζ_7	2	-2	0	0	1	-1	$-\sqrt{2}$	$\sqrt{2}$
ζ_8	4	-4	0	0	-1	1	0	0

Notice that once the character ζ_6 is determined, since it is not self-associate, we can immediately write down its associate character ζ_7. Then ζ_8 can be obtained using column orthogonality. The character ζ_6 is actually the character of the spin representation discussed in Theorem 2.8. This representation maps the generators t_i to the matrices T_i, where

$$T_1 = \begin{pmatrix} 0 & i \\ i & 0 \end{pmatrix}; \qquad T_2 = \frac{1}{2}\begin{pmatrix} 0 & -\sqrt{3}-i \\ \sqrt{3}-i & 0 \end{pmatrix};$$

and

$$T_3 = \frac{1}{\sqrt{3}}\begin{pmatrix} i\sqrt{2} & 1 \\ -1 & -i\sqrt{2} \end{pmatrix}.$$

Under the projection to S_4, these generators t_1, t_2 and t_3 map onto the transpositions (1 2), (2 3) and (3 4) respectively. Thus t_1t_2 and $t_1t_2t_3$ map onto a 3-cycle and 4-cycle respectively. Since

$$T_1T_2 = \frac{1}{2}\begin{pmatrix} 1+i\sqrt{3} & 0 \\ 0 & 1-i\sqrt{3} \end{pmatrix},$$

which has trace 1, and

$$T_1T_2T_3 = \frac{-1}{2\sqrt{3}}\begin{pmatrix} \sqrt{6}-i\sqrt{2} & -1-i\sqrt{3} \\ 1-i\sqrt{3} & \sqrt{6}+i\sqrt{2} \end{pmatrix},$$

which has trace $-\sqrt{2}$, we see that the character of this representation is ζ_6 as given in the table.

Remark. In the notation which will be used (and explained) later, ζ_6 will be denoted by <4>, ζ_7 by $<4>^a$, and ζ_8 by <3 1>.

To obtain the table of $\tilde{A}_4$, first note that A_4 has a quotient group which is cyclic of order 3, and so has three irreducible complex characters obtained by mapping the generator of the quotient group to either 1, ω or ω^2, where $\omega = e^{2\pi i/3}$. The remaining character of A_4 may be obtained by orthogonality, or by restricting the character θ_4 of S_4. As for the negative characters, restricting ζ_6 gives a self-conjugate character χ_5 of degree 2. Multiplying χ_5 by χ_2 and χ_3 produces χ_6 and χ_7.

Table 4.8 The character table of $\tilde{A}_4$.

	$(1^4)'$	$(1^4)''$	(2^2)	$(3\ 1)'_1$	$(3\ 1)''_1$	$(3\ 1)'_2$	$(3\ 1)''_2$
χ_1	1	1	1	1	1	1	1
χ_2	1	1	1	ω	ω	ω^2	ω^2
χ_3	1	1	1	ω^2	ω^2	ω	ω
χ_4	3	3	-1	0	0	0	0
χ_5	2	-2	0	1	-1	1	-1
χ_6	2	-2	0	ω	$-\omega$	ω^2	$-\omega^2$
χ_7	2	-2	0	ω^2	$-\omega^2$	ω	$-\omega$

Notes

1. The theory of group characters was developed by Frobenius at the turn of the century, the very special case of abelian groups having already had a long history of applications to number theory and probability; see Mackey (1980). A great deal of information about the structure of a finite group can be deduced from its character table. For example, the number of irreducible characters of degree 1 is equal to the index of the commutator subgroup. Thus a group is perfect if and only if the only character of degree 1 is the trivial character. The ATLAS of finite groups, Conway *et al.* (1985), includes character tables for many finite simple groups.

2. When we consider representations over an algebraically closed field of characteristic p, the situation is very much more complicated than that outlined at the beginning of Chapter 4, if p divides $|G|$. For example, in this p-modular context, two representations with the same composition factors need not be isomorphic. This is one reason why it is not true that the trace of such a p-modular representation determines the representation. A simpler way to see this is to note that the trace of a direct sum of p copies of any representation is zero everywhere. Brauer has a very elegant resolution of this second difficulty. The Brauer character of a p-modular representation, Brauer (1941), is a complex-valued class function defined on the conjugacy classes of p-regular elements of G (that is, on the elements of G of order coprime to p). It may be shown that two p-modular representations of G have the same Brauer character if and only if the representations have the same composition factors. It can also be shown that the number of irreducible p-modular representations is equal to the number of p-regular conjugacy classes of G. However, there is no straightforward test (such as calculating an inner product) for irreducibility of p-modular representations. Thus the determination of the table of irreducible Brauer characters is often very difficult. See Isaacs (1976) or Serre (1977) for an introduction to this theory.

5

A CONSTRUCTION FOR NEGATIVE REPRESENTATIONS

In this chapter, we shall give an operation which enables the irreducible negative representations of $G \,\tilde{\mathbf{Y}}\, H$ to be built from those of given objects G and H in $\mathcal{G}$. It is more convenient to express this construction in terms of modules than of representations. We therefore first recall the basic facts about G-modules and tensor products. The irreducible representations of a direct product $G \times H$ are then constructed from those of G and H. In the case of $G \,\tilde{\mathbf{Y}}\, H$, we define a twisted action on the tensor product, which is used to construct the negative modules. Its character values are then determined in terms of those of its factors.

Definition. Given a finite group G, a G-module is a finite dimensional (complex) vector space M with an action $G \times M \to M$, written as multiplication, such that, for g and h in G, for u and v in M, and for λ in $\mathbb{C}$:

(a) $g(u+v) = gu+gv$;
(b) $g(\lambda v) = \lambda(gv)$;
(c) $1v = v$; and
(d) $(gh)v = g(hv)$.

There is a natural $1-1$ correspondence between G-modules and representations of the group G. If M is a G-module, a linear endomorphism $R(g)$ of M is associated to the group element g by defining

$$R(g)(v) = gv \tag{5.1}$$

for all v in M. Conversely, given a representation R of G on M, formula (5.1) may be used to define an action of G on M.

Two G-modules M_1 and M_2 are *isomorphic* if and only if there is a vector space isomorphism θ from M_1 to M_2 such that, for all g in G and v in M_1,

$$\theta(gv) = g\theta(v) .$$

Such a map θ is called a G-module isomorphism. It is clear that M_1 and M_2 are isomorphic if and only if the representations associated with them are equivalent.

We next describe tensor products, and show how these determine the representations of a direct product. Given complex vector spaces U and V, their tensor product $U \otimes V$ is a vector space containing elements of the form $u \otimes v$ for u in U and v in V, although not all the elements of $U \otimes V$ can be written in this way. The function λ from $U \times V$ to $U \otimes V$, sending (u, v) to $u \otimes v$, is bilinear, and universal in the sense that, given any vector space W and any bilinear function $\beta : U \times V \to W$, there is a unique linear map $\gamma : U \otimes V \to W$ such that β is the composite $\gamma \circ \lambda$. It is easy to see that $U \otimes V$ is determined up to isomorphism by this property. Now we give an explicit construction for $U \otimes V$, in order to show that it exists, and also to provide a concrete object for purposes of calculation. Choose bases $\{u_1, u_2, ..., u_r\}$ and $\{v_1, v_2, ..., v_s\}$ of U and V respectively. Then $U \otimes V$ is the vector space with basis the rs symbols $u_i \otimes v_j$. Elements u in U and v in V have the forms $\sum \beta_i u_i$ and $\sum \gamma_j v_j$, respectively. We then define $u \otimes v$ by

$$u \otimes v = \sum \beta_i \gamma_j (u_i \otimes v_j) .$$

Note that a general element of $U \otimes V$ is a *sum* of elements of the form $u \otimes v$.

Now let M be a G-module and N be an H-module. The tensor product $M \otimes N$ of the vector spaces M and N may be given a $(G \times H)$-module structure by defining

$$(g, h)(u \otimes v) = gu \otimes hv.$$

This extends linearly to an action of $G \times H$ on all of $M \otimes N$. To check that this is well-defined, note that the map sending (u, v) to $gu \otimes hv$ is bilinear, and we have simply defined its unique linear extension. Taking the ordered basis for $M \otimes N$ to be

$$\{u_1 \otimes v_1, u_1 \otimes v_2, ..., u_1 \otimes v_s, u_2 \otimes v_1, ..., u_r \otimes v_s\} .$$

it follows that, if $R \otimes S$ is the representation associated with $M \otimes N$, then $(R \otimes S)(g, h)$ is the Kronecker product $R(g) \otimes S(h)$ (as occurs in Chapter 2). Thus, if χ and ξ are the characters of R and S respectively, then the character of $R \otimes S$ on (g, h) is $\chi(g)\xi(h)$. The following result is well-known (see V.10.3 of Huppert (1967) or Serre (1977), Theorem 10, p. 27). We include a proof to help motivate later developments.

Theorem 5.2. *Let G and H be finite groups. If M and N are irreducible modules for G and H respectively, then $M \otimes N$ is an irreducible $G \times H$-module, and every irreducible $G \times H$-module may be produced in this way. Furthermore, $M \otimes N$ is not isomorphic to $M' \otimes N'$, unless M is isomorphic to M' and N is isomorphic to N'.*

Proof. Let χ and ξ be the characters of M and N, so that the character of $\chi\otimes\xi$ is given by

$$(\chi\otimes\xi)(g, h) = \chi(g)\xi(h)\,.$$

For two such pairs,

$$\begin{aligned}
<\chi'\otimes\xi', \chi\otimes\xi> &= \frac{1}{|G\times H|}\sum(\chi'\otimes\xi')(g,h)\overline{(\chi\otimes\xi)}(g,h) \\
&= \frac{1}{|G\times H|}\sum\chi'(g)\bar{\chi}(g)\xi'(h)\bar{\xi}(h) \\
&= \left(\frac{1}{|G|}\sum\chi'(g)\bar{\chi}(g)\right)\left(\frac{1}{|H|}\sum\xi'(h)\bar{\xi}(h)\right) \\
&= <\chi', \chi><\xi', \xi>\,.
\end{aligned}$$

This is zero unless $\chi' = \chi$ and $\xi' = \xi$. In that case it follows that $<\chi\otimes\xi, \chi\otimes\xi> = 1$ and so $M\otimes N$ is irreducible. In general, it follows that $M'\otimes N'$ and $M\otimes N$ are distinct, since they have different characters, unless $M' \cong M$ and $N' \cong N$.

Finally, we show that all the irreducible characters of $G \times H$ have the form $\chi\otimes\xi$. Now by (C4) of Chapter 4,

$$\sum_{\chi}\chi(1)^2 = |G|\,,$$

where χ ranges over Irr(G), the set of all irreducible characters of G. Thus, letting ξ range over Irr(H),

$$\begin{aligned}
\sum_{\chi,\xi}(\chi\otimes\xi)(1)^2 &= \sum_{\chi}\chi(1)^2\sum_{\xi}\xi(1)^2 \\
&= |G|\,|H| \\
&= |G\times H|\,.
\end{aligned}$$

This completes the proof.

Next, let G and H be objects in $\mathcal{G}$ with σ being the zero map on G. Then the group $G \tilde{\times} H$ is simply the direct product $G \times H$, with which we have just dealt. In an obvious analogue of the ideas in Chapter 4, a module M for an object G in $\mathcal{G}$ is said to be *negative* if the element z acts by taking each v to $-v$.

Corollary 5.3. *Let G and H be objects in $\mathcal{G}$ with σ being the zero map on G. The irreducible $G \tilde{\times} H$-modules are all of the form $M\otimes N$, where M and N are irreducible modules for G and H respectively. The irreducible $G\, \tilde{\curlyvee}\, H$-modules are* *precisely those irreducible $G \tilde{\times} H$-modules which contain (z, z) in their kernels; that is, those of the form $M \otimes N$, where M and N are either both positive or both negative irreducible modules. The module $M \otimes N$ is negative provided both M and N are negative. Whether regarded as a module for $G \tilde{\times} H$ or for $G \,\tilde{\mathbf{Y}}\, H$, the module $M \otimes N$ is not isomorphic to $M' \otimes N'$, unless M is isomorphic to M' and N is isomorphic to N'.*

Proof. Since $G \tilde{\times} H = G \times H$, the only part remaining to be checked is the assertion about the modules for $G \,\tilde{\mathbf{Y}}\, H$. These are clearly the modules for $G \tilde{\times} H$ on which (z, z) acts as the identity. Thus M and N are either both negative or both positive. The negative modules for $G \,\tilde{\mathbf{Y}}\, H$ are the subset of these in which $(z, 1)Z$ acts as the negative of the identity map, and so M and N must both be negative. A $G \,\tilde{\mathbf{Y}}\, H$-module isomorphism is necessarily a $G \tilde{\times} H$-module isomorphism, proving the assertion about distinctness.

Now let G be an object in $\mathcal{G}$ in which σ is not the zero map, and let g_1 be an element of G with $\sigma(g_1) \neq 0$. Given a representation R of $\ker\sigma$, the induced representation $R \uparrow G$, as defined in (C4) of Chapter 4, takes the following form:

$$(R \uparrow G)(x) = \begin{pmatrix} \dot{R}(x) & \dot{R}(xg_1) \\ \dot{R}(g_1^{-1}x) & \dot{R}(g_1^{-1}xg_1) \end{pmatrix},$$

where $\dot{R}(g) = R(g)$ if g is in $\ker\sigma$, and $\dot{R}(g)$ is zero otherwise. To see this, one need only check that this is a representation which has the correct character, as given in (C4). Thus, if x is in $\ker\sigma$,

$$(R \uparrow G)(x) = \begin{pmatrix} R(x) & 0 \\ 0 & R(g_1^{-1}xg_1) \end{pmatrix},$$

and if x is not in $\ker\sigma$,

$$(R \uparrow G)(x) = \begin{pmatrix} 0 & R(xg_1) \\ R(g_1^{-1}x) & 0 \end{pmatrix}.$$

If χ is the character of R, then the character $R \uparrow G$ as given above has value on x given by

$$\dot{\chi}(x) + \dot{\chi}(g_1^{-1}xg_1),$$

where, by definition, $\dot{\chi}(g)$ is zero unless g is in $\ker\sigma$. This is the formula for $\chi \uparrow G$ given in (C4), as required. Note that this formula does not depend on the choice of g_1, since any other element of G not in $\ker\sigma$ may be written as $g_1 s$, for some s in $\ker\sigma$, and since

$$\dot{\chi}(g_1^{-1}xg_1) = \dot{\chi}(s^{-1}g_1^{-1}xg_1s)\,.$$

It will be convenient to have a description of inducing from $\ker\sigma$ to G in terms of modules. Let L be the $\ker\sigma$-module corresponding to R. A description of $L \uparrow G$, which avoids the need to choose g_1, can be given as follows. The module $L \uparrow G$ can be any G-module V for which there exists a vector space decomposition

$$V = L_0 \oplus L_1 \tag{5.4}$$

such that:

(i) L_0 and L_1 are $\ker\sigma$-submodules, with action determined by the restriction of the action of G on V, such that L_0 is isomorphic to L (as a $\ker\sigma$-module); and

(ii) if g_1 is an element of G which is not in $\ker\sigma$, then $g_1L_i \subseteq L_{i+1}$ (subscripts taken modulo 2).

This description even applies when σ is zero, provided that we take L_1 to be $\{0\}$. When σ is non-zero, it follows that L_1 is isomorphic to the conjugate of L as a $\ker\sigma$-module.

Let M and N be irreducible negative modules for given objects G and H respectively, in $\mathcal{G}$. The next task will be to define a module $M \mathbin{\widetilde{\otimes}} N$ for the group $G \mathbin{\tilde{\times}} H$. There will be some choices involved in this definition. In fact, $M \mathbin{\widetilde{\otimes}} N$ will only be well-defined up to replacement by its associate or by an isomorphic module. It will not be altered by replacing either of M or N by its associate or by an isomorphic module. Since both $(1, z)$ and $(z, 1)$ will act as multiplication by -1 on $M \mathbin{\widetilde{\otimes}} N$, the element (z, z) acts as the identity on $M \mathbin{\widetilde{\otimes}} N$. Hence $M \mathbin{\widetilde{\otimes}} N$ will be a negative $G \mathbin{\tilde{\mathbf{Y}}} H$-module. The operation $\widetilde{\otimes}$ will be commutative to the extent that, up to associates, the $H \mathbin{\tilde{\mathbf{Y}}} G$-module $N \mathbin{\widetilde{\otimes}} M$, when made into a $G \mathbin{\tilde{\mathbf{Y}}} H$-module using the isomorphism of Theorem 3.3 (b), will be isomorphic to $M \mathbin{\widetilde{\otimes}} N$. Although not strictly necessary, we shall restrict the definition to the case when M and N are both irreducible.

Definition. With G, H, M and N as above, let L be an irreducible constituent of the restriction of M to $\ker\sigma$. Let $L_0 \oplus L_1$ be a decomposition for $L \uparrow G$ as in (5.4). In all cases except when M is not self-associate and N is self-associate, define the twisted tensor product $M \mathbin{\widetilde{\otimes}} N$ to be the space $(L_0 \oplus L_1) \otimes N$ with action of $G \mathbin{\tilde{\times}} H$ as follows:

$$(g, h)(v \otimes w) = (-1)^{j\sigma(h)} gv \otimes hw,$$

for all v in L_j and w in N. Symmetrically, let K be an irreducible constituent of the restriction of N to $\ker\sigma$, and let $K_0 \oplus K_1$ be a decomposition for $K \uparrow H$. In all cases except when M is self-associate and N is not self-associate, define $M \mathbin{\tilde{\otimes}} N$ to be $M \otimes (K_0 \oplus K_1)$ with action determined by

$$(g, h)(v \otimes w) = (-1)^{j\sigma(g) + \sigma(g)\sigma(h)} gv \otimes hw,$$

for all v in M and w in K_j.

Remarks. (1) If σ_G is the zero map, then $L = L_0 = M$ and $L_1 = \{0\}$. Thus $M \mathbin{\tilde{\otimes}} N$ reduces to the usual tensor product of representations. A similar remark applies when σ_H is zero.

(2) When σ_G (or σ_H) is non-zero, there are at most two choices for L (or K), and these are conjugates of each other. We shall see later how these choices affect the isomorphism class of $M \mathbin{\tilde{\otimes}} N$. If M is not self-associate, both M and its associate M^a restrict to L, and so the modules $M \mathbin{\tilde{\otimes}} N$ and $M^a \mathbin{\tilde{\otimes}} N$ are isomorphic.

(3) When M and N are both self-associate or when neither is self-associate, we have given two definitions of $M \mathbin{\tilde{\otimes}} N$. The character computations given after the proof of Theorem 5.6 below show that these definitions agree up to associates and isomorphism.

Proposition 5.5. *With G, H, M and N as above, $M \mathbin{\tilde{\otimes}} N$ is a $G \mathbin{\tilde{\times}} H$-module. If M and N are both negative modules, then (z, z) is in the kernel of $M \mathbin{\tilde{\otimes}} N$, so that $M \mathbin{\tilde{\otimes}} N$ is also a $G \mathbin{\tilde{\Upsilon}} H$-module.*

Proof. The case when either σ_G or σ_H is the zero map is covered by Corollary 5.3. When both are non-zero, proceed as follows. The formulae for the action of (g, h) are linear in both v and w, so extend uniquely to all of $M \mathbin{\tilde{\otimes}} N$. In the first case of the definition, take v in $L_0 \cup L_1$. Let $\tau(v)$ be 0 or 1 (modulo 2) according as v is in L_0 or L_1 respectively. It follows by (i) and (ii) after (5.4) that

$$\tau(gv) = \sigma(g) + \tau(v) .$$

Now for elements g, g' in G and h, h' in H,

$$(g, h)\big((g', h')(v \otimes w)\big) = (g, h)\big((-1)^{\tau(v)\sigma(h')} g'v \otimes h'w\big)$$
$$= (-1)^{\tau(v)\sigma(h') + \tau(g'v)\sigma(h)} gg'v \otimes hh'w ,$$

and

$$\big((g, h)(g', h')\big)(v \otimes w) = (z^{\sigma(g')\sigma(h)} gg', hh')(v \otimes w)$$

$$= (-1)^{\sigma(g')\sigma(h)+\tau(v)\sigma(hh')} gg'v \otimes hh'w \,.$$

It remains only to note that

$$\tau(v)\sigma(h')+\tau(g'v)\sigma(h) = \sigma(g')\sigma(h)+\tau(v)\sigma(hh') \,.$$

The proof in the second case of the definition is similar and is left to the reader. When M and N are both negative, $(z, 1)$ and $(1, z)$ both act as -1 whereas $(1, 1)$ and (z, z) both act as $+1$.

The reader may ask why it is necessary to define $M \mathbin{\tilde{\otimes}} N$ in two cases. The answer is contained in the next result, where we show that the module $M \mathbin{\tilde{\otimes}} N$ is always irreducible. If either of the two formulae for $M \mathbin{\tilde{\otimes}} N$ had been used exclusively as the definition, the module $M \mathbin{\tilde{\otimes}} N$ would sometimes be reducible.

Theorem 5.6. *Let M and N be irreducible negative modules for objects G and H, respectively. Then the module $M \mathbin{\tilde{\otimes}} N$ is irreducible as a representation of either $G \mathbin{\tilde{\curlyvee}} H$ or of $G \mathbin{\tilde{\times}} H$.*

Proof. It suffices to prove irreducibility with respect to $G \mathbin{\tilde{\times}} H$. We shall do so for the first case of the definition of $\tilde{\otimes}$, the other case being exactly analogous. If σ_G is zero, then $M \mathbin{\tilde{\otimes}} N$ is the usual tensor product of modules, and so is irreducible by Proposition 5.3. Assume therefore that σ_G is non-zero.

If $\sigma(g) \neq 0$, then $gL_j \subset L_{j+1}$. It follows from the definition of $M \mathbin{\tilde{\otimes}} N$ that the matrix representing (g, h) has two zero matrix blocks down its main diagonal. Thus, when $\sigma(g) \neq 0$,

$$\chi_{M \tilde{\otimes} N}(g, h) = 0 \qquad (1)$$

(where $\chi_V(x)$ denotes the value at x of the character associated with the G-module V).

Let S be the matrix representation of $\ker\sigma$ corresponding to the module L in the definition of $\tilde{\otimes}$. Fix an element g_1 of G with $\sigma(g_1) = 1$. Then

$$(S \uparrow G)(g) = \begin{pmatrix} \dot{S}(g) & \dot{S}(gg_1) \\ \dot{S}(g_1^{-1}g) & \dot{S}(g_1^{-1}gg_1) \end{pmatrix}.$$

If $\sigma(g) = 0$, the fact that

$$(g, h)(v \otimes w) = (-1)^{\tau(v)\sigma(h)} gv \otimes hw$$

means that the matrix representing (g, h) has the form

$$\begin{pmatrix} S(g) \otimes R(h) & 0 \\ 0 & (-1)^{\sigma(h)} S(g_1^{-1} g g_1) \otimes R(h) \end{pmatrix}.$$

Thus, when $\sigma(g) = 0$,

$$\chi_{M \tilde{\otimes} N}(g, h) = \left(\chi_L(g) + (-1)^{\sigma(h)} \chi_L(g_1^{-1} g g_1)\right) \chi_N(h) . \tag{2}$$

Combining (1) and (2),

$$<\chi_{M \tilde{\otimes} N}, \chi_{M \tilde{\otimes} N}> = |G \tilde{\times} H|^{-1} \sum_{(g,\, h) \in G \tilde{\times} H} |\chi_{M \tilde{\otimes} N}(g,h)|^2$$

$$= (2|\ker\sigma_G \times H|)^{-1} \sum_{(g,h) \in \ker\sigma \times H} |\chi_L(g) + (-1)^{\sigma(h)} \chi_L(g_1^{-1} g g_1)|^2 \, |\chi_N(h)|^2$$

$$= \left\{ (2|\ker\sigma_G|)^{-1} \sum_{\sigma(g)=0} |\chi_L(g) + \chi_L(g_1^{-1} g g_1)|^2 \right\} \left\{ |H|^{-1} \sum_{\sigma(h)=0} |\chi_N(h)|^2 \right\}$$

$$+ \left\{ (2|\ker\sigma_G|)^{-1} \sum_{\sigma(g)=0} |\chi_L(g) - \chi_L(g_1^{-1} g g_1)|^2 \right\} \left\{ |H|^{-1} \sum_{\sigma(h)=1} |\chi_N(h)|^2 \right\}$$

$$= AB + CD$$

say, where A, B, C and D are the factors inside the brackets { }.

We must prove that $AB + CD = 1$. This divides into three cases.

For the first two cases, assume that M is self-associate. By Theorem 4.2, the restriction of M to $\ker\sigma$ has two irreducible constituents, namely L and its conjugate L^c. These two are distinct. Furthermore,

$$\chi_L(g_1^{-1} g g_1) = \chi_L^c(g) .$$

Thus

$$A = 2^{-1} <\chi_L + \chi_L^c, \chi_L + \chi_L^c> = 1 ,$$

since χ_L and χ_L^c are distinct irreducible characters. Similarly,

$$C = 2^{-1} <\chi_L - \chi_L^c, \chi_L - \chi_L^c> = 1 .$$

For the first case, assume that N is self-associate (as well as M). Then $\chi_N(h) = 0$ if $\sigma(h) = 1$. Thus $D = 0$, and the summation in B may be taken over all h in H, yielding

$$B = <\chi_N, \chi_N>_H = 1 .$$

Thus $AB + CD = 1{\cdot}1 + 1{\cdot}0 = 1$, as required.

For the second case, assume N is not self-associate (and M is self-associate). Then the restriction of N to $\ker\sigma_H$ is irreducible, so

$$\begin{aligned} B &= 2^{-1}|\ker\sigma_H|^{-1}\sum_{\ker\sigma} |\chi_N(h)|^2 \\ &= 2^{-1}<\chi_N\downarrow\ker\sigma, \chi_N\downarrow\ker\sigma> \\ &= 1/2\,. \end{aligned}$$

This also shows

$$\begin{aligned} \sum_{\sigma(h)=1} |\chi_N(h)|^2 &= \sum_H |\chi_N(h)|^2 - \sum_{\ker\sigma} |\chi_N(h)|^2 = |H| - |\ker\sigma| \\ &= |H|/2\,, \end{aligned}$$

yielding $D = 1/2$. Thus

$$AB + CD = (1)(1/2) + (1)(1/2) = 1\,,$$

as required.

For the final case, M is not self-associate, and so N is also not self-associate, since we are dealing with the first case of the definition of $\widetilde{\otimes}$. Here L is the restriction of M to $\ker\sigma$, and $\chi_L(g_1^{-1}gg_1) = \chi_L(g)$. Thus $C = 0$, and

$$\begin{aligned} A &= (2|\ker\sigma_G|)^{-1}\sum_{\sigma(g)=0} |2\chi_L(g)|^2 \\ &= 2|\ker\sigma_G|^{-1}\sum_{\ker\sigma} |\chi_L(g)|^2 \\ &= 2<\chi_L, \chi_L>_{\ker\sigma} \\ &= 2\,. \end{aligned}$$

Since N is not self-associate, the previous paragraph shows that $B = 1/2$. Thus $AB + CD = (2)(1/2) + 0 = 1$, as required.

This completes the proof.

Remark. This is a suitable point at which to summarize the values for the character of $M \,\widetilde{\otimes}\, N$. These are recorded Table 5.7, in which SA and NSA are abbreviations for "self-associate" and "not self-associate" respectively. Also χ_L and χ_L^c are the irreducible constituents of the restriction of χ_M to $\ker\sigma$ when M is self-associate, with χ_K and χ_K^c those for χ_N, when N is self-associate. This applies in the case of non-zero σ. When σ_G is zero, $\chi_L = \chi_M$ and $\chi_L^c := 0$; similarly for (H, N, K). The entries in the first, second and fourth columns are gleaned from the calculations in the proof of Theorem 5.6. By carrying out the analogous calculations for the second case in the definition of $\widetilde{\otimes}$, the entries in the third column can be found, as well as confirming the first and fourth columns. Alternatively, apply the isomorphism $G \mathbin{\tilde{\curlyvee}} H \to H \mathbin{\tilde{\curlyvee}} G$ of 3.3(b).

Table 5.7 Table of values of the character $\chi_{M \widetilde{\otimes} N}$.

	M	SA	SA	NSA	NSA
	N	SA	NSA	SA	NSA
	$M \widetilde{\otimes} N$	SA	NSA	NSA	SA
	$\dim(M \widetilde{\otimes} N)$	$\dim(M)\dim(N)$	$\dim(M)\dim(N)$	$\dim(M)\dim(N)$	$2\dim(M)\dim(N)$
$\chi_{M \widetilde{\otimes} N}(g,h)$	$\sigma(g)=0=\sigma(h)$	$\chi_M(g)\chi_N(h)$	$\chi_M(g)\chi_N(h)$	$\chi_M(g)\chi_N(h)$	$2\chi_M(g)\chi_N(h)$
	$\sigma(g)=0\neq\sigma(h)$	0	$(\chi_L(g)-\chi_{L}^{c}(g))\chi_N(h)$	0	0
	$\sigma(g)\neq 0=\sigma(h)$	0	0	$\chi_M(g)(\chi_K(h)-\chi_{K}^{c}(h))$	0
	$\sigma(g)=1=\sigma(h)$	0	0	0	0

Remarks. (1) Note that $M \widetilde{\otimes} N$ is self-associate if and only if $\chi_{M \widetilde{\otimes} N}(g, h) = 0$ for all g and h such that $\sigma(g) \neq \sigma(h)$ (as indicated in the first and last columns). This is in agreement with remark (ii) before Theorem 4.2.

(2) If σ is the zero map on G, we take L to be M and L^c to be the zero subspace. In Table 5.7 only the two columns apply in which M is self-associate, and only the rows with $\sigma(g) = 0$. (Similarly if σ is the zero map on H.) The table collapses to one row and one column if both σ_G and σ_H are zero.

(3) When defining $M \widetilde{\otimes} N$, choices were made for the constituent L of the restriction of M to $\ker\sigma$, and also for K. The table shows that changing these choices would at most replace the constructed module by its associate.

(4) The definition of $M \widetilde{\otimes} N$ was given in two parts. Table 5.7 shows that the constructed modules have the same character (and are therefore isomorphic up to associates) in the overlapping cases when M and N are both self-associate and when neither is self-associate. The two parts of the definition are therefore compatible.

Definition. Let G be an object in $\mathcal{G}$. Denote by $\mathcal{C}(G)$ the set of self-associate irreducible negative characters of G. Denote by $\mathcal{D}(G)$ the set of unordered pairs $\{\chi, \chi^a\}$ of associated irreducible negative characters of G for which $\chi \neq \chi^a$. Then, as in Chapter 4, let the cardinalities of $\mathcal{C}(G)$ and $\mathcal{D}(G)$ be $c(G)$ and $d(G)$ respectively.

Note that $d=0$ if σ is the zero map, since then every representation is self-associate. The following is a rephrasing of Proposition 4.1.

Proposition 5.8. *With the above notation,*

$$|G|/2 = \sum_{\chi \in \mathscr{C}(G)} \chi(1)^2 + 2 \sum_{\{\chi, \chi^a\} \in \mathscr{D}(G)} \chi(1)^2 .$$

We can now describe the irreducible negative representations of $G \tilde{\curlyvee} H$.

Theorem 5.9. *Let G and H be objects in $\mathscr{G}$. Then all the irreducible negative representations of $G \tilde{\curlyvee} H$ are constructed by using the twisted tensor product $\tilde{\otimes}$. We have two distinct (associate) irreducible modules $M \tilde{\otimes} N$ when exactly one of M or N is self-associate, and one such module otherwise. Thus*

$$c(G \tilde{\curlyvee} H) = c(G)c(H) + d(G)d(H) ,$$

and

$$d(G \tilde{\curlyvee} H) = c(G)d(H) + d(G)c(H) .$$

Proof. If σ is zero on either G or H, then $G \tilde{\times} H$ is just the direct product $G \times H$ and the result is identical to Theorem 5.3. We may therefore suppose that σ is non-zero on both G and H. Suppose that $M_1 \tilde{\otimes} N_1$ and $M_2 \tilde{\otimes} N_2$ are isomorphic $G \tilde{\curlyvee} H$-modules, up to associates. Choose an element h_0 in $\ker \sigma_H$ such that $\chi_{N_1}(h_0) \neq 0$. By inspecting the row labelled "$\sigma(g) = 0 = \sigma(h)$" in Table 5.7, it follows that there is a number s such that, for all $g \in \ker \sigma_G$,

$$\chi_{M_1}(g) = s\chi_{M_2}(g) ,$$

namely

$$s = \chi_{N_1}(h_0)^{-1} \chi_{N_2}(h_0) .$$

Setting $g = 1$ yields that s is a positive rational number. Since M_1 and M_2 are irreducible, $s = 1$, so M_1 and M_2 restrict to isomorphic modules for $\ker \sigma$. It follows that they are isomorphic up to associates. Symmetrically, N_1 and N_2 are also isomorphic up to associates. We therefore obtain

$$c(G)c(H) + d(G)d(H)$$

self-associate irreducible characters for $G \tilde{\curlyvee} H$, and

$$c(G)d(H) + d(G)c(H)$$

pairs which are not self-associate.

It only remains to show that these are all the irreducibles for $G \tilde{\curlyvee} H$. Let the degrees of the representations in $\mathscr{C}(G)$, $\mathscr{D}(G)$, $\mathscr{C}(H)$, and $\mathscr{D}(H)$ be denoted respectively by e_i, e_i', f_i and f_i'. Thus, by Proposition 5.8,

$$|G|/2 = \sum_{1 \le j \le c(G)} (e_j)^2 + 2 \sum_{1 \le j \le d(G)} (e_j')^2 ,$$

and similarly for H. The dimension over $\mathbb{C}$ of $M \,\tilde{\otimes}\, N$ is $(\dim M)(\dim N)$ except when neither is self-associate, when its dimension is $2(\dim M)(\dim N)$. Calculating the sums of the squares of the dimensions of the modules $M \,\tilde{\otimes}\, N$ gives, with summations over $1 \le j \le c(G)$, $1 \le k \le c(H)$, $1 \le \ell \le d(G)$, $1 \le m \le d(H)$:

$$\begin{aligned}
&\sum (e_j f_k)^2 + \sum (2e_\ell' f_m')^2 + 2\sum (e_j f_m')^2 + 2\sum (e_\ell' f_k)^2 \\
&= \left(\sum (e_j)^2 + 2\sum (e_\ell')^2\right)\left(\sum (f_k)^2 + 2\sum (f_m')^2\right) \\
&= (|G|/2)(|H|/2) \\
&= |G \,\tilde{\curlyvee}\, H|/2 .
\end{aligned}$$

By Proposition 5.8 the $M \,\tilde{\otimes}\, N$ are all the irreducible negative representations of $G \,\tilde{\curlyvee}\, H$.

Remarks. (1) Notice that the description of the irreducible representations of $G \,\tilde{\curlyvee}\, H$ requires a knowledge of the irreducible representations of $\ker \sigma_G$ and $\ker \sigma_H$ as well as those of G and H. Equivalently, one needs to know what the associates of all the irreducibles for G and H are.

(2) As we have already noted, when $N \,\tilde{\otimes}\, M$ is converted from an $H \,\tilde{\curlyvee}\, G$-module to a $G \,\tilde{\curlyvee}\, H$-module by defining

$$(g, h)u = (-1)^{\sigma(g)\sigma(h)}(h, g)u ,$$

it is isomorphic to $M \,\tilde{\otimes}\, N$. There is also an "associative" law for $\tilde{\otimes}$. Let P be a third negative module, for an object F in $\mathcal{G}$. By iteration from Table 5.7 one can calculate the character values of $(M \,\tilde{\otimes}\, N) \,\tilde{\otimes}\, P$ and of $M \,\tilde{\otimes}\, (N \,\tilde{\otimes}\, P)$, for the objects $(G \,\tilde{\curlyvee}\, H) \,\tilde{\curlyvee}\, F$ and $G \,\tilde{\curlyvee}\, (H \,\tilde{\curlyvee}\, F)$. The character values at $((g, h), f)$ agree with those at $(g, (h, f))$. Thus these two modules are isomorphic up to associates if we identify the groups using the natural isomorphism given in part (c) of Theorem 3.3. In future, brackets will be omitted from both $\tilde{\curlyvee}$-products and $\tilde{\otimes}$-products with more than two factors.

In the final result of this chapter, the effect on characters of restriction from $G \,\tilde{\curlyvee}\, H$ to $\ker \sigma$ will be determined.

Theorem 5.10. *Assume that σ is non-zero on both G and H. Then the restriction of $M \,\tilde{\otimes}\, N$ to* $\ker \sigma$ *is irreducible if one of M or N is self-associate and the other is not. Otherwise,*

$$(\chi_M \mathbin{\widetilde{\otimes}} \chi_N) \downarrow \ker\sigma = \beta + \beta^c$$

for an irreducible β *such that:*

(*i*) *if M and N are both self-associate, then*

$$(\beta - \beta^c)(g, h) = \begin{cases} \pm(\chi_L - \chi_L^c)(g)(\chi_K - \chi_K^c)(h) & \text{if } \sigma(g) = 0 = \sigma(h); \\ 0 & \text{if } \sigma(g) = 1 = \sigma(h); \end{cases}$$

(*ii*) *if neither M nor N is self-associate, then*

$$(\beta - \beta^c)(g, h) = \begin{cases} 0 & \text{if } \sigma(g) = 0 = \sigma(h); \\ \pm 2i\chi_M(g)\chi_N(h) & \text{if } \sigma(g) = 1 = \sigma(h). \end{cases}$$

(*Combined with Table 5.7, these determine* β *and* β^c, *up to switching their names.*)

Proof. By Theorem 4.2, if $M \mathbin{\widetilde{\otimes}} N$ is not self-associate, its restriction to $\ker\sigma$ is irreducible. This restriction is a sum of two irreducibles when $M \mathbin{\widetilde{\otimes}} N$ is self-associate. If M and N are both self-associate, it is easy to check that the subspaces

$$(L \otimes K) \oplus (L' \otimes K') \text{ and } (L \otimes K') \oplus (L' \otimes K)$$

of $M \mathbin{\widetilde{\otimes}} N$ are invariant under the action by those elements (g, h) with $\sigma_G(g) = \sigma_H(h)$, i.e. those in the kernel of $\sigma_{G \curlyvee H}$. It follows that these modules must be the irreducible constituents of the restriction. This proves all but the character formula in (ii).

When neither M nor N is self-associate, their restrictions to the appropriate index two subgroups are irreducible. Let

$$L_0 = \{(1, m) : m \in M\},$$

and

$$L_1 = \{(-1, m) : m \in M\}.$$

Define the action of an element $g \in G$ on $L_0 \oplus L_1$ by

$$g\big((-1)^k, m\big) = \big((-1)^{k+\sigma(g)}, gm\big).$$

It is clear that $L_0 \oplus L_1$ is a G-module; that L_0 and L_1 are $\ker\sigma$-submodules; that L_0 is isomorphic to the restriction of M to $\ker\sigma$; and that the action of elements outside $\ker\sigma$ interchanges L_0 and L_1. It follows by (5.4) that $L_0 \oplus L_1$ is isomorphic to $L \uparrow G$, where $L = M \downarrow \ker\sigma$. The definition of $\widetilde{\otimes}$ gives the action on $(L_0 \oplus L_1) \mathbin{\widetilde{\otimes}} N$ as

$$(g, h)\{((-1)^k, m) \otimes n\} = (-1)^{k\sigma(h)}((-1)^{k+\sigma(g)}, gm)\otimes hn\,.$$

Consider the subspace V of $M \tilde{\otimes} N$ spanned by the vectors of the form

$$v(m, n) = (1, m)\otimes n + i(-1, m)\otimes n$$

where $i^2 = -1$. If $\sigma(g) = 0 = \sigma(h)$, then

$$\begin{aligned}(g, h)v(m, n) &= (g, h)[(1, m)\otimes n + i(-1, m)\otimes n]\\ &= (1, gm)\otimes hn + i(-1, gm)\otimes hn\\ &= v(gm, hn)\,.\end{aligned}$$

If $\sigma(g) = 1 = \sigma(h)$, then

$$\begin{aligned}(g, h)v(m, n) &= (g, h)[(1, m)\otimes n + i(-1, m)\otimes n]\\ &= (-1, gm)\otimes hn - i(1, gm)\otimes hn\\ &= -iv(gm, hn)\,.\end{aligned}$$

Thus the space V is invariant under $\ker\sigma$. The same holds for the space W spanned by vectors of the form $(1, m)\otimes n - i(-1, m)\otimes n$. It follows that V and W are the irreducible $\ker\sigma$-submodules of $M \tilde{\otimes} N$. Thus β and β^c are χ_V and χ_W. If now A is the matrix for the action of g on M, and B is that for the action of h on N, then the above calculations show that the matrix for the action of (g, h) on the subspace V above is

$$(-i)^{\sigma(g)\sigma(h)}A\otimes B\,.$$

Thus

$$\chi_V(g, h) = (-i)^{\sigma(g)\sigma(h)}\chi_M(g)\chi_N(h)\,.$$

Similarly,

$$\chi_W(g, h) = i^{\sigma(g)\sigma(h)}\chi_M(g)\chi_N(h)\,.$$

Subtracting completes the proof.

Example. As an illustration of Theorems 5.9 and 5.10, we calculate the irreducible characters of $\tilde{S}_4 \,\tilde{\curlyvee}\, \tilde{S}_4$, and of its index 2 subgroup $\ker\sigma$. Recall from Chapter 4 that $\tilde{S}_4$ has three irreducible negative representations, of which one is self-associate and the other pair are associates. The irreducible negative characters are determined by the following table:

	1	8	6
	1	3	4
<4>	4	-1	0
<3 1>	2	1	$\sqrt{2}$
<3 1>^{a}	2	1	$-\sqrt{2}$

Here, the integers in the first row represent the numbers of elements in certain conjugacy classes, and those in the second row the orders of those elements. The rows give the values of the irreducible negative characters on half of the set of those conjugacy classes for which at least one of these characters takes a non-zero value. The missing values are precisely the negatives of those given. Using this notation, the values of the irreducible negative characters of $\tilde{A}_4$ are as follows, where <4> $\downarrow \tilde{A}_4 = \beta + \beta^c$:

	1	4	4
	1	3	3
β	2	ω	ω^2
β^c	2	ω^2	ω
<3,1> $\downarrow$	2	1	1

where ω denotes a primitive cube root of unity. Using Table 5.7 and Theorem 5.9, the table of irreducible negative characters of $\tilde{S}_4 \,\tilde{\mathbf{Y}}\, \tilde{S}_4$ is as shown:

	1	8	8	64	48	48
	1	3	3	3	12	12
<4> $\tilde{\otimes}$ <4>	16	-4	-4	1	0	0
<3,1> $\tilde{\otimes}$ <3,1>	8	4	4	2	0	0
<4> $\tilde{\otimes}$ <3,1>	8	4	-2	-1	$(\omega-\omega^2)\sqrt{2}$	0
(<4> $\tilde{\otimes}$ <3,1>)^{a}	8	4	-2	-1	$(\omega^2-\omega)\sqrt{2}$	0
<3,1> $\tilde{\otimes}$ <4>	8	-2	4	-1	0	$(\omega-\omega^2)\sqrt{2}$
(<3,1> $\tilde{\otimes}$ <4>)^{a}	8	-2	4	-1	0	$(\omega^2-\omega)\sqrt{2}$

Finally, we use Theorem 5.10 to obtain the characters of the index two subgroup, $\ker\sigma$, of $\tilde{S}_4 \,\tilde{\mathbf{Y}}\, \tilde{S}_4$:

	1	8	8	32	32	36
	1	3	3	3	3	4
β_1	4	2	2	1	1	$2i$
β_1^c	4	2	2	1	1	$-2i$
β_2	8	-2	-2	2	-1	0
β_2^c	8	-2	-2	-1	2	0
$(<4> \tilde{\otimes} <3,1>)\downarrow$	8	4	-2	-1	-1	0
$(<3,1> \tilde{\otimes} <4>)\downarrow$	8	-2	4	-1	-1	0

where

$$(<4> \tilde{\otimes} <4>)\downarrow \ker\sigma = \beta_2 + \beta_2^c;$$

$$(<3,1> \tilde{\otimes} <3,1>)\downarrow \ker\sigma = \beta_1 + \beta_1^c.$$

Notes

The operation $\widetilde{\otimes}$ is based on an operation given in Hoffman and Humphreys (1986), and defined in Appendix 8 of this book, where it is denoted $\boxtimes^-$. By dealing with $\mathbb{Z}/2$-graded modules, $\boxtimes^-$ recovers the interdeterminacy up to associates which is built into $\widetilde{\otimes}$. An earlier version of $\widetilde{\otimes}$ was defined in Humphreys (1988) over algebraically closed fields of arbitrary characteristic, and more generally for p-fold covering groups. It was used to study blocks of certain modular representations. Theorems 5.6 and 5.9 are given as Theorem 2.24 in Hoffman and Humphreys (1986). Essentially equivalent results are (5.1), (5.2), (5.3) and (5.4) in Jozefiak (1989). Independently, Stembridge (1989, Theorem 4.3) proves this result when G and H both have the form $\tilde{S}_n$, so that $\tilde{\mathbf{Y}}$ (and $\widetilde{\otimes}$) can be obtained by manipulations within Clifford algebras (and Clifford modules, respectively).

6

THE BASIC REPRESENTATION

In this chapter we return to the basic negative representation of $\tilde{S}_n$, defined in the proof of Theorem 2.8. This representation is shown to be irreducible and its character is determined. It is then used, together with the operation $\tilde{\otimes}$ of Chapter 5, to produce an irreducible negative representation of each object $\tilde{S}_a \tilde{\curlyvee} \tilde{S}_b \tilde{\curlyvee} \ldots$. The character values and restriction to $\ker\sigma$ of the latter are then calculated.

The technicalities of these calculations have a certain appeal, but, on first reading, some may prefer to look at only what will be used later. This consists of the initial part of the chapter up to Theorem 6.2, the last two theorem statements, 6.11 and 6.12, together with the definitions preceding them and the examples after. The notes at the end refer to other methods of calculation, both in this book and in the literature.

For each integer $m \geq 0$, we produced, in the proof of Theorem 2.8, matrices $M_1, \ldots, M_{2m+1}$ of degree 2^m. Their definitions will be recalled in the proof of the next result, but subsequently it will be only the relations stated in the proposition which will be needed. For each subset J of $\{1, 2, \ldots, 2m\}$, let $M_J = M_{j_1} \cdots M_{j_t}$, where $j_1, \ldots, j_t$ are the elements of J in increasing order. Thus, for the empty set $\varnothing$, we have $M_\varnothing = I$, the identity matrix of degree 2^m. Denote the trace of a matrix A as $\operatorname{tr} A$.

Proposition 6.1. *For fixed m, we have*:

(*i*) $M_j^2 = -I$ *and* $M_k M_j = -M_j M_k$ *if* $j \neq k$;

(*ii*) $M_J M_K = \pm M_{(J \setminus K) \cup (K \setminus J)}$ *for subsets J and K of* $\{1, \ldots, 2m\}$;

(*iii*) $M_1 M_2 \cdots M_{2m+1} = i^{m+1} I$;

(*iv*) *if J is not empty, then* $\operatorname{tr} M_J = 0$;

(*v*) *the* M_J, *as J varies over subsets of* $\{1, \ldots, 2m\}$, *form a basis for the space of all complex matrices of degree* 2^m.

Proof. To recall the definition of M_k, let

$$A := \begin{pmatrix} 0 & i \\ i & 0 \end{pmatrix}, \quad B := \begin{pmatrix} 0 & -1 \\ 1 & 0 \end{pmatrix}, \quad C := \begin{pmatrix} 1 & 0 \\ 0 & -1 \end{pmatrix},$$

where $i^2 = -1$. In Theorem 2.8, M_k was defined as a Kronecker product

$$M_k = M_{k1} \otimes M_{k2} \otimes \cdots \otimes M_{km}$$

where M_{kj} is the (k, j)th matrix in the following array:

$$\begin{array}{cccccccccc}
C & C & C & C & \ldots & C & C & C & A \\
C & C & C & C & \ldots & C & C & C & B \\
C & C & C & C & \ldots & C & C & A & I \\
C & C & C & C & & C & C & B & I \\
C & C & C & C & & C & A & I & I \\
\cdot & \cdot & & & & & & & \cdot \\
\cdot & \cdot & & & & & & & \cdot \\
\cdot & \cdot & & & & & & & \cdot \\
C & A & & & & & & & \cdot \\
C & B & I & I & & I & I & I & I \\
A & I & I & I & & I & I & I & I \\
B & I & I & I & \ldots & I & I & I & I \\
iC & C & C & C & \ldots & C & C & C & C
\end{array}$$

There are m columns and $2m+1$ rows. Multiplying two such iterated Kronecker products is done by iterating the rule $(S \otimes T)(S' \otimes T') = (SS') \otimes (TT')$. We have

$$A^2 = B^2 = -I; \quad C^2 = I; \quad AB = iC = -BA;$$
$$AC = -CA; \quad BC = -CB.$$

(i) The relation $M_k^2 = -I$ follows because each row has exactly one entry squaring to $-I$, the others giving I. The relation $M_k M_j = -M_j M_k$ follows because, for each pair of rows, one pair of corresponding entries anticommute and all other pairs commute.

(ii) This is an easy exercise using (i).

(iii) Multiplying the entire pth column (from the right) from top to bottom produces $iC^{2p} = iI$, except that the leftmost column produces $i^2 I$. Now use the fact that the Kronecker product of identity matrices is an identity matrix.

(iv) Here we use the fact that $\operatorname{tr}(S \otimes T) = \operatorname{tr} S \operatorname{tr} T$. If

$$J = \{j_1, \ldots, j_t\} \subset \{1, \ldots, 2m\},$$

we obtain M_J by using the rows at levels j_s, multiplying all the entries in each column, and tensoring the resulting products. Consider the column furthest to the right in which one of rows $j_1, ..., j_t$ contains something other than I. This column produces one of A, or B, or AB, each of which has trace zero.

(v) The number of M_J is the dimension of the space of matrices, so we must prove linear independence. Suppose $\sum s_J M_J = 0$. By (ii), $M_J M_K = \pm M_L$, where $L \neq \emptyset$ if $K \neq J$. Thus, for each K, by (iv)

$$0 = \mathrm{tr}\left(\sum s_J M_J M_K\right) = \pm s_K \mathrm{tr} M_\emptyset = \pm 2^m s_K .$$

Definition. Given $n > 2$, the *basic* representation R_n of $\tilde{S}_n$ is the complex representation determined by writing $n = 2m+1$ or $2m+2$ for $m \geq 1$, and defining

$$R_n(t_k) = (2k)^{-1/2}\left[(k+1)^{1/2} M_k - (k-1)^{1/2} M_{k-1}\right]$$

for $1 \leq k < n$, where M_k is the previous matrix of degree 2^m. (Obviously, M_0 can be anything.) The matrix $R_n(t_k)$ was called T_k in the proof of Theorem 2.8.

Using only 6.1 (i), this is easily shown to give a representation (see the proof of Theorem 2.8). Its degree is $2^{\lceil (n-1)/2 \rceil}$ where $\lceil \ \rceil$ is the "rounding down" (or "integer part") function. For $n = 1$ and 2, these formulae are interpreted as reducing to the unique negative one-dimensional representations for which, when $n = 2$, the generator t_1 maps to i.

Theorem 6.2. *The basic representation R_n is an irreducible representation of $\tilde{S}_n$.*

Proof. Given $m \geq 1$, we shall prove this simultaneously for $n = 2m+1$ and $2m+2$. Let T be the group generated by $M_1, ..., M_{2m}$. By 6.1 (i) and (v), each element of T is uniquely of the form $\pm M_J$ for some $J \subset \{1, ..., 2m\}$. Thus T has order 2^{2m+1}. Since T is a group of matrices, it is a representation of itself, the "tautological" representation. Its character has value $\pm 2^m$ at $\pm I$, and zero elsewhere, by 6.1 (iv). Thus, its inner product with itself is

$$\left[(2^m)^2 + (-2^m)^2\right]/2^{2m+1} = 1 ,$$

and so the tautological representation is irreducible. Each matrix $M_1, ..., M_{2m}$ is a linear combination of $\{R_{2m+1}(t_1), ..., R_{2m+1}(t_{2m})\}$. Thus R_{2m+1} is also irreducible, since $\{t_1, ..., t_{2m}\}$ generates $\tilde{S}_{2m+1}$. Clearly, the restriction of R_{2m+2} to $\tilde{S}_{2m+1}$ is R_{2m+1}, so R_{2m+2} is an irreducible representation of $\tilde{S}_{2m+2}$.

The next task is to compute the character of R_n.

Lemma 6.3. *For any matrices*

$$C = \sum c_J M_J \quad \text{and} \quad D = \sum d_J M_J ,$$

with summations over $J \subset \{1, ..., 2m\}$ and coefficients c_J and d_J in $\mathbb{C}$, suppose that C and D are disjoint in the sense that $c_J d_J = 0$ for all non-empty J. Then

$$\operatorname{tr}(CD) = 2^{-m} \operatorname{tr} C \operatorname{tr} D .$$

Proof. By Proposition 6.1 (iv),

$$\operatorname{tr} C = 2^m c_\varnothing \text{ and } \operatorname{tr} D = 2^m d_\varnothing .$$

By disjointness and 6.1 (ii),

$$CD = \sum_{J,K} c_J d_K M_J M_K = c_\varnothing d_\varnothing I + \sum_{\varnothing \neq J \neq K \neq \varnothing} \pm c_J d_K M_{(J \setminus K) \cup (K \setminus J)} .$$

Using 6.1 (iv) again,

$$\operatorname{tr}(CD) = c_\varnothing d_\varnothing \operatorname{tr}(I) = 2^m c_\varnothing d_\varnothing = 2^{-m} \operatorname{tr} C \operatorname{tr} D .$$

Definition. For $1 \le a \le b \le 2m+2$, define

$$t(a, b) = t_a t_{a+1} \cdots t_{b-1} \in \tilde{S}_{2m+2} ,$$

so that $t(a, a) = 1$ and $t(a, b) \in \tilde{S}_{2m+1}$ for $b < 2m+2$. The projection of $t(a, b)$ in S_{2m+2} is the cycle $(a\; a+1 \ldots b)$ of length $b-a+1$, so $\sigma t(a, b) \equiv b-a \pmod 2$.

Proposition 6.4. *For $1 \le a < b \le 2m+1$, we have (with $J \subset \{1, ..., 2m\}$, so that there is no J-th term in the summation for those J containing 0 when $a = 1$),*

$$R_{2m+2}[t(a, b)] = \sum_{J \subset \{a-1, ..., b-1\}} s_J M_J ,$$

for coefficients s_J which satisfy:

(*i*) $s_{\{a-1,...,b-1\}} = 0$ *if* $a > 1$;

(*ii*) $s_{\{a,...,b-1\}} = 2^{(a-b)/2} (b/a)^{1/2}$;

(*iii*) *if $b-a$ is even, then $s_J = 0$ for odd $|J|$, and $s_\varnothing = 2^{(a-b)/2}$*;

(*iv*) *if $b-a$ is odd, then $s_J = 0$ for even $|J|$, and $s_{\{b-1\}} = 2^{(a-b+1)/2} b^{1/2} (2b-2)^{-1/2}$.*

Proof. This is a straightforward calculation by induction on $b-a$. When $b = a+1$, it is immediate from the definition of $R_n(t_a)$. For the inductive step,

$$R_n[t(a, b)] = R_n[t(a, b-1)]R_n(t_{b-1})$$
$$= \sum_{K\subset\{a-1,...,b-2\}} r_K M_K (2b-2)^{-1/2}\left[b^{1/2}M_{b-1} - (b-2)^{1/2}M_{b-2}\right],$$

where the r_K satisfy (i) to (iv) with (b, s, J) replaced by $(b-1, r, K)$. By 6.1 (ii), both $M_K M_{b-1}$ and $M_K M_{b-2}$ have the form $\pm M_J$ for $J \subset \{a-1, ..., b-1\}$. To prove (i), note that in fact $J = (K\backslash\{b-1\}) \cup (\{b-1\}\backslash K)$ in the case of $M_K M_{b-1}$, so $J \neq \{a-1, ..., b-1\}$ if $K \neq \{a-1, ..., b-2\}$. The case of $M_K M_{b-2}$ is even easier. As for (ii), the coefficient of $M_{\{a, ..., b-1\}}$ is obtained from $M_K M_{b-1}$ for $K = \{a,...,b-2\}$ and is

$$2^{(a-b+1)/2}[(b-1)/a]^{1/2}(2b-2)^{-1/2}b^{1/2} = 2^{(a-b)/2}(b/a)^{1/2},$$

as required. To prove (iii), assume $b-a$ is even. Then $r_K = 0$ for even $|K|$, and if $|K|$ is odd, then $|(K\backslash\{x\}) \cup (\{x\}\backslash K)|$ is even for all x, proving the first part. Since $(K\backslash\{x\}) \cup (\{x\}\backslash K)$ is non-empty for $K \subset \{a-1, ..., b-2\}$ and $x = b-1$ or $b-2$, except for $K = \{b-2\}$ and $x = b-2$, we obtain

$$s_\varnothing = (2b-2)^{-1/2}(-1)(b-2)^{1/2}(-1)2^{(a-b+2)/2}(b-1)^{1/2}(2b-4)^{-1/2} = 2^{(a-b)/2},$$

as required. The proof of (iv) is similar but simpler.

Lemma 6.5. *If* $1 \leq a_1 < b_1 < a_2 < \ldots < b_\ell \leq 2m+2 = n$, *then*

$$R_n[t(a_1, b_1)t(a_2, b_2)\ldots t(a_{\ell-1}, b_{\ell-1})] \quad \textit{and} \quad R_n[t(a_\ell, b_\ell)]$$

are disjoint, as defined in 6.3.

Proof. Let Q and S respectively be these two matrices. Since $b_{\ell-1} < 2m-1$, we can multiply the expressions given in Proposition 6.4 for $R_n[t(a_i, b_i)]$ for $1 \leq i \leq \ell-1$, to obtain, for some r_K,

$$Q = \sum_{K\subset\{a_1-1,...,b_{\ell-1}-1\}} r_K M_K .$$

First assume $b_\ell < 2m+2$. Then 6.4 also applies to S, giving the result, since $\{a_1-1, ..., b_{\ell-1}-1\} \cap \{a_\ell-1, ..., b_\ell-1\}$ is empty. On the other hand, if $b_\ell = 2m+2$, using 6.4,

$$S = \sum_{J \subset \{a_\ell - 1, \ldots, 2m\}} s_J M_J (\alpha M_{2m+1} + \beta M_{2m})$$

for certain α and β given by $R_n(t_{2m+1})$. The products $M_J M_{2m}$ give M_L for sets L disjoint from those K occurring in Q. The products $M_J M_{2m+1}$ are multiples of $M_1 M_2 \ldots M_{2m} M_J = M_L$ (say), by 6.1 (iii). We can assume $J \neq \{a_\ell - 1, \ldots, 2m\}$ by 6.4 (i). Then if $a_\ell - 1 \leq i \leq 2m$ and $i \in J$, we have $i \in L$. Thus $L \neq K$ for any K occurring in Q. This completes the proof.

Definition. The basic character χ_n is the character of R_n.

Proposition 6.6. *Given a sequence* $1 \leq a_1 < b_1 < a_2 < \ldots < b_\ell \leq 2m+2$, *where* $\ell \geq 1$, *and where* $(a_1, b_1) \neq (1, 2m+2)$ *if* $\ell = 1$, *we have*

$$\chi_{2m+2}[t(a_1, b_1) \ldots t(a_\ell, b_\ell)] = \begin{cases} 2^{m - \sum (b_i - a_i)/2} & \textit{if all } b_i - a_i \textit{ are even;} \\ 0 & \textit{otherwise.} \end{cases}.$$

Proof. Proceed by induction on ℓ. The inductive step is immediate from 6.5 and 6.3: since

$$\chi_{2m+2}[t(a_1, b_1) \ldots t(a_\ell, b_\ell)] = \operatorname{tr}(QS) = 2^{-m} \operatorname{tr} Q \operatorname{tr} S$$

with Q and S as given in 6.5, one applies the inductive hypothesis to $\operatorname{tr} Q$. The initial case is divided in two.

For $\ell = 1$ and $b_1 < 2m+2$, by 6.4 and 6.1 (iv), $\chi_n[t(a, b)] = s_\varnothing \operatorname{tr}(I)$, which is zero for $b-a$ odd by 6.4 (iv), and, for $b-a$ even, by 6.4 (iii), is

$$s_\varnothing 2^m = 2^{(a-b)/2} 2^m = 2^{m-(b-a)/2},$$

as required.

For $\ell = 1$ and $b_1 = 2m+2$, we have $a_1 > 1$ by assumption. Then, by 6.4,

$$\begin{aligned} & R_{2m+2}[t(a_1, b_1)] \\ &= R_{2m+2}[t(a_1, 2m+1)] R_{2m+2}(t_{2m+2}) \\ &= \sum_{J \subset \{a_1 - 1, \ldots, 2m\}} s_J M_J (4m+2)^{-1/2} \left[(2m+2)^{1/2} M_{2m+1} - (2m)^{1/2} M_{2m} \right]. \end{aligned}$$

By 6.1 (iv), we need to calculate the coefficient of $M_\varnothing = I$ after multiplying. But $M_J M_{2m+1} \neq \pm I$ for any J in the summation (since $s_J = 0$ when $a_1 = 2$ and

$J = \{1, \ldots, 2m\}$ by 6.4 (i)); and $M_J M_{2m} = \pm I$ only for $J = \{2m\}$. If a_1 is odd, then $s_{\{2m\}} = 0$ by 6.4 (iii). If a_1 is even,

$$s_{\{2m\}} = 2^{(a_1 - 2m)/2}(2m+1)^{1/2}(4m)^{-1/2} .$$

Multiplying this by $(4m+2)^{-1/2}(2m)^{1/2}\operatorname{tr} I$ yields $2^{(a_1/2)-1}$, as required.

The case omitted in 6.6 is given next.

Proposition 6.7. *For all m,*

$$\chi_{2m+2}[t(1, 2m+2)] = i^{m+1}(m+1)^{1/2} .$$

Proof. Using 6.4 as in the last part of the previous proof,

$$\begin{aligned} &\chi_{2m+2}[t(1, 2m+2)] \\ &= \operatorname{tr} \sum_{J \subset \{1,\ldots,2m\}} s_J M_J (4m+2)^{-1/2}\Big[(2m+2)^{1/2} M_{2m+1} - (2m)^{1/2} M_{2m}\Big] , \end{aligned}$$

where $s_J = 0$ for odd $|J|$, and $s_{\{1,\ldots,2m\}} = 2^{-m}(2m+1)^{1/2}$. For even $|J|$, we have $M_J M_{2m} = \pm M_K$ for some $K \neq \varnothing$; and $M_J M_{2m+1}$ is a multiple of $M_J M_1 \ldots M_{2m} = \pm M_K$, where $K \neq \varnothing$ unless $J = \{1, 2, \ldots, 2m\}$. We obtain

$$\begin{aligned} &\operatorname{tr}\Big[2^{-m}(2m+1)^{1/2}(4m+2)^{-1/2}(2m+2)^{1/2} M_1 M_2 \ldots M_{2m} M_{2m+1}\Big] \\ &\quad = 2^{-m} 2^{-1/2}(2m+2)^{1/2} \operatorname{tr}(i^{m+1} I) \\ &\quad = i^{m+1}(m+1)^{1/2} , \end{aligned}$$

as required.

From this point onwards, it will be necessary to be somewhat more formal and technical concerning sequences and partitions.

Definitions. By a *partition*, we mean a sequence $\lambda \in \mathbb{Z}^\ell$ for some $\ell \geq 0$ such that λ is weakly decreasing and has positive terms or *parts*; i.e.

$$\lambda_1 \geq \lambda_2 \geq \ldots \geq \lambda_\ell > 0 .$$

The *weight*, $|\lambda|$, of λ is $\lambda_1 + \lambda_2 + \ldots + \lambda_\ell$, so we say that λ is a partition of $|\lambda|$. The *length* of λ is ℓ, sometimes denoted $\ell(\lambda)$. The unique partition of length zero also has weight zero. It is referred to as the empty partition, and often denoted by $\varnothing$. In a few instances, it will be convenient to identify the partition λ

with any other (finite or infinite) sequence of the form $\lambda_1,\ldots,\lambda_\ell,0,0,\ldots$. For this reason, we shall refrain from referring to the length of a *sequence* α, but instead use the phrases $\alpha \in \mathbb{Z}^k$ or $\alpha \in \mathbb{N}^k$. Define

$$\begin{aligned}
\mathcal{P} &:= \{\lambda : \lambda \text{ is a partition}\} ;\\
\mathcal{P}(n) &:= \{\lambda \in \mathcal{P} : |\lambda| = n\} ;\\
\mathcal{P}^0 &:= \{\lambda \in \mathcal{P} : \text{each } \lambda_i \text{ is odd}\} .
\end{aligned}$$

The partition λ is *odd* if and only if the number of even parts in λ is odd, and is *even* if and only if it is not odd. Thus, the parity of a permutation agrees with the parity of its cycle type. The parity of λ is also the parity of the integer $|\lambda| + \ell(\lambda)$. Define

$$\begin{aligned}
\mathcal{P}^0(n) &:= \mathcal{P}(n) \cap \mathcal{P}^0 ;\\
\mathcal{D} &:= \{\lambda \in \mathcal{P} : \lambda_i \neq \lambda_j \text{ if } i \neq j\} .
\end{aligned}$$

The partition λ is *strict* if and only if $\lambda \in \mathcal{D}$; that is, its parts are distinct.

$$\mathcal{D}(n) := \mathcal{P}(n) \cap \mathcal{D} .$$

Except when otherwise noted, whenever an order relation $<$ on $\mathcal{P}(n)$ and its subsets $\mathcal{D}(n)$ and $\mathcal{P}^0(n)$ is referred to, we mean the *reverse lexicographic order*. Thus $\lambda < \mu$ means $\lambda_m > \mu_m$ for

$$m = \min\{i : \lambda_i \neq \mu_i\} .$$

The following theorem is a recording of the character calculations in 6.6 and 6.7, weakened by not specifying signs. This is all we need for starting the induction to obtain the later generalization, parts (i) and (ii) of 6.11. Part (iii) of 6.11 contains the precise statement concerning signs which is needed later, and that is where the more accurate work in 6.6 will be employed.

Theorem 6.8. *Let $g \in \tilde{S}_n$ project to cycle type μ in S_n, where $\mu \in \mathcal{P}(n)$ has length ℓ.*

(*i*) *If n is odd, then χ_n is self-associate, and*

$$\chi_n(g) = \begin{cases} \pm 2^{(\ell-1)/2} & \text{if } \mu \in \mathcal{P}^0(n); \\ 0 & \text{otherwise.} \end{cases}$$

(*ii*) *If n is even, then χ_n is not self-associate, and*

$$\chi_n(g) = \begin{cases} \pm 2^{(\ell/2)-1} & \textit{if } \mu \in \mathcal{P}^0(n); \\ \pm i^{n/2}(n/2)^{1/2} & \textit{if } \mu = (n); \\ 0 & \textit{otherwise.} \end{cases}$$

Proof. Let n be $2m+1$ or $2m+2$. Then we need only calculate $\chi_{2m+2}(g)$, where

$$g = t(1, \mu_1)t(\mu_1+1, \mu_1+\mu_2)t(\mu_1+\mu_2+1, \mu_1+\mu_2+\mu_3)\ldots t(n-\mu_\ell+1, n)\,.$$

This is done in 6.6 and 6.7, giving the required values. The other statements are immediate from the fact that a character is self-associate if and only if its value on all odd elements is zero. See Remark (ii) before Theorem 4.2.

By Theorem 4.2, when n is even, the restriction of R_n to the index two subgroup $\tilde{A}_n$ is irreducible. When n is odd, R_n restricts to $\tilde{A}_n$ as a sum of two irreducible conjugate representations. Their characters are given by the following theorem.

Theorem 6.9. *If $n = 2m+1$ with $m > 0$, then the restriction of the basic representation R_n to $\tilde{A}_n$ is a sum of two irreducible representations. Their characters agree except on elements of $\tilde{A}_n$ which project to n-cycles. On these two conjugacy classes, the difference of their values is $\pm i^m(2m+1)^{1/2}$. (Since χ_n, their sum, is given by the previous theorem, this determines them.)*

Proof. Let $H = iM_{2m+1}$, so that $H^2 = I$, and $HM_k = -M_kH$ for $1 \le k \le 2m$. Since the minimal polynomial of H is x^2-1, the eigenvalues of H are 1 and -1. Since H has trace zero, the multiplicities of 1 and -1 as eigenvalues are equal. Let C be the change of basis matrix given by the eigenvectors of H, so that

$$H' = C^{-1}HC = \begin{pmatrix} I & 0 \\ 0 & -I \end{pmatrix},$$

where I denotes the identity matrix of degree 2^{m-1}.

For any matrix A of degree 2^m, let $A' = C^{-1}AC$. Then for $1 \le j \le 2m$,

$$H'R_n(t_j)' = -R_n(t_j)'H'\,.$$

Thus any product of an even number of the matrices $R_n(t_j)'$ commutes with H'. It can be easily checked that the matrices which commute with H' have the form

$$\begin{pmatrix} X & 0 \\ 0 & Y \end{pmatrix},$$

where X and Y are matrices of degree 2^{m-1}. Thus

$$R_n(d)' = \begin{pmatrix} X(d) & 0 \\ 0 & Y(d) \end{pmatrix}$$

for all elements d in $\tilde{A}_n$. Taking traces, we see that

$$\chi_n(d) = \beta(d) + \beta^c(d) ,$$

where β and β^c are the traces of X and Y. Thus

$$\begin{aligned}(\beta - \beta^c)(d) &= \operatorname{tr}\begin{pmatrix} X(d) & 0 \\ 0 & -Y(d) \end{pmatrix} \\ &= \operatorname{tr}\left[H' R_n(d)'\right] = \operatorname{tr}\left[iM_{2m+1} R_n(d)\right].\end{aligned}$$

Let $d_0 = t(1,2m+1) = t_1 t_2 \ldots t_{2m}$, which projects to an n-cycle in S_n. By 6.4,

$$R_n(d_0) = \sum_{J \subset \{1,\ldots,2m\}} s_J M_J ,$$

where $s_{\{1,\ldots,2m\}} = 2^{-m}(2m+1)^{1/2}$. Since $M_1 M_2 \ldots M_{2m}$ is a multiple of M_{2m+1} by 6.1 (iii), unless $J = \{1, \ldots, 2m\}$, the product $M_{2m+1} M_J$ is a multiple of M_K for some non-empty K, and $\operatorname{tr} M_K = 0$ for non-empty K, by 6.1 (iv). Thus for $d_0 = t(1, 2m+1)$,

$$\begin{aligned}(\beta - \beta^c)(d_0) &= \operatorname{tr}\left[iM_{2m+1} 2^{-m}(2m+1)^{1/2} M_1 \ldots M_{2m}\right] \\ &= i2^{-m}(2m+1)^{1/2} \operatorname{tr}(M_1 \ldots M_{2m} M_{2m+1}) \\ &= i2^{-m}(2m+1)^{1/2} \operatorname{tr}(i^{m+1} I) \\ &= -i^m (2m+1)^{1/2} .\end{aligned}$$

It remains to prove that β^c agrees with β on all those d in $\tilde{A}_n$ which do not project to an n-cycle. Now

$$\begin{aligned}&(n!)^{-1} \sum_{d \in \tilde{A}_n} |\beta(d) - \beta^c(d)|^2 \\ &= \langle\beta, \beta\rangle - \langle\beta, \beta^c\rangle - \langle\beta^c, \beta\rangle + \langle\beta^c, \beta^c\rangle = 2 ,\end{aligned}$$

since β and β^c are distinct irreducible characters of $\tilde{A}_n$. But the number of n-cycles in A_n is clearly $(n-1)!$, so there are $(n-1)!2$ elements d for which

$$|\beta(d)-\beta^c(d)|^2 = |\pm i^m(2m+1)^{1/2}|^2 = 2m+1 = n\,.$$

Summing over only these elements above yields $(n!)^{-1}(n-1)!2n = 2$, and so $\beta(d) = \beta^c(d)$ for all other elements d, as required, completing the proof.

We shall now use the characters χ_n and the operation $\widetilde{\otimes}$ of Chapter 5 to produce new characters. For example, for any positive k and ℓ, neither χ_{2k} nor $\chi_{2\ell}$ is self-associate, so $\chi_{2k} \widetilde{\otimes} \chi_{2\ell}$ is a self-associate irreducible character of $\widetilde{S}_{2k} \,\tilde{\curlyvee}\, \widetilde{S}_{2\ell}$. Its restriction to $\ker\sigma$, a group which we shall call $\widetilde{A}_{2k,\,2\ell}$, is therefore the sum of two conjugate irreducible constituents.

Lemma 6.10. *The constituents of* $(\chi_{2k} \widetilde{\otimes} \chi_{2\ell}) \downarrow \widetilde{A}_{2k,2\ell}$ *agree on all classes, except those of elements* (g, h), *where* g *and* h *project to* $(2k)$- *and* (2ℓ)-*cycles respectively. On such elements, their difference is* $\pm 2i^{k+\ell+1}(k\ell)^{1/2}$.

Proof. This is immediate from Theorem 5.10 (ii), substituting the character value given in 6.8 (ii), middle case.

Definition. Normally λ will denote a strict partition, but for convenience in the next proof, it will denote any finite sequence of *positive* integers, not necessarily distinct, say $\lambda = (\lambda_1,...,\lambda_r) \in \mathbb{Z}^r$. Its weight, $|\lambda|$, and its parity are defined exactly as for partitions. A *refinement*, μ^*, of λ is a sequence $(\mu^{(1)},...,\mu^{(r)})$ such that each $\mu^{(i)}$ is a partition with $|\mu^{(i)}| = \lambda_i$. Define the object

$$\widetilde{S}_\lambda := \widetilde{S}_{\lambda_1} \,\tilde{\curlyvee} \dots \tilde{\curlyvee}\, \widetilde{S}_{\lambda_r}$$

of $\mathcal{G}$. By Theorem 3.4, $\widetilde{S}_\lambda$ has a natural embedding as a sub-object of $\widetilde{S}_{|\lambda|}$. The quotient group $\widetilde{S}_\lambda/\{1,z\}$ is isomorphic to $S_\lambda = S_{\lambda_1} \times \dots \times S_{\lambda_r}$, a so-called Young subgroup of $S_{|\lambda|}$. The conjugacy classes in S_λ are in $1-1$ correspondence with refinements μ^* of λ. Such a conjugacy class in S_λ determines in $S_{|\lambda|}$ the conjugacy class of cycle type μ, where μ is the partition obtained by rearranging the juxtaposed sequence $\mu^{(1)}\mu^{(2)}\dots\mu^{(r)}$. The set of all refinements μ^* of λ which produce a given μ is denoted $\mathrm{Ref}(\lambda, \mu)$.

Definition. For each $\lambda \in \mathbb{Z}^r$ with all $\lambda_i > 0$, define

$$\chi_\lambda := \chi_{\lambda_1} \widetilde{\otimes} \chi_{\lambda_2} \widetilde{\otimes} \dots \widetilde{\otimes} \chi_{\lambda_r}\,,$$

an irreducible character of $\widetilde{S}_\lambda$. By the remarks before the definition of $\widetilde{\otimes}$ and after Theorem 5.9, χ_λ is well-defined up to replacement by its associate, and is independent of the order of the terms in λ, modulo the identification of the different groups involved via the isomorphism of Theorem 3.3 (b). By the iteration of Theorem 3.3 (c), $\tilde{S}_\lambda$ is well-defined without using brackets when $r > 2$, and can be identified with $\tilde{S}_{\lambda_1,\ldots,\lambda_i} \tilde{\curlyvee} \tilde{S}_{\lambda_{i+1},\ldots,\lambda_r}$ for each i. In other words, if $\mu \sqcup \nu$ denotes the juxtaposition of sequences μ and ν, we can write $\tilde{S}_{\mu\sqcup\nu} = S_\mu \tilde{\curlyvee} \tilde{S}_\nu$. The associativity of $\tilde{\otimes}$ in Remark (2) after Theorem 5.9 can then be written as $\chi_{\mu\sqcup\nu} = \chi_\mu \tilde{\otimes} \chi_\nu$.

Using Table 5.7, the character χ_λ is self-associate if and only if an even number of χ_{λ_i} are not self-associate. By Theorem 6.8 this is true when an even number of λ_i are even integers; that is, when the partition obtained by rearranging λ is even. When $\ell(\lambda) > 1$, the character χ_λ is only well-defined up to replacement by its associate. Thus its values on even elements are well defined (this is relevant to part (iii) of the theorem below), but its values on odd elements are only defined up to a change of sign.

Theorem 6.11. *Let $\lambda \in \mathbb{Z}^r$ be a sequence of positive integers, and let μ^* be a refinement of λ. Let $g \in \tilde{S}_\lambda$ project to an element in S_λ whose associated conjugacy class corresponds to μ^*. Let $\mu \in \mathcal{P}(|\lambda|)$ be the partition obtained by rearranging the juxtaposed sequence $\mu^{(1)}\mu^{(2)}\ldots\mu^{(r)}$ into weakly decreasing order. Suppose μ has length s.*

(i) If λ is even (i.e. $|\lambda|+r$ is an even integer), then χ_λ is self-associate, and

$$\chi_\lambda(g) = \begin{cases} \pm 2^{(s-r)/2} & \textit{if } \mu \in \mathcal{P}^0(|\lambda|); \\ 0 & \textit{otherwise.} \end{cases}$$

(ii) If λ is odd, then χ_λ is not self-associate, and

$$\chi_\lambda(g) = \begin{cases} \pm 2^{(s-r-1)/2} & \textit{if } \mu \in \mathcal{P}^0(|\lambda|); \\ \pm i^{(|\lambda|-r+1)/2}(\lambda_1\lambda_2\ldots\lambda_r/2)^{1/2} & \textit{if } \mu = \lambda; \\ 0 & \textit{otherwise.} \end{cases}$$

(iii) When $\mu \in \mathcal{P}^0(|\lambda|)$, the real number $\chi_\lambda(g)$ has the same sign as $\chi_{|\lambda|}(g)$.

Proof. The proofs of (i) and (ii) are by simultaneous induction on r. When $r = 1$, they reduce to Theorem 6.8. The inductive steps will make repeated use of Table 5.7.

(i) Consider first the case in which some rearrangement of λ is the juxtaposition of two proper subsequences λ' and λ'' which are both even. Then, using the commutative property of $\tilde{\otimes}$, we can replace λ by this rearrangement, and write $\chi_\lambda = \chi_{\lambda'} \tilde{\otimes} \chi_{\lambda''}$. Since $\chi_{\lambda'}$ and $\chi_{\lambda''}$ are both self-associate, the first column of Table 5.7 applies. The same rearrangement applied to μ^* produces refinements of λ' and λ'' with underlying partitions μ' and μ'', say. The element g may be written as $(g', g'') \in S_{\lambda'} \tilde{\curlyvee} S_{\lambda''}$, and, by Table 5.7, $\chi_\lambda(g) = 0$ unless $\sigma(g')$ and $\sigma(g'')$ are zero. When g' and g'' are both even, $\chi_\lambda(g)$ equals $\chi_{\lambda'}(g')\chi_{\lambda''}(g'')$. Applying the inductive hypothesis now yields the result. The required arithmetic arises when μ' and μ'' are both in $\mathcal{P}^0$. It is as follows, with r', r'', s', s'' denoting the lengths of $\lambda', \lambda'', \mu', \mu''$ respectively:

$$2^{(s'-r')/2}\,2^{(s''-r'')/2} = 2^{(s-r)/2},$$

since $r'+r'' = r$ and $s'+s'' = s$.

Alternatively, suppose no rearrangement of λ is a juxtaposition of two even proper subsequences. Since a single odd part or a pair of even parts give even sequences, λ must have the form $(2k, 2\ell)$, and $\chi_\lambda = \chi_{2k} \,\widetilde{\otimes}\, \chi_{2\ell}$. Since neither χ_{2k} nor $\chi_{2\ell}$ is self-associate, the last column of Table 5.7 applies. As in the case above, it remains to deal with $\chi_\lambda(g', g'')$ where g' and g'' project to cycle types μ' and μ'' in $\mathcal{P}^0(2k)$ and $\mathcal{P}^0(2\ell)$ of lengths s' and s'', respectively. Using Table 5.7 and the inductive hypothesis relevant to part (ii),

$$\begin{aligned}\chi_\lambda(g) &= 2\chi_{2k}(g')\chi_{2\ell}(g'') \\ &= 2[\pm 2^{(s'-2)/2}][\pm 2^{(s''-2)/2)}], \\ &= \pm 2^{(s-2)/2}\end{aligned}$$

as required, since $s = s'+s''$.

(ii) Again we divide into two cases, the first being the case when λ has at least one odd part, say $2k+1$. We may take $\lambda_r = 2k+1$ and $\lambda' = (\lambda_1,...,\lambda_{r-1})$, and write $\chi_\lambda = \chi_{\lambda'} \,\widetilde{\otimes}\, \chi_{2k+1}$, so that χ_{2k+1} is self-associative and $\chi_{\lambda'}$ is not. The element g has the form $(g', g'') \in \tilde{S}_{\lambda'} \tilde{\curlyvee} \tilde{S}_{2k+1}$. Let g' and g'' project to elements whose conjugacy classes in $S_{\lambda'}$ and S_{2k+1} give partitions μ' and μ'' respectively. The third column of Table 5.7 applies to $\chi_{\lambda'} \,\widetilde{\otimes}\, \chi_{2k+1}$. Suppose that μ is even and $\chi_\lambda(g) \neq 0$. By the table, μ' and μ'' are both even and

$$\chi_\lambda(g) = \chi_{\lambda'}(g')\chi_{2k+1}(g'').$$

By the inductive hypotheses for (ii) and (i) respectively, both μ' and μ'' are in $\mathcal{P}^0$. If their lengths are s' and s'' respectively, induction gives

$$\chi_\lambda(g) = \{\pm 2^{[s'-(r-1)-1]/2}\}\{\pm 2^{(s''-1)/2}\} = \pm 2^{(s-r-1)/2}$$

as required, since $s'+s'' = s$. Now suppose instead that μ is odd and $\chi_\lambda(g) \neq 0$. Table 5.7 implies that μ' is odd and μ'' is even, and that

$$\chi_\lambda(g) = \chi_{\lambda'}(g')(\beta - \beta^c)(g''),$$

where $\chi_{2k+1} \downarrow \tilde{A}_{2k+1}$ splits as $\beta + \beta^c$. Since $\chi_{\lambda'}(g') \neq 0$, the inductive hypothesis for (ii) implies that $\mu' = \lambda'$ and that

$$\begin{aligned}\chi_{\lambda'}(g') &= \pm i^{(|\lambda'| - (r-1)+1)/2}(\lambda_1 \ldots \lambda_{r-1}/2)^{1/2} \\ &= \pm i^{(|\lambda| - r - 2k - 1)/2}(\lambda_1 \ldots \lambda_{r-1}/2)^{1/2}.\end{aligned}$$

Since $(\beta - \beta^c)(g'') \neq 0$, Theorem 6.9 implies that $\mu'' = (2k+1)$ and

$$(\beta - \beta^c)(g'') = \pm i^k(2k+1)^{1/2} = \pm i^k \lambda_r^{1/2}.$$

Multiplying gives the required value, completing the induction when λ has an odd part.

Finally, suppose λ consists entirely of even parts, of which there are an odd number. Here we write

$$\chi_\lambda = \chi_{\lambda'} \tilde{\otimes} \chi_{2k,2\ell}; \quad g = (g', g'') \in \tilde{S}_{\lambda'} \curlyvee \tilde{S}_{2k,2\ell},$$

and so $\chi_{\lambda'}$ is not self-associate, and again the third column of Table 5.7 applies. Suppose that $\chi_\lambda(g) \neq 0$ and μ is even. As before, the table implies that μ' and μ'' are even, and by the inductive hypothesis they are both in $\mathcal{P}^0$. The relevant arithmetic is

$$2^{[s'-(r-2)-1]/2}\, 2^{(s''-2)/2} = 2^{(s-r-1)/2}.$$

Suppose instead that $\chi_\lambda(g) \neq 0$ and μ is odd. Again, in parallel with the previous paragraph, μ' is odd, μ'' is even, and

$$\chi_\lambda(g) = \chi_{\lambda'}(g')[\beta - \beta^c](g''),$$

where $\chi_{2k,2\ell} \downarrow \tilde{A}_{2k,2\ell}$ splits as $\beta + \beta^c$. By induction, $\mu' = \lambda' = (\lambda_1, \ldots, \lambda_{r-2})$ and

$$\chi_{\lambda'}(g') = \pm i^{(|\lambda'| - r + 3)/2}(\lambda_1 \ldots \lambda_{r-2}/2)^{1/2}.$$

By Lemma 6.10, $\mu'' = (2k, 2\ell) = (\lambda_{r-1}, \lambda_r)$ and

$$(\beta - \beta^c)(g'') = \pm 2 i^{k+\ell+1}(k\ell)^{1/2}.$$

Multiplying yields the required result.

(iii) Given $\mu \in \mathcal{P}^0(|\lambda|)$ and $\mu^* \in \mathrm{Ref}(\lambda, \mu)$, it suffices to find a particular $g \in \tilde{S}_\lambda$ in the conjugacy class indexed by μ^* for which both $\chi_{|\lambda|}(g)$ and $\chi_\lambda(g)$ are positive. Let $\mu^* = (\mu^{(1)}, \ldots, \mu^{(r)})$ with

$$\mu^{(i)} = (\mu_1^{(i)},...,\mu_{\ell_i}^{(i)}) \in \mathcal{P}(|\lambda_i|)\,.$$

Let

$$g_i = t(1, \mu_1^{(i)})t(\mu_1^{(i)}+1, \mu_1^{(i)}+\mu_2^{(i)})\ldots \in \tilde{S}_{\lambda_i}\,,$$

and

$$g = (g_1, ..., g_\ell) \in \tilde{S}_{\lambda_1}\, \tilde{\mathbf{Y}} \ldots \tilde{\mathbf{Y}}\, \tilde{S}_{\lambda_r} = \tilde{S}_\lambda \subset \tilde{S}_{|\lambda|}\,.$$

The image of g in $\tilde{S}_{|\lambda|}$ has the form $t(1, b_1)t(a_2, b_2)\ldots$ with $1 = a_1 < b_1 < a_2 < b_2 \ldots$. By Proposition 6.6, $\chi_{|\lambda|}(g)$ is a certain power of 2 and in particular is positive, as required. For the same reason, $\chi_{\lambda_i}(g_i)$ is positive for each i. But then

$$\chi_\lambda(g) = \chi_{\lambda_1}(g_1)\chi_{\lambda_2}(g_2)\ldots > 0\,,$$

as required, by the iteration of the top entry in the first column of Table 5.7.

Remarks. (1) Note that in all cases, $\chi_\lambda(g)$ depends up to sign only on μ, not on μ^*; that is, it depends only on the conjugacy class of g in $\tilde{S}_{|\lambda|}$. This is perhaps unexpected.

(2) Clause (iii) of the theorem, but no other more exact accounting of signs, will be needed later. It is stated in such a way as to be independent of choices made for splitting conjugacy classes or choices between characters and their associates.

Example. We illustrate the calculation of the basic irreducible characters when $n = 5$. The only non-zero values of the basic character χ_5 are given by

$$\chi_5(1^5) = 4\,,\ \chi_5(3\ 1^2) = 2 \text{ and } \chi_5(5) = 1\,,$$

where, as in Chapter 4, we have denoted one choice of the conjugacy classes of elements of $\tilde{S}_5$ by the cycle type of the corresponding element of S_5. We shall restrict use of this notation to cycle types in $\mathcal{P}^0$. The non-zero values of $\chi_{4,1}$ are given as follows, where g is any element projecting to cycle type (4,1):

$$\chi_{4,1}(1^5) = 2\,,\ \chi_{4,1}(3\ 1^2) = 1 \text{ and } \chi_{4,1}(g) = \pm\sqrt{2}\,.$$

Notice that $\chi_{4,1}$ is not self-associate. The character of its associate coincides with it except on g, where $\chi^a_{4,1}(g) = -\chi_{4,1}(g)$. The non-zero values of $\chi_{3,2}$ are given by

$$\chi_{3,2}(1^5) = 2, \ \chi_{3,2}(3\ 1^2) = 1 \ \text{ and } \chi_{3,2}(h) = \pm\sqrt{3},$$

where h projects to cycle type (3,2). Its associate has the same character, except that

$$\chi^a_{3,2}(h) = -\chi_{3,2}(h) .$$

Definition. Let $\tilde{A}_\lambda$ be the kernel of

$$\sigma : \tilde{S}_\lambda \to \mathbb{Z}/2 .$$

Theorem 6.12. *Let $\lambda \in \mathbb{Z}^r$ be a sequence of positive integers. Then the restriction of χ_λ to $\tilde{A}_\lambda$ is irreducible if and only if λ is odd. When λ is even,*

$$\chi_\lambda \downarrow \tilde{A}_\lambda = \beta + \beta^c$$

for an irreducible character β such that

$$(\beta - \beta^c)(g) = \begin{cases} \pm i^{(|\lambda| - r)/2}(\lambda_1\lambda_2 \ldots \lambda_r)^{1/2} & \text{if } g \text{ has cycle type } \lambda \text{ in } A_{|\lambda|}\,; \\ 0 & \text{otherwise.} \end{cases}$$

Remark. Together with the values given for χ_λ in Theorem 6.11, this determines β and β^c up to interchanging their names. Note that 6.9 and 6.10 are special cases of 6.12.

Proof. The first assertion follows from Theorem 6.11, since $\chi_\lambda \downarrow \tilde{A}_\lambda$ is irreducible if and only if χ_λ is not self-associate. The character calculation proceeds by induction on r. When $r = 1$, it becomes the formula in Theorem 6.9. For the inductive step, let $\lambda_r = k$ and $\lambda' = (\lambda_1,\ldots,\lambda_{r-1})$, so that we can write $\chi_\lambda = \chi_{\lambda'} \tilde{\otimes} \chi_k$ and $g = (g', g'') \in \tilde{S}_{\lambda'} \curlyvee \tilde{S}_k$, and apply Theorem 5.10.

If k is even, then neither $\chi_{\lambda'}$ nor χ_k is self-associate. Suppose that $(\beta - \beta^c)(g) \neq 0$. By 5.10 (ii), g' and g'' are both odd, and

$$(\beta - \beta^c)(g) = \pm 2i\chi_{\lambda'}(g')\chi_k(g'') .$$

Since this is not zero, Theorem 6.11 (ii) yields that g' and g'' correspond to cycle types λ' and (k), with

$$\chi_{\lambda'}(g') = \pm i^{(|\lambda'| - r+2)/2}(\lambda_1 \ldots \lambda_{r-1}/2)^{1/2}$$

and

$$\chi_k(g'') = \pm i^{k/2}(k/2)^{1/2}.$$

Thus

$$\begin{aligned}(\beta-\beta^c)(g) &= \pm i^{(|\lambda'|+k-r+4)/2}\, 2(\lambda_1 \ldots \lambda_r/4)^{1/2} \\ &= \mp i^{(|\lambda|-r)/2}(\lambda_1 \ldots \lambda_r)^{1/2},\end{aligned}$$

as required.

If k is odd, then both $\chi_{\lambda'}$ and χ_k are self-associate. Suppose that $(\beta-\beta^c)(g) \neq 0$. By 5.10 (i), both g' and g'' are even, and

$$(\beta-\beta^c)(g) = \pm(\beta_{\lambda'}-\beta^c_{\lambda'})(g')(\beta_k-\beta^c_k)(g''),$$

where $\beta_{\lambda'}$ and β_k are constituents of the restrictions of $\chi_{\lambda'}$ and χ_k to $\tilde{A}_{\lambda'}$ and $\tilde{A}_k$, respectively. By the inductive hypothesis, this is zero unless g' and g'' correspond to cycle types λ' and (k), in which case it is

$$\begin{aligned}&\pm i^{(|\lambda'|-r+1)/2}(\lambda_1 \ldots \lambda_{r-1})^{1/2} i^{(k-1)/2} k^{1/2} \\ &= \pm i^{(|\lambda|-r)/2}(\lambda_1 \ldots \lambda_r)^{1/2},\end{aligned}$$

as required.

Example. The restrictions $\chi_{4,1} \downarrow \tilde{A}_{4,1}$ and $\chi_{3,2} \downarrow \tilde{A}_{3,2}$ are irreducible. The restriction of $\chi_5 = <5>$ to $\tilde{A}_5$ is a sum of two characters β and β^c with

$$\begin{aligned}&\beta(1^5) = \beta^c(1^5) = 2, \quad \beta(3\ 1^2) = \beta^c(3\ 1^2) = 1, \\ &\beta(5) = (1+\sqrt{5})/2 \ \text{ and } \beta^c(5) = (1-\sqrt{5})/2.\end{aligned}$$

We can now give a preview of how the basic negative character of $\tilde{S}_\lambda$ is used in the same way as the trivial character of S_λ is used for the linear representations of S_n. The basic characters χ_λ of $\tilde{S}_\lambda$ will be induced to $\tilde{S}_n$ to give a set of negative characters ξ_λ. It will be proved in Chapter 8 that all the irreducible negative characters of $\tilde{S}_n$ are $\mathbb{Z}$-linear combinations of $\{\xi_\lambda : \lambda \in \mathcal{D}(n)\}$.

We can illustrate the process when $n = 5$. The computation of the negative characters of $\tilde{S}_5$ induced from $\tilde{S}_\lambda$ is quite straightforward. In general, see the proof of Proposition 8.3 for calculating ξ_λ. In this case, for $\lambda = (4,1)$ or $(3,2)$, and for s and t projecting to the same cycle type $(3,1,1)$ or $(1,1,1,1,1)$, it happens that s and t being conjugate in $\tilde{S}_5$ is enough to make them conjugate in $\tilde{S}_\lambda$. It follows that, with $C_G(g)$ denoting the centralizer of g in G,

$$\xi_\lambda(t) = (\chi_\lambda \uparrow \tilde{S}_5)(t) = \chi_\lambda(t)|C_{S_5}(\tau)| \,/\, |C_{S_\lambda}(\tau)|\;,$$

where τ is the image in S_5 of t.

Inducing the characters given in the example after Theorem 6.11 yields the following array of negative characters of $\tilde{S}_5$, written according to our usual conventions:

	(5)	(31^2)	(1^5)	(4 1)	(3 2)
ξ_5	4	2	1	0	0
$\xi_{4,1}$	10	2	0	$\sqrt{2}$	0
$\xi_{4,1}^a$	10	2	0	$-\sqrt{2}$	0
$\xi_{3,2}$	20	1	0	0	$\sqrt{3}$
$\xi_{3,2}^a$	20	1	0	0	$-\sqrt{3}$

The first character, ξ_5, which is also denoted <5>, is irreducible. The second and third are each the sum of two irreducibles, as may be seen by computing inner products. Since <5> is a constituent of each of these (by computing inner products), these second and third characters may be denoted <5>+<4,1> and <5>+<4,1>^{a}. We can therefore calculate the values of <4,1> and its associate by subtracting the values of <5> from the characters $\xi_{4,1}$ and $\xi_{4,1}^a$. The fourth and fifth characters are of the form <5> + <4,1> + <4,1>^{a} + <3,2> and <5> + <4,1> + <4,1>^{a} + <3,2>^{a}, respectively. Subtracting <5> + <4,1> + <4,1>^{a} from $\xi_{3,2}$ and $\xi_{3,2}^a$, we can see that the "projective character" table of S_5 is as shown.

	1	20	24	30	20
	1	3	5	4	6
<5>	4	2	1	0	0
<4,1>	6	0	−1	$\sqrt{2}$	0
<4,1>^{a}	6	0	−1	$-\sqrt{2}$	0
<3,2>	4	−1	1	0	$\sqrt{3}$
<3,2>^{a}	4	−1	1	0	$-\sqrt{3}$

The first of these characters is the basic character and is self-associate. The other characters are pairs of distinct associate characters. Theorems 4.2 and 6.9 may be applied to see that the "projective character table" of A_5 is as shown.

	1	20	12	12
	1	3	5	5
α_5	2	1	$\frac{1+\sqrt{5}}{2}$	$\frac{1-\sqrt{5}}{2}$
α_5^c	2	1	$\frac{1-\sqrt{5}}{2}$	$\frac{1+\sqrt{5}}{2}$
$<4,1>\downarrow=\alpha_{4,1}$	6	0	-1	-1
$<3,2>\downarrow=\alpha_{3,2}$	4	-1	1	1

where $<5> \downarrow \tilde{A}_5 = \alpha_5+\alpha_5^c$. The notation α_λ will be used later in Theorem 8.7.

Notes

The character calculations in this chapter are based on those in Schur (1911). An alternative approach, using trigometric identities, is given in Morris (1962). Proposition 6.1 exhibits the Clifford algebra, implicit in Schur's calculations. The Clifford algebra is used explicitly in the analogous character calculations in Stembridge (1989), and in Jozefiak (1989). The first paper on $\tilde{S}_n$ after Schur's was Morris (1962), where the fact that it was a subgroup of Pin (n) was noted. The connection of Pin to Clifford algebras goes back at least to Brauer and Weyl (1930). In the following appendix we make three trace calculations in $\mathbb{Z}/2$-graded Clifford modules. In Appendix 12, these will be used to recalculate the characters of $\tilde{S}_n$ by an inductive method, based on knowledge of its irreducible representations as linear combinations of $\chi_\lambda \uparrow \tilde{S}_n$, which is obtained in Appendix 8 independently of the character calculations in Chapter 6.

Appendix 6

A trace of Clifford modules

In this appendix, we present some additional facts relating to Clifford algebras and $\tilde{S}_n$. These are not needed for the main development in the book. However, this is the first of three appendices (to Chapters 6, 8 and 12) which develop the theory in a way which is further removed from Schur's original arguments than is that in Chapters 1 to 8.

We shall study only the *complex* Clifford algebras, denoted by $C(V, f)$ or C_n. They are defined exactly as was done after Theorem 2.8, except that the algebra is defined over $\mathbb{C}$ rather than $\mathbb{R}$. Thus, for a real vector space V with basis $e_1, \ldots, e_n$, and with symmetric bilinear form f for which $f(e_i, e_j) = -\delta_{ij}$, the algebra $C(V, f)$ is the quotient of the free associative $\mathbb{C}$-algebra with identity generated by $\{e_1, \ldots, e_n\}$ by the two sided ideal generated by the elements e_k^2+1 and $e_je_k+e_ke_j$ for $j \neq k$. As a complex vector space, $C(V, f)$ has dimension 2^n. Note that the signature of f is now irrelevant, since, within the algebra, replacing e_j by ie_j changes the sign of e_j^2. For our purposes, it is simplest to keep V as a real vector space and f as a definite form. Then any subspace W of V will generate a subalgebra of $C(V, f)$ which is also a Clifford algebra for a definite form, namely $C(W, g)$, where g is the restriction of f to $W \times W$. We shall be particularly interested in the case when $V = \mathbb{R}^n$ and W is the subspace

$$\left\{ \sum \alpha_j e_j : \sum \alpha_j = 0 \right\}$$

of codimension one. In that case denote the subalgebra as B_{n-1}, contained in $C(V, f)$, which we call C_n.

The reason for interest in B_{n-1} is that it has a basis $\{t'_1, \ldots, t'_{n-1}\}$ (not orthonormal) consisting of invertible elements $t'_j := 2^{-1/2}(e_j - e_{j+1})$. The group generated by this basis is isomorphic to $\tilde{S}_n$, using the map which sends the generator t_j to t'_j and z to -1. To see this, one need only check that the defining relations for $\tilde{S}_n$ also hold for the t'_j. The relations $t'_k t'_j = -t'_j t'_k$ for $k > j+1$ and $t'^2_j = -1$ are easy to check. A small calculation yields

$$(e_j - e_{j+1})(e_{j+1} - e_{j+2})(e_j - e_{j+1}) = 2(e_j - e_{j+2}).$$

Thus

$$t'_j t'_{j+1} t'_j = 2^{-1/2}(e_j - e_{j+2}),$$

and, by interchanging t'_j and t'_{j+1},

$$t'_{j+1} t'_j t'_{j+1} = -2^{-1/2}(e_{j+2} - e_j).$$

Hence

$$(t'_j t'_{j+1})^3 = 2^{-1}(e_j - e_{j+2})^2 = -1,$$

as required. Note that this gives a shorter proof of Theorem 2.8 (essentially, that $z \neq 1$ in $\tilde{S}_n$). Whereas in the earlier proof we used a somewhat complicated irreducible representation which is needed later, here we have used the reducible representation C_n. An irreducible module for B_{n-1} will restrict to an irreducible representation of $\tilde{S}_n$, since $\{t'_1, .., t'_{n-1}\}$ generates B_{n-1} as an algebra. The irreducible representation in Theorem 2.8 will reappear below disguised as a module over the Clifford algebra.

We shall denote the element $e_1 e_2 ... e_n$ of C_n as ℓ_{C_n}. It can be shown to be independent, up to sign, of the choice of ordered orthonormal basis $\{e_1, ..., e_n\}$ for V. In fact, a change of basis will change the sign of ℓ_C if and only if the determinant of the change of basis matrix is -1. A proof of this may be found in Greub (1978, p. 239, (10.5)). It is analogous to the relation between determinants and the Grassman (or exterior) algebra, Hoffman and Kunze (1971, 5.7). The latter would be an example of a Clifford algebra if we allowed singular forms such as the identically zero form.

Proposition A6.1. *We have* (*up to sign*)

$$\ell_{B_{n-1}} = n^{-1/2} \sum_{j=1}^{n} (-1)^j e_1 ... \hat{e}_j ... e_n,$$

where $\hat{e}_j$ means "omit e_j".

Proof. It is easy to check that $\{d_1, ..., d_{n-1}\}$ is an orthonormal basis for W^{n-1}, where

$$d_j := (j^2 + j)^{-1/2}(je_{j+1} - e_1 - e_2 - \dots - e_j).$$

The given formula then follows by induction on n, since, when it is multiplied by $d_n \in B_n \subset C_{n+1}$, it becomes the "same" formula with n changed to $n+1$.

Note that C_n is a direct sum $C_n^{(0)} \oplus C_n^{(1)}$, where $C_n^{(k)}$ is spanned by those products $e_{j_1} \ldots e_{j_\ell}$ for which $\ell \equiv k \pmod 2$. Furthermore, multiplication maps $C_n^{(k)} \times C_n^{(\ell)}$ into $C_n^{(k+\ell)}$; that is, C_n is a $\mathbb{Z}/2$-*graded algebra*. Physicists and others also use the word *superalgebra* for such a structure. It is clear that the $\mathbb{Z}/2$-grading is independent of the choice of orthonormal basis. Each $C_n^{(k)}$ has complex dimension 2^{n-1}.

We have, for x in C_n,

$$x\ell_{C_n} = \begin{cases} \ell_{C_n} x & \text{if } n \text{ is odd;} \\ (-1)^k \ell_{C_n} x & \text{if } x \in C_n^{(k)} \text{ and } n \text{ is even.} \end{cases}$$

This is clear when $x = e_j$, from which it follows for general x.

The structure of C_n has been explained in an elegant manner in Atiyah, Bott and Shapiro (1964). We summarize this as follows. Let $\mathbb{C}^{k\times k}$ be the algebra of $k\times k$ complex matrices. Then the Kronecker product of matrices defines an isomorphism

$$\mathbb{C}^{k\times k} \otimes_{\mathbb{C}} \mathbb{C}^{\ell\times\ell} \to \mathbb{C}^{k\ell\times k\ell}$$

of $\mathbb{C}$-algebras. Simple calculations show that the following determine algebra isomorphisms:

$$C_1 \cong \mathbb{C} \oplus \mathbb{C}; \qquad C_2 \cong \mathbb{C}^{2\times 2};$$

$$e_1 \mapsto (i,-i); \qquad e_1 \mapsto \begin{pmatrix} 0 & 1 \\ -1 & 0 \end{pmatrix};$$

$$e_2 \mapsto \begin{pmatrix} i & 0 \\ 0 & -i \end{pmatrix}.$$

An algebra isomorphism

$$C_{n+2} \to C_n \otimes_{\mathbb{C}} C_2$$

is determined by mapping e_1 to $1 \otimes e_1$; e_2 to $1 \otimes e_2$; and e_j to $ie_{j-2} \otimes e_1e_2$ for $j > 2$. Combining these, we inductively obtain

Theorem A6.2. *As* $\mathbb{C}$*-algebras,* $C_{2n} \cong \mathbb{C}^{2^n \times 2^n}$ *and* $C_{2n+1} \cong C_{2n} \oplus C_{2n}$.

Since the modules for matrix algebras and for direct sums of algebras are well known (see Lang, 1965), we now have

Corollary A6.3. *For all integers* $n \geq 0$*, there are left modules:* M^*_{2n} *for* C_{2n}*; and non-isomorphic* M_{2n+1}, M'_{2n+1} *for* C_{2n+1}*; such that every finite dimensional left module for these Clifford algebras is isomorphic to a unique direct sum of these modules. All three modules have dimension* 2^n *over* $\mathbb{C}$.

All the above modules will be referred to as *Clifford modules.*

A $\mathbb{Z}/2$*-graded module* M over a $\mathbb{Z}/2$-graded algebra A is a module together with a given direct sum decomposition $M = M^{(0)} \oplus M^{(1)}$ of vector spaces, such that the module action maps $A^{(j)} \times M^{(k)}$ into $M^{(j+k)}$. An isomorphism of $\mathbb{Z}/2$-graded modules is a module isomorphism which preserves the given direct sum decomposition. One moral of these appendices will be that treating both graded and ungraded modules simultaneously tends to result in more symmetrical and memorable formulae and proofs. The $\mathbb{Z}/2$-graded Clifford modules can be classified as with the ungraded modules above by developing $\mathbb{Z}/2$-analogues of classical techniques in the theory of algebras; see Wall (1964) and Jozefiak (1988). However, we shall proceed directly since the situation is transparent in the case of the Clifford algebra.

Since $(\ell_{C_{2n}})^2 = (-1)^n$, as is easily checked, any C_{2n}-module splits as a vector space into the $\pm i^n$-eigenspaces of the action of $\ell_{C_{2n}}$. We can use these two subspaces to give a $\mathbb{Z}/2$-grading to the module. This is verified using the relation

$$x\ell_{C_{2n}} = (-1)^k \ell_{C_{2n}} x \text{ for } x \in C^{(k)}_{2n}.$$

Applying this to the module M^*_{2n}, we obtain two $\mathbb{Z}/2$-graded modules M_{2n} and M'_{2n}, depending on which eigenspace is assigned grading zero. It will be unnecessary to be more specific about which is called M_{2n} and which M'_{2n}. They are "reverses" of each other, and are not isomorphic as $\mathbb{Z}/2$-graded modules, since any module endormorphism of the irreducible module M^*_{2n} is a scalar multiplication, and so cannot reverse $\mathbb{Z}/2$-gradings.

On the other hand, the non-isomorphic C_{2n+1}-modules M_{2n+1} and M'_{2n+1} are "associates" of each other; that is, they may be regarded as being the same vector space with the two actions of any element in $C^{(1)}_{2n+1}$ differing by a sign, and the actions of any element in $C^{(0)}_{2n+1}$ agreeing. Now let M^*_{2n+1} be the $\mathbb{Z}/2$-graded module which is $M_{2n+1} \oplus M'_{2n+1}$ as a module and with $\mathbb{Z}/2$-grading defined by

$$M^{*(j)}_{2n+1} := \{(m, (-1)^j m) : m \in M_{2n+1}\}.$$

We shall use the following facts in the proof below of Proposition A6.5 (iii). Any ungraded C_{2n+1}-module which is self-associate is isomorphic to a direct sum of equal numbers of copies of M_{2n+1} and M'_{2n+1}. Equivalently, it is a direct sum of copies of M^*_{2n+1}, after "forgetting" the graded structure of the latter. Also, since $(\ell_{C_{2n+1}})^2 = (-1)^{n+1}$, and left multiplication by $\ell_{C_{2n+1}}$ on any C_{2n+1}-module is a module morphism, this left multiplication is scalar multiplication by i^{n+1} on one of M_{2n+1} or M'_{2n+1}, and by $-i^{n+1}$ on the other.

Since any of the Clifford algebras of this appendix is isomorphic to C_n for some n, the classification of modules above is applicable, for example, to B_k. We shall denote its irreducible modules as N_k, N'_k and N^*_k, so that N_{2n}, N'_{2n} and N^*_{2n+1} are graded, whereas N_{2n+1}, N'_{2n+1} and N^*_{2n} are not.

If M is an A-module and is finite dimensional as a vector space, denote the trace of the action of $a \in A$ on M by $\mathrm{tr}(a \mid M)$.

Proposition A6.4. *For sequences $\alpha = (i_1, \ldots, i_\ell)$ with $1 \le i_1 < \ldots < i_\ell \le n$, let e_α be the basis element $e_{i_1} \ldots e_{i_\ell}$ of C_n, and let $k_\alpha \in \mathbb{C}$. (Thus $e_\varnothing = 1$ for the empty sequence $\varnothing$.) Then*

$$\mathrm{tr}\left(\sum_\alpha k_\alpha e_\alpha \mid C_n\right) = 2^n k_\phi.$$

Proof. This is clear since $e_\alpha e_\beta$ is always $\pm e_\gamma$ for some γ, and $\gamma \neq \beta$ unless α is the empty sequence.

Let $s_n = t'_1 t'_2 \ldots t'_{n-1}$, an element which projects to the n-cycle $(1\ 2 \ldots n)$ in S_n.

Proposition A6.5. *We have the following trace calculations*:

(*i*) $\mathrm{tr}(s_{2n+1} \mid N^*_{2n}) = 1$;

(*ii*) $\mathrm{tr}(s_{2n+1} \mid N_{2n}) - \mathrm{tr}(s_{2n+1} \mid N'_{2n}) = \pm i^n \sqrt{2n+1}$;

(*iii*) $\mathrm{tr}(s_{2n} \mid N_{2n-1}) = \pm i^n \sqrt{n}$.

Proof. In (i) and (ii), we use the fact that C_{2n+1}, as a B_{2n}-module, is a direct sum of 2^{n+1} copies of N^*_{2n}. This is clear by counting dimensions.

(i) Using Proposition A6.4, the left side is

$$\mathrm{tr}\left[(2^{-1/2})^{2n}(e_1 - e_2) \ldots (e_{2n} - e_{2n+1}) \mid N^*_{2n}\right]$$

$$= 2^{-(n+1)}\mathrm{tr}(2^{-n}(e_1 - e_2)\ldots(e_{2n} - e_{2n+1}) \mid C_{2n+1})$$

$$= 2^{-(n+1)}2^{-n}2^{2n+1}(0-e_2)(e_2-0)(0-e_4)(e_4-0)\ldots(e_{2n}-0) = 1.$$

(ii) Note that if a linear operator $T: V_+ \oplus V_- \to V_+ \oplus V_-$ has invariant subspaces $V_\pm$, then $\mathrm{tr}(T \mid V_+) - \mathrm{tr}(T \mid V_-) = \mathrm{tr}(TJ)$, where $J \mid V_\pm = \pm I$. Since $\ell_{B_{2n}}$ acts on N_{2n} (respectively N'_{2n}) as $\pm i^n$ (resp. $\mp i^n$), using Propositions A6.1 and A6.4, the left side of (ii) becomes

$$\pm\mathrm{tr}(i^n s_{2n+1}\ell_{B_{2n}} \mid N^*_{2n})$$

$$= \pm 2^{-(n+1)}\mathrm{tr}(i^n(2^{-1/2})^{2n}(e_1-e_2)\ldots(e_{2n}-e_{2n+1})(2n+1)^{-1/2} \times$$

$$\times \sum_{i=1}^{2n+1}(-1)^i e_1\ldots\hat{e}_i\ldots e_{2n+1} \mid C_{2n+1})$$

$$= \pm 2^{-(n+1)}i^n 2^{-n}2^{2n+1} \times$$

$$\times \sum_{i=1}^{2n+1}(e_1-0)\ldots(e_{i-1}-0)(0-e_{i+1})\ldots(0-e_{2n+1})(-1)^i e_1\ldots\hat{e}_i\ldots e_{2n+1}(2n+1)^{\cdot}$$

$$= \pm i^n(2n+1)^{-1/2}\sum_{i=1}^{2n+1}(-1)^{2n+1-i}(-1)^i(e_1\ldots\hat{e}_i\ldots e_{2n+1})^2$$

$$= \pm i^n(2n+1)^{-1/2}(\pm(2n+1))$$

$$= \pm i^n\sqrt{2n+1}.$$

(iii) If $S: V \to V$ is any linear operator, and we define operators on $V \oplus V$ by $T := S \oplus (-S)$ and $K(x, y) := (x, -y)$, then $\mathrm{tr}\, S = \mathrm{tr}(TK)/2$. Now take $V = N_{2n-1}$, but regard $V \oplus V$ as being $N_{2n-1} \oplus N'_{2n-1}$ as a module. Note that, as a B_{2n-1}-module, C_{2n} is the direct sum of 2^n copies of N^*_{2n-1}. Using this and Propositions A6.1 and A6.4, the left side becomes

$$
\begin{aligned}
&\pm \mathrm{tr}(i^n s_{2n} \ell_{B_{2n-1}} \mid N^*_{2n-1})/2 \\
&= \pm 2^{-(n+1)} i^n \mathrm{tr}((2^{-1/2})^{2n-1}(e_1 - e_2)\ldots(e_{2n-1} - e_{2n})(2n)^{-1/2} \times \\
&\times \sum_{i=1}^{2n} (-1)^i e_1 \ldots \hat{e}_i \ldots e_{2n} \mid C_{2n}) \\
&= \pm i^n 2^{-1/2}(2n)^{-1/2} \sum_{i=1}^{2n} (e_1 - 0)\ldots(e_{i-1} - 0)(0 - e_{i+1})\ldots(0 - e_{2n})(-1)^i e_1 \ldots \hat{e}_i \ldots e_2 \\
&= \pm 2^{-1} i^n n^{-1/2} \sum_{i=1}^{2n} (-1)^{2n-i}(-1)^i (e_1 \ldots \hat{e}_i \ldots e_{2n})^2 \\
&= \pm i^n \sqrt{n}.
\end{aligned}
$$

We have obtained these trace formulae, without having to find an explicit basis for the irreducible Clifford modules, by passing to the module C_{n+1} over B_n, where trace calculations are easy.

To conclude, we describe the C_{2n}-module M^*_{2n}, both as a submodule and as a quotient module of C_{2n}. The proof also provides a $\mathbb{C}$-basis for the module. Constructions for the other graded and ungraded Clifford modules (and for the basic irreducible $\tilde{S}_n$-modules) are then easily made, using the relations between these modules given in this appendix.

Proposition A6.6. *For* $1 \leq j \leq n$, *let* $f_j = e_{2j-1} + ie_{2j}$. *Then* M^*_{2n} *is isomorphic both to the submodule of* C_{2n} *generated by the product* $f_1 f_2 \ldots f_n$, *and to the quotient module* C_{2n}/I_{2n}, *where* I_{2n} *is the left ideal generated by* $\{f_1, \ldots, f_n\}$.

Proof. For $1 \leq j \leq n$, define $g_j = e_{2j} + ie_{2j-1}$. Clearly $\{f_1, \ldots, f_n, g_1, \ldots, g_n\}$ spans the vector space over $\mathbb{C}$ with basis $\{e_1, \ldots, e_{2n}\}$, and therefore it generates C_{2n} as an algebra over $\mathbb{C}$. The relations

$$f_j g_k + g_k f_j = -4i\delta_{kj}$$

allow one to write any product of f's and g's as a linear combination of such products in which all the g's come first (on the left). Hence the set

$$\{g_{i_1} \ldots g_{i_k} : 1 \leq i_1 < \ldots < i_k \leq n\}$$

spans C_{2n}/I_{2n} over $\mathbb{C}$. Since $f_j^2 = 0$, we obtain $xf_1 \ldots f_n = 0$ for all x in I_{2n}. But $f_1 \ldots f_n \neq 0$, so C_{2n}/I_{2n} is non-zero and has dimension at most 2^n over $\mathbb{C}$. By

Corollary A6.3, it is isomorphic to M^*_{2n}.

Now let F_{2n} be the submodule of C_{2n} generated by $f_1 f_2 ... f_n$. By the above paragraph, the kernel of the module surjection $C_{2n} \to F_{2n}$ taking 1 to $f_1 ... f_n$ contains I_{2n}. Thus F_{2n} is isomorphic to a quotient of C_{2n}/I_{2n}, and so to a quotient of M^*_{2n}. But F_{2n} is non-zero, so $F_{2n} \cong M^*_{2n}$.

The material in this appendix will be used in Appendices 8 and 12. In Appendix 8, we study an algebra made up of projective representations. The Clifford modules will be used to define certain generators for that algebra. In Appendix 12, we use the trace calculation in Proposition A6.5 to recalculate some characters, based on the algebra defined in Appendix 8.

7

THE Q-FUNCTIONS

In this chapter, we shall define the Q-functions. These symmetric functions were introduced by Schur. It will be convenient to use a different definition. Schur's defining formula will appear in Chapter 9. First we give some basic facts about symmetric functions.

The symmetric group S_n acts on the ring $\mathbb{Z}[x_1, ..., x_n]$, of polynomials with integer coefficients, by permuting the variables. A polynomial f in $\mathbb{Z}[x_1, ..., x_n]$ is said to be symmetric if f is invariant under the action of S_n. It can be easily seen that the set Λ_n of symmetric polynomials is a subring of $\mathbb{Z}[x_1, ..., x_n]$. For $k \geq 0$, let $\Lambda_n^{(k)}$ consist of the homogeneous symmetric polynomials of degree k, including the zero polynomial. The abelian groups $\Lambda_n^{(k)}$ $(k \geq 0)$ give Λ_n the structure of a graded ring

$$\Lambda_n = \bigoplus_{k \geq 0} \Lambda_n^{(k)} .$$

Let λ be any partition. The symmetric polynomial $m_\lambda(x_1, x_2, ..., x_n)$ is defined by

$$m_\lambda(x_1, ..., x_n) = \sum x_1^{\alpha_1} \cdots x_n^{\alpha_n} ,$$

where the sum is over all sequences $(\alpha_1, ..., \alpha_n)$ of non-negative integers whose non-zero terms are exactly the parts of λ in any order. Thus if $\lambda = (3, 2, 2)$,

$$m_\lambda(x_1, x_2, x_3) = x_1^3 x_2^2 x_3^2 + x_1^2 x_2^3 x_3^2 + x_1^2 x_2^2 x_3^3 .$$

Furthermore, $m_\lambda(x_1, x_2, ..., x_n)$ is zero if λ is a partition of length greater than n. It is clear that the $m_\lambda(x_1, ..., x_n)$, as λ runs through partitions of length at most n, form a $\mathbb{Z}$-basis for Λ_n. In particular, if $\mathcal{P}(k)$ denotes the set of partitions of k, and $\ell(\lambda)$ denotes the length of λ, then

$$\{m_\lambda(x_1, ..., x_n) : \lambda \in \mathcal{P}(k) \text{ and } \ell(\lambda) \leq n\}$$

is a basis for $\Lambda_n^{(k)}$. Thus for $n \geq k$, a basis for $\Lambda_n^{(k)}$ is $\{m_\lambda(x_1, ..., x_n) : \lambda \in \mathcal{P}(k)\}$, a set whose size is independent of n.

The number of variables in a symmetric polynomial is often unimportant to the discussion, so it is convenient to work in the ring Λ of symmetric functions in infinitely many variables. To give a precise definition of Λ, suppose that $m \geq n$ and consider the map

$$\rho_{m,n} : \Lambda_m \to \Lambda_n$$

obtained by setting each x_i, for $i > n$, equal to zero. Thus ρ is a ring homomorphism and

$$\rho_{m,n}\bigl(m_\lambda(x_1, \ldots, x_m)\bigr) = m_\lambda(x_1, \ldots, x_n) .$$

Hence $\rho_{m,n}$ is surjective, and gives rise (by restriction) to a homomorphism

$$\rho_{m,n}^{(k)} : \Lambda_m^{(k)} \to \Lambda_n^{(k)}$$

which is surjective and, in fact, bijective if $m \geq n \geq k$.

Now define the abelian group $\Lambda^{(k)}$ to be the inverse limit

$$\Lambda^{(k)} = \lim_{\substack{\leftarrow \\ n}} \Lambda_n^{(k)}$$

relative to the homomorphisms $\rho_{m,n}^{(k)}$. An element of $\Lambda^{(k)}$ is therefore a sequence $f = (f_n)_{n \geq 1}$ of homogeneous symmetric polynomials of degree k such that

$$f_m(x_1, \ldots, x_n, 0, \ldots, 0) = f_n(x_1, \ldots, x_n)$$

whenever $m \geq n$. If $m \geq n \geq k$, the map

$$\rho_{m,n}^{(k)} : \Lambda_m^{(k)} \to \Lambda_n^{(k)}$$

is an isomorphism, and so the projection

$$\rho_n^{(k)} : \Lambda^{(k)} \to \Lambda_n^{(k)} ,$$

sending f to f_n, is also an isomorphism if $n \geq k$. Now, for a partition λ of k, let the monomial symmetric function m_λ be defined by

$$\rho_n^{(k)}(m_\lambda) = m_\lambda(x_1, \ldots, x_n) .$$

It follows that $\Lambda^{(k)}$ has a $\mathbb{Z}$-basis consisting of the monomial symmetric functions m_λ for $\lambda \in \mathcal{P}(k)$, and that the graded ring Λ of symmetric functions, defined by

$$\Lambda = \bigoplus_{k \geq 0} \Lambda^{(k)} ,$$

is freely generated as a $\mathbb{Z}$-module by the m_λ. There are surjective homomorphisms

$$\rho_n = \bigoplus_{k\geq 0} \rho_n^{(k)} : \Lambda \to \Lambda_n$$

such that $\rho_n^{(k)}$ is an isomorphism in degree $k \leq n$.

The elements of Λ are certain formal infinite sums of monomials. Note that Λ is not the inverse limit of the Λ_n in the category of rings. For example, this inverse limit contains the infinite product $\prod(1+x_i)$ which does not belong to Λ. However, Λ is the inverse limit of the Λ_n in the category of graded rings. Although the elements of Λ are neither functions nor polynomials (they resemble polynomials more than functions!), we shall continue the custom of calling them symmetric functions.

It is possible to describe Λ without using inverse limits. Let $X = \{x_1, x_2, ...\}$ be a countable set of variables. Let S_∞ be the group under composition of all bijective functions from X to itself which fix all but a finite number of the x_i. Let $\mathcal{S}$ be the set of sequences of non-negative integers such that all but a finite number of terms of the sequence are zero. Given a sequence $\omega = (\omega_1, \omega_2, ...)$ in $\mathcal{S}$, let x^ω be the formal product $x_1^{\omega_1} x_2^{\omega_2} x_3^{\omega_3}$.... Denote by M the ring of all formal infinite sums

$$\sum_{\omega \in \mathcal{S}} n_\omega x^\omega ,$$

where n_ω is an integer which is zero if $\sum \omega_i$ is sufficiently large. We can obviously identify X with a subset of M. The action of S_∞ then extends in an obvious way to an action on M by ring homomorphisms. Furthermore, the subring of elements fixed by S_∞ is isomorphic to Λ.

There is a simple way to produce elements of Λ. Let $F(x) = 1 + \sum_{i>0} a_i x^i$ be a formal power series with integer coefficients. The equation

$$\prod_{i=1}^{n} F(x_i) = \sum_k f_k(x_1, ..., x_n) ,$$

where f_k is homogeneous of degree k, defines symmetric polynomials f_k, since the left side is clearly invariant under the action of S_n. The above equation takes place in $\prod_k (\mathbb{Z}[x_1, ..., x_n])^{(k)}$. One either deduces, or adopts the convention, that

$f_0 = 1$ and $f_k = 0$ for $k < 0$. Since

$$\sum_k f_k(x_1, ..., x_n, 0) = F(0)\prod_{i=1}^{n} F(x_i) = \sum_k f_k(x_1, ..., x_n),$$

the elements f_k may be regarded as being in Λ. It will sometimes be convenient to rewrite the defining equation in one of the forms:

$$\prod_{x \in X} F(x) = \sum_k f_k(X),$$

where X is the set of variables; or

$$\prod_{i=1}^{n} F(x_i\theta) = \sum_k f_k(x_1, ..., x_n)\theta^k,$$

where θ is another variable whose chief role is as a reminder of homogeneous degree.

Definition. Define symmetric functions q_k of degree k by

$$\prod_{x \in X} (1+x)/(1-x) = \sum_k q_k(X);$$

that is, take $F(x)$ to be

$$(1+x)/(1-x) = 1+2\sum_{i>0} x^i$$

in the discussion above. Multiplying, we find

$$q_k = \sum_{\lambda \in \mathcal{P}(k)} 2^{\ell(\lambda)} m_\lambda .$$

Proposition 7.1. *For each r (using the Kronecker delta)*

$$\sum_i (-1)^i q_i q_{r-i} = \delta_{r,0} .$$

When r is odd, this relation is trivial. When r $= 2k > 0$, *it yields*

$$q_k^2 = (-1)^{k+1}\, 2 \sum_{i=o}^{k-1} (-1)^i q_i q_{2k-i} . \tag{7.1}$$

Proof. Since q_i has degree i,

$$q_i(-x_1, \ldots, -x_n) = (-1)^i q_i(x_1, \ldots, x_n)\,.$$

Thus

$$1 = \prod_x 1 = \prod_x F(-x)F(x) = \prod_x F(-x)\prod_x F(x)$$

$$= \sum_i q_i(-X)\sum_j q_j(X) = \sum_r \sum_{i+j=r} (-1)^i q_i(X) q_j(X)\,,$$

as required.

Definition. Let Δ be the subring of Λ generated by $\{q_i : i \in \mathbb{Z}\}$. For any finite sequence $\alpha = (\alpha_1, \ldots, \alpha_\ell)$ of integers, let

$$q_\alpha = q_{\alpha_1} q_{\alpha_2} \cdots q_{\alpha_\ell}$$

(so $q_\alpha = 0$ if any α_i is negative and $q_\varnothing = 1$ for the empty partition $\varnothing$).

Since any power q_k^s for $s > 1$ can be reduced inductively using the relation in Proposition 7.1, we obtain

Corollary 7.2. *The set* $\{q_\lambda : \lambda \in \mathcal{D}$, *the set of strict partitions*$\}$ *spans* Δ *as a* $\mathbb{Z}$*-module.*

For disjoint sets X and Y of variables, any symmetric function in all the variables can be written in terms of the two sets separately; that is

$$f(X \cup Y) = \sum_\alpha f_\alpha(X) \bar{f}_\alpha(Y)$$

with each f_α and $\bar{f}_\alpha$ in Λ, where α ranges over some finite indexing set. This decomposition is not unique, but it is easily seen that the element $\sum_\alpha f_\alpha \otimes \bar{f}_\alpha$ of $\Lambda \otimes_{\mathbb{Z}} \Lambda$ depends only on f. This is immediate from the isomorphism

$$\mathbb{Z}[x_1, \ldots, x_n, y_1, \ldots, y_m] \cong \mathbb{Z}[x_1, \ldots, x_n] \otimes_{\mathbb{Z}} \mathbb{Z}[y_1, \ldots, y_m]\,.$$

Definition. The function from Λ to $\Lambda \otimes_{\mathbb{Z}} \Lambda$ which takes f to $\sum_\alpha f_\alpha \otimes \bar{f}_\alpha$ will be referred to as the *coproduct*.

Since $(fg)(X \cup Y) = f(X \cup Y)g(X \cup Y)$, we immediately see that, under the coproduct, if g maps to $\sum_{\beta} g_{\beta} \otimes \bar{g}_{\beta}$, then fg maps to $\sum_{\alpha,\beta} (f_{\alpha}\, g_{\beta}) \otimes (\bar{f}_{\alpha}\, \bar{g}_{\beta})$. This is called the *Hopf algebra property*. Hopf algebra theory will not be needed here. If preferred, the reader can forget about $\otimes$ and replace every occurrence of $g \otimes h$ by $g(X)h(Y)$.

Proposition 7.3. *The coproduct maps Δ into $\Delta \otimes \Delta$. In particular, q_r maps to $\sum_i q_i \otimes q_{r-i}$.*

Proof. The second assertion follows because q_r arises from a power series F (independently of which F):

$$\sum_r q_r(X \cup Y) = \prod_{z \in X \cup Y} F(z) = \prod_{x \in X} F(x) \prod_{y \in Y} F(y)$$

$$= \sum_i q_i(X) \sum_j q_j(Y) = \sum_r \sum_{i+j=r} q_i(X) q_j(Y)\,,$$

as required. Now the first assertion follows immediately, using the definition of Δ and the Hopf algebra property.

Definition. For $r > 0$, let $p_r \in \Lambda$ be the rth power sum function; that is, $p_r(X) = \sum_{x \in X} x^r$. Let $p_{\alpha} = p_{\alpha_1} p_{\alpha_2} \ldots$, for a finite sequence α of positive integers.

Proposition 7.4. *For all $k \geq 0$, we have $2p_{2k+1} \in \Delta$. In fact, for all n,*

$$nq_n = 2 \sum_{\text{odd } j > 0} p_j q_{n-j}\,.$$

Thus the $\mathbb{Q}$-algebra $\Delta \otimes_{\mathbb{Z}} \mathbb{Q}$ (which is denoted $\Delta_{\mathbb{Q}}$ and identified with a subalgebra of symmetric functions with rational coefficients) is spanned over $\mathbb{Q}$ by $\{p_{\lambda} : \lambda \in \mathcal{P}^0\}$.

Proof. The first and third sentences follow from the second one, using that equation to inductively write q_n as a polynomial in the odd power sums (see also (7.9)). To prove the second sentence, we use standard identities in the ring of formal power series in the variable θ (Goulden and Jackson, 1983):

$$\Big(\sum_i q_i\theta^i\Big)^{-1}\Big(\sum_n nq_n\theta^{n-1}\Big) = \frac{d}{d\theta}\log\sum_n q_n\theta^n$$

$$= \frac{d}{d\theta}\log\prod_x (1+x\theta)/(1-x\theta)$$

$$= \frac{d}{d\theta}\sum_x [\log(1+x\theta) - \log(1-x\theta)]$$

$$= \frac{d}{d\theta}\sum_x \sum_{\text{odd } j>0} 2x^j\theta^j/j$$

$$= 2\sum_{\text{odd } j>0} p_j(X)\theta^{j-1}\,.$$

Now multiply by $\sum_i q_i\theta^i$ to obtain the result (which could be called a "Newton formula").

Proposition 7.5. *The set* $\{p_r : r \geq 1\}$ *of power sum functions is algebraically independent.*

Proof. There is a conventional argument which examines monomials; we shall use the coproduct instead. It maps p_r to $1\otimes p_r + p_r\otimes 1$, since $p_r(X\cup Y) = p_r(X)+p_r(Y)$. By the Hopf algebra property,

$$p_\lambda \mapsto \sum p_\alpha \otimes p_{\lambda\backslash\alpha}\,,$$

where α varies over all subsequences of λ, and $\lambda\backslash\alpha$ is the complementary subsequence. Now suppose, for a contradiction, that $\sum_{\lambda\in\mathcal{P}(n)} c_\lambda p_\lambda = 0$ for $c_\lambda \in \mathbb{Q}$, not all zero, and with n minimal. This minimality implies that the set

$$X = \{p_\nu \otimes p_\eta : \nu\in\mathcal{P}(i), \eta\in\mathcal{P}(j), i>0, j>0, i+j=n\}$$

is linearly independent. But each such $p_\nu \otimes p_\eta$ occurs with non-zero coefficient in the coproduct of p_λ for exactly one $\lambda\in\mathcal{P}(n)$, with λ being obtained by rearranging the juxtaposition $\nu\eta$, and the coefficient being the number of subsequences of λ which agree with ν as sequences. Applying the coproduct to the above relation contradicts the linear independence of X.

Remark. This is a standard Hopfian argument: linearly independent primitives (elements p with $p \mapsto 1 \otimes p + p \otimes 1$) tend to be algebraically independent.

Corollary 7.6. *We have*:

(*i*) $\Delta_{\mathbb{Q}} = \mathbb{Q}[p_1, p_3, p_5, \ldots]$;

(*ii*) $\Delta = \mathbb{Z}[q_1, q_2, q_3, \ldots]/I$, *where I is the ideal generated by*

$$\left\{ \sum_i (-1)^i q_i q_{r-i} : r > 0 \right\};$$

(*iii*) $\{q_\lambda : \lambda \in \mathcal{D}\}$ *is a $\mathbb{Z}$-basis for Δ.*

Proof. The first assertion is immediate from Propositions 7.4 and 7.5. Thus $\Delta_{\mathbb{Q}}^{(k)}$ has dimension $|\mathcal{P}^0(k)|$, and so $\Delta^{(k)}$ is a free abelian group of rank $|\mathcal{P}^0(k)|$, which equals $|\mathcal{D}(k)|$. Since $\{q_\lambda : \lambda \in \mathcal{D}(k)\}$ spans $\Delta^{(k)}$, it is therefore a $\mathbb{Z}$-basis, implying that the ideal I of all relations satisfied by $\{q_i\}$ is generated as given.

The next task is to define an inner product $< , >$ mapping $\Delta \times \Delta$ to $\mathbb{Z}$.

Theorem 7.7. *There is a symmetric $\mathbb{Z}$-valued inner product on Δ which is uniquely characterized by any one of* (*i*), (*ii*) *or* (*iii*) *below*:

(*i*) *Each f in Δ is given by*

$$f = \sum_{\lambda \in \mathcal{P}} <f, q_\lambda> m_\lambda;$$

that is, $<f, q_\alpha>$ is the coefficient of the monomial X^α in $f(X)$.

(*ii*) *We have*

(*a*) $<1, 1> = 1$;

(*b*) $<q_i, q_j> = 2\delta_{ij}$ *for positive i and j*;

(*c*) *for all f,g and h in Δ,*

$$<f, gh> = \sum_\alpha <f_\alpha, g> <\bar{f}_\alpha, h>$$

where f maps to $\sum f_\alpha \otimes \bar{f}_\alpha$ under the coproduct.

(*iii*) *Extend $< , >$ to a $\mathbb{Q}$ bilinear inner product on $\Delta_{\mathbb{Q}}$. Let*

$$z_\lambda = \prod_{i>0} i^{f_i}(f_i!)$$

where f_i is the frequency of occurrence of i in λ; that is, $f_i = |\{j : \lambda_j = i\}|$. Then *for all* λ *and* μ *in* $\mathcal{P}^0$

$$<p_\lambda, p_\mu> = \delta_{\lambda\mu} 2^{-\ell(\lambda)} z_\lambda ,$$

where $\ell(\lambda)$ *is the length of* λ.

Remark. Property (ii) (c) says that the product and coproduct are duals of each other with respect to $< , >$.

Proof. Uniqueness is easy in each case. Equation (i) determines $<f, q_\lambda>$ for all f and λ, and, therefore, by linearity, it determines $<f, g>$ for all f and g in Δ, since the q_λ span Δ over $\mathbb{Z}$. Equation (ii) (c) and Proposition 7.3 allow one to calculate $<q_\lambda, q_\mu>$ by double induction on $\ell(\lambda)$ and $\ell(\mu)$, using (ii) (a) and (b) to start the induction. (The answer is $\sum_A 2^{\eta(A)}$, where A ranges over all $\mathbb{N}$-matrices with row sums λ and column sums μ, and where $\eta(A)$ is the number of non-zero entries.) Since all $<q_\lambda, q_\mu>$ are determined, this fixes $< , >$ by linearity.

By Corollary 7.6, equation (iii) uniquely determines a $\mathbb{Q}$-valued inner product on $\Delta_\mathbb{Q}$, so we shall use it to define $< , >$. It remains only to prove (i) and (ii) for this inner product (with f,g and h in $\Delta_\mathbb{Q}$), since either (i) or (ii) clearly implies that the inner product maps the subset $\Delta \times \Delta$ of $\Delta_\mathbb{Q} \times \Delta_\mathbb{Q}$ into $\mathbb{Z}$.

The integers z_λ arise naturally in the reversal of the calculation in the proof of Proposition 7.4:

$$\begin{aligned}\sum_n q_n(X)\theta^n &= \prod_{x\in X}(1+x\theta)/(1-x\theta)\\ &= \exp \int 2 \sum_{\text{odd}\, j>0} p_j(X)\theta^{j-1} d\theta\\ &= \exp \sum_{\text{odd}\, j>0} 2p_j(X)\theta^j/j\\ &= \sum_m \sum_{\lambda\in\mathcal{P}^0(m)} 2^{\ell(\lambda)} z_\lambda^{-1} p_\lambda(X)\theta^m .\end{aligned}$$

Eliminating θ, we have

$$\prod_x (1+x)/(1-x) = \sum_{\lambda\in\mathcal{P}^0} 2^{\ell(\lambda)} z_\lambda^{-1} p_\lambda(X) \qquad (7.8)$$

and

$$q_m = \sum_{\lambda \in \mathcal{P}^0(m)} 2^{\ell(\lambda)} z_\lambda^{-1} p_\lambda . \tag{7.9}$$

Now let $X{\cdot}Y$ be the set of products $\{xy : x \in X, y \in Y\}$. Then $p_n(X)p_n(Y) = p_n(X{\cdot}Y)$, so $p_\lambda(X)p_\lambda(Y) = p_\lambda(X{\cdot}Y)$. Replacing X by $X{\cdot}Y$ in (7.8) yields

$$\prod_{X \times Y} (1+xy)/(1-xy) = \sum_{\lambda \in \mathcal{P}^0} 2^{\ell(\lambda)} z_\lambda^{-1} p_\lambda(X) p_\lambda(Y) . \tag{7.10}$$

On the other hand,

$$\begin{aligned}
\prod_{X \times Y} (1+xy)/(1-xy) &= \prod_{y \in Y} \prod_{x \in X} (1+xy)/(1-xy) \\
&= \prod_{y \in Y} \left(\sum_{n_y} q_{n_y}(X) y^{n_y} \right) \\
&= \sum q_\alpha(X) Y^\alpha \qquad (7.11) \\
&= \sum_{\lambda \in \mathcal{P}} q_\lambda(X) m_\lambda(Y) .
\end{aligned}$$

The summation in (7.11) is over all sequences $\alpha \in \mathbb{N}^{|Y|}$, and Y^α means $y_1^{\alpha_1} y_2^{\alpha_2} \ldots$. Thus

$$\sum_{\lambda \in \mathcal{P}} q_\lambda(X) m_\lambda(Y) = \sum_{\lambda \in \mathcal{P}^0} 2^{\ell(\lambda)} z_\lambda^{-1} p_\lambda(X) p_\lambda(Y) . \tag{7.12}$$

Rewritten using $\otimes$ rather than disjoint variables,

$$\sum_{\lambda \in \mathcal{P}} q_\lambda \otimes m_\lambda = \sum_{\lambda \in \mathcal{P}^0} 2^{\ell(\lambda)} z_\lambda^{-1} p_\lambda \otimes p_\lambda . \tag{7.12$'$}$$

Let T be either side in (7.12)′ (which we call the fundamental tensor of $<\ ,\ >$). Consider the following composition of two linear maps:

$$\Delta_{\mathbb{Q}} \to \Delta_{\mathbb{Q}} \otimes \Delta_{\mathbb{Q}} \otimes \Delta_{\mathbb{Q}} \to \Delta_{\mathbb{Q}},$$
$$f \to f \otimes T;\ f \otimes g \otimes h \to <f, g> h.$$

This composite map is the identity, since, using the right side of (7.12)′, it sends p_μ to itself for all $\mu \in \mathcal{P}^0$. But, using the left side of (7.12)′, it sends f to $\sum_{\lambda \in \mathcal{P}} <f, q_\lambda> m_\lambda$. This proves property (i).

Taking $f = q_n$ in (i) yields (ii) (a) and (b). To prove (ii) (c), note that both sides are trilinear functions of (f, g, h), so we may take f, g and h in some basis for $\Delta_{\mathbb{Q}}$. Use the basis $\{p_\lambda : \lambda \in \mathcal{P}^0\}$. Let $f = p_\lambda$, $g = p_\mu$ and $h = p_\nu$. The left side of (ii) (c) becomes

$$<p_\lambda, p_\mu p_\nu> \; = \; <p_\lambda, p_{\mu \cup \nu}> \; = 2^{-\ell(\lambda)} z_\lambda \delta_{\lambda, \mu \cup \nu} ,$$

where $\mu \cup \nu$ is obtained by juxtaposing μ and ν, then shuffling into weakly decreasing order. As explained in the proof of 7.5, the coproduct maps p_λ to $\sum_\alpha p_\alpha \otimes p_{\lambda \backslash \alpha}$, (summation over all subsequences of α of λ, with $\lambda \backslash \alpha$ being the complementary subsequence). The right side of (ii) (c) becomes

$$\sum_\alpha <p_\alpha, p_\mu> \; <p_{\lambda \backslash \alpha}, p_\nu> .$$

This is zero unless μ occurs as a subsequence of λ, with complementary subsequence ν (that is, unless $\lambda = \mu \cup \nu$), as required. On the other hand, if $\lambda = \mu \cup \nu$, the right side of (ii) (c) becomes

$$2^{-\ell(\mu)} z_\mu \, 2^{-\ell(\nu)} z_\nu \, | Y(\lambda, \mu) | ,$$

where $Y(\lambda, \mu)$ is the set of subsequences of λ which, as sequences, agree with μ. The proof is completed by observing that

$$z_{\mu \cup \nu} = z_\mu z_\nu | Y(\mu \cup \nu, \mu) | .$$

(Together with $z_{(k)} = k$, this is a useful inductive characterization of z_λ.)

This completes the proof of Theorem 7.7.

At this point, the Q-functions of the title can be defined.

Definition. Let $\{Q_\lambda : \lambda \in \mathcal{D}(n)\}$ be the unique orthogonal basis for $\Delta_{\mathbb{R}}^{(n)} (= \Delta^{(n)} \otimes \mathbb{R})$ such that $Q_\lambda - q_\lambda$ is a real linear combination of $\{q_\mu : \mu \in \mathcal{D}(n), \mu < \lambda\}$.

Thus in each homogeneous degree, the transition matrix between $\{Q_\lambda\}$ and $\{q_\lambda\}$ is unitriangular, where the indices are ordered by reverse lexicographic order. The existence of $\{Q_\lambda\}$ is clear from the Gram–Schmidt process, which applies to $(\Delta_{\mathbb{R}}^n, <,>)$ since $\{p_\lambda : \lambda \in \mathcal{P}^0(n)\}$ is an orthogonal basis with $<p_\lambda, p_\lambda>$ positive. Clearly $Q_{(n)} = q_n$, since (n) is lexicographically maximal. For the main application of Q-functions to projective representations, we need only their definition and the following proposition, whose proof occupies the last lines of this chapter.

Proposition 7.13. (*i*) *For all* λ *in* $\mathcal{D}$,

$$<Q_\lambda, Q_\lambda> = 2^{\ell(\lambda)}.$$

(*ii*) *If* $\lambda \in \mathcal{D}(m)$ *and* $Q_\lambda = \sum_{\mu \in \mathcal{D}(m)} a_{\lambda\mu} q_\mu$, *then*:

(*a*) $a_{\lambda\mu} = 0$ *if either* $\mu > \lambda$ *or* $\ell(\mu) > \ell(\lambda)$;
(*b*) $a_{\lambda\lambda} = 1$, *and* $a_{\lambda\mu} \in 2\mathbb{Z}$ *for all* $\mu \neq \lambda$;
(*c*) $a_{\lambda\mu} \in 2^{\ell(\lambda)-\ell(\mu)}\mathbb{Z}$ *for all* μ.

Remarks. By (ii), Q_λ is in Δ, so it has integer coefficients. We shall see later that its coefficients are divisible by $2^{\ell(\lambda)}$.

There are two corollaries of (i) which we shall discuss before proceeding to the material needed for the proof of 7.13.

The following is a standard fact from linear algebra.

Lemma 7.14. *Suppose that* V *is a finite dimensional vector space with a symmetric inner product* $< , >$ *for which each of* $(\{v_i\}, \{v'_i\})$ *and* $(\{w_j\}, \{w'_j\})$ *is a pair of dual bases; that is*

$$<v_i, v'_k> = \delta_{ik} = <w_i, w'_k>.$$

Then

$$\sum_i v_i \otimes v'_i = \sum_j w_j \otimes w'_j,$$

an element which we refer to as the "fundamental tensor" of $< , >$.

Proof. Let

$$w_j = \sum A_{ij} v_i \text{ and } w'_k = \sum B_{\ell k} v'_\ell .$$

Then

$$\delta_{jk} = <w_j, w'_k> = \sum_{i,\ell} A_{ij} B_{\ell k} <v_i, v'_\ell> = \sum_i A_{ij} B_{ik}.$$

Thus $A^{tr}B = I$, so $AB^{tr} = I$. Hence

$$\begin{aligned}\sum_j w_j \otimes w'_j &= \sum_j \left(\sum_i A_{ij} v_i\right) \otimes \left(\sum_\ell B_{\ell j} v'_\ell\right)\\ &= \sum_{i,\ell} \sum_j A_{ij} B_{\ell j} v_i \otimes v'_\ell\\ &= \sum_{i,\ell} \delta_{i\ell} v_i \otimes v'_\ell\\ &= \sum_i v_i \otimes v'_i\,.\end{aligned}$$

Corollary 7.15. $\sum_{\lambda\in\mathcal{D}} 2^{-\ell(\lambda)} Q_\lambda \otimes Q_\lambda = \sum_{\lambda\in\mathcal{P}^0} 2^{\ell(\lambda)} z_\lambda^{-1} p_\lambda \otimes p_\lambda.$

Remarks. The equation in 7.15 takes place in the "strong" direct product of $\Delta^n_{\mathbb{R}} \otimes \Delta^n_{\mathbb{R}}$ for all n; alternatively, restrict the summations to $\mathcal{D}(n)$ and $\mathcal{P}^0(n)$. Note that, by (7.12), the fundamental tensor is also $\sum_{\lambda\in\mathcal{P}} q_\lambda \otimes m_\lambda$. However, $\{q_\lambda : \lambda\in\mathcal{P}\}$ and $\{m_\lambda : \lambda\in\mathcal{P}\}$ are not a dual pair of bases for Δ: the former is linearly dependent, and the latter is not a subset of Δ, being almost disjoint from Δ.

Corollary 7.16 (Schur). *For any disjoint sets X and Y of variables,*

$$\prod_{(x,\,y)\in X\times Y} (1+xy)/(1-xy) = \sum_{\lambda\in\mathcal{D}} 2^{-\ell(\lambda)} Q_\lambda(X) Q_\lambda(Y)\,.$$

Proof. Combine 7.15 with 7.10.

This is a fundamental identity. We have chosen definitions to make it almost trivial, postponing to Chapter 9 several explicit formulae for Q_λ which put flesh on its bones. Such formulae are not needed for the formulation and proof in Chapter 8 of the basic theory connecting Q-functions to projective representations. They will be needed for many of the applications.

Before proving 7.13, we shall give an operator formula for Q_λ which will also lead to the more explicit formulae in Chapters 9 and 11.

Since $\Delta_{\mathbb{R}}$ has an orthonormal basis (certain multiples of p_λ), the following equation uniquely defines a linear map $f^\perp : \Delta_{\mathbb{R}} \to \Delta_{\mathbb{R}}$, which decreases degree by the degree of the given element f of $\Delta_{\mathbb{R}}$:

$$<f^{\perp}(g),h> \;=\; <g,fh> . \tag{7.17}$$

Clearly $1^{\perp}$ is the identity, and

$$(fg)^{\perp} = g^{\perp} \circ f^{\perp} = f^{\perp} \circ g^{\perp} . \tag{7.18}$$

Then 7.3 and 7.7 (ii) (c) immediately yield

$$q_n^{\perp}(fg) = \sum_i q_i^{\perp}(f) q_{n-i}^{\perp}(g) . \tag{7.19}$$

Furthermore, for any f in $\Delta_{\mathbb{R}}$,

$$\begin{aligned} <q_n^{\perp}(q_i),f> &= <q_i, q_n f> \\ &= \sum_r <q_r, q_n> <q_{i-r}, f> \quad \text{by 7.7 (ii) (c)} \\ &= <q_n, q_n> <q_{i-n}, f> . \end{aligned}$$

Thus, for $n > 0$,

$$q_n^{\perp}(q_i) = 2q_{i-n} . \tag{7.20}$$

Combining (7.18), (7.19), (7.20) and the definition of Δ yields $q_\lambda^{\perp}(\Delta) \subset \Delta$. Since the map $f \mapsto f^{\perp}$ is itself linear, we find $f^{\perp}(\Delta) \subset \Delta$ if $f \in \Delta$. Thus, for f in Δ, we shall regard $f^{\perp}$ as an endomorphism of the subgroup Δ of $\Delta_{\mathbb{R}}$.

Definition. For each integer n, define a group homomorphism $\mathcal{B}_n : \Delta \to \Delta$ by

$$\mathcal{B}_n(f) = \sum_i (-1)^i q_{n+i} \, q_i^{\perp}(f) .$$

The summation is effectively finite; that is, $0 \le i \le \text{degree}(f)$. The operator $\mathcal{B}_n$ increases degree by n (decreases it by $|n|$, if $n<0$).

Theorem 7.21. *For each* $\lambda = (\lambda_1, \ldots, \lambda_\ell)$ *in* $\mathcal{D}$,

$$Q_\lambda = \mathcal{B}_{\lambda_1} \circ \ldots \circ \mathcal{B}_{\lambda_\ell}(1) .$$

Proof. As a temporary notation, define b_λ, for $\lambda \in \mathcal{D}$ of length ℓ, to be $\mathcal{B}_{\lambda_1} \circ \mathcal{B}_{\lambda_2} \circ \ldots \circ \mathcal{B}_{\lambda_\ell}(1)$. We must show:

(i) $b_\lambda - q_\lambda$ is a linear combination of $\{q_\mu : \mu < \lambda\}$; and
(ii) $\{b_\lambda\}$ is orthogonal.

The proof of (i) is by induction on the length ℓ of λ. When $\ell = 1$,

$$b_r = \mathcal{B}_r(1) = \sum_i (-1)^i q_{r+i} q_i^\perp(1) = q_r \,.$$

For the inductive step, let $\lambda_1 = r$ and let λ' be the partition obtained by deleting r from λ. Thus

$$b_\lambda = \mathcal{B}_r(b_{\lambda'}) = \sum_i (-1)^i q_{r+i} q_i^\perp(b_{\lambda'}) \,.$$

By the inductive hypothesis, we may write

$$b_{\lambda'} = q_{\lambda'} + \sum_{\nu<\lambda'} n_\nu q_\nu \,.$$

Also, for each i,

$$q_i^\perp(b_{\lambda'}) = \sum_\rho k_\rho q_\rho \,,$$

for some k_ρ. Thus

$$\begin{aligned} b_\lambda &= q_r b_{\lambda'} + \sum_{i>0} (-1)^i q_{r+i} q_i^\perp(b_{\lambda'}) \\ &= q_r q_{\lambda'} + \sum_{\nu<\lambda'} n_\nu q_r q_\nu + \sum_{i>0} (-1)^i q_{r+i} q_i^\perp(b_{\lambda'}) \\ &= q_\lambda + \sum_{\nu<\lambda'} n_\nu q_r q_\nu + \sum_{i>0} \sum_\rho (-1)^i k_\rho q_{r+i} q_\rho \,. \end{aligned}$$

Since $(r) \cup \nu$ (for $\nu < \lambda'$) and $(r+i) \cup \rho$ are both less than λ, the induction is completed.

In order to verify (ii), we first show, again by induction on ℓ, that for $k \geq 0$,

$$q_{r+k}^\perp(b_\lambda) = 2\delta_{k,0} b_{\lambda'} \,. \tag{7.22}$$

To see this note that

$$\begin{aligned}
q^{\perp}_{r+k}(b_\lambda) &= q^{\perp}_{r+k}[\mathcal{B}_r(b_{\lambda'})] \\
&= q^{\perp}_{r+k}\left(\sum_i (-1)^i q_{r+i} q_i^{\perp}(b_{\lambda'})\right), \quad \text{by definition of } \mathcal{B}_r \\
&= \sum_i \sum_j (-1)^i q_j^{\perp}(q_{r+i}) q^{\perp}_{r+k-j}\left(q_i^{\perp}(b_{\lambda'})\right), \quad \text{by (7.19)} \\
&= \sum_i \sum_{j>0} (-1)^i 2q_{r+i-j} q_i^{\perp}\left(q^{\perp}_{r+k-j}(b_{\lambda'})\right) + \sum_i (-1)^i q_{r+i} q_i^{\perp}\left(q^{\perp}_{r+k}(b_{\lambda'})\right),
\end{aligned}$$

by (7.18) and (7.20).

But $r+k > \lambda_2$, the first part of λ', so $q^{\perp}_{r+k}(b_{\lambda'}) = 0$ by the inductive hypothesis. Thus the second term above is zero. Writing $s = r+i+k-j$, we obtain

$$q^{\perp}_{r+k}(b_\lambda) = \sum_s 2q_{s-k}\left(\sum_{i>s-r-k} (-1)^i q_i^{\perp} q^{\perp}_{s-i}(b_{\lambda'})\right).$$

If $i \leq s-r-k$, then $s-i \geq r+k$, so that $q^{\perp}_{s-i}(b_{\lambda'}) = 0$ by induction. Thus

$$\begin{aligned}
q^{\perp}_{r+k}(b_\lambda) &= \sum_s 2q_{s-k}\left(\sum_i (-1)^i q_i^{\perp} q^{\perp}_{s-i}(b_{\lambda'})\right) \\
&= 2q_{-k} b_{\lambda'}, \quad \text{by Proposition 7.1} \\
&= 2\delta_{k,0} b_{\lambda'}
\end{aligned}$$

as required.

To complete the proof of (ii), assume that μ and λ are partitions of the same integer and that $\mu < \lambda$, where $\mu_j = \lambda_j$ for all $j < i$, but $\mu_i > \lambda_i$. Using (7.22),

$$\begin{aligned}
<q_\mu, b_\lambda> &= q_\mu^{\perp}(b_\lambda) \\
&= \ldots q^{\perp}_{\mu_3} q^{\perp}_{\mu_2} q^{\perp}_{\mu_1}(b_{\lambda_1,\lambda_2,\lambda_3}\ldots) \\
&= 2\delta_{\mu_1,\lambda_1}[\ldots q^{\perp}_{\mu_3} q^{\perp}_{\mu_2}(b_{\lambda_2,\lambda_3}\ldots)] \\
&= 2^2\delta_{\mu_1,\lambda_1}\delta_{\mu_2,\lambda_2}[\ldots q^{\perp}_{\mu_3}(b_{\lambda_3}\ldots)] = \text{etc.}
\end{aligned}$$

At the ith step, this gives zero. Now write

$$b_\mu = q_\mu + \sum_{\nu<\mu} n_\nu q_\nu .$$

Then, if $\mu < \lambda$,

$$<b_\mu, b_\lambda> = <q_\mu, b_\lambda> + \sum_{\nu<\mu} n_\nu <q_\nu, b_\lambda> = 0$$

as required.

Proof of Proposition 7.13. Part (i) follows by repeating the last few lines of the proof of Theorem 7.21, but with $\mu = \lambda$ rather than $\mu < \lambda$:

$$\begin{aligned} <q_\lambda, Q_\lambda> &= \ldots q_{\lambda_3}^\perp q_{\lambda_2}^\perp q_{\lambda_1}^\perp (Q_{\lambda_1, \lambda_2, \ldots}) \\ &= 2[\ldots q_{\lambda_3}^\perp q_{\lambda_2}^\perp (Q_{\lambda_2, \ldots})] \\ &= 2^{\ell(\lambda)}, \end{aligned}$$

by induction on $\ell(\lambda)$. Thus

$$\begin{aligned} <Q_\lambda, Q_\lambda> &= <q_\lambda, Q_\lambda> + \sum_{\nu<\lambda} n_\nu <q_\nu, Q_\lambda> \\ &= <q_\lambda, Q_\lambda> \\ &= 2^{\ell(\lambda)}. \end{aligned}$$

As for part (ii), that $a_{\lambda\lambda} = 1$ and $a_{\lambda\mu} = 0$ for $\mu > \lambda$ are part of the definition of Q_λ. The other assertions are proved by induction on $\ell(\lambda)$. To start the induction there is nothing to prove, since $Q_{(m)} = q_m$. Now let $\lambda = (\lambda_1, \ldots, \lambda_\ell)$, $\lambda' = (\lambda_2, \ldots, \lambda_\ell)$ and

$$Q_{\lambda'} = \sum_\mu a_{\lambda'\mu} q_\mu .$$

For a sequence $\alpha \in \mathbb{N}^s$, let $\eta(\alpha)$ be the number of i for which $\alpha_i > 0$. It is immediate by induction on s using (7.19) and (7.20) that, for $\mu = (\mu_1, \ldots, \mu_s)$,

$$q_i^\perp(q_\mu) = \sum_{\alpha \in \mathbb{N}^s} 2^{\eta(\alpha)} q_{\mu-\alpha}$$

where $\mu-\alpha$ means $(\mu_1-\alpha_1, \ldots, \mu_s-\alpha_s)$. Thus

$$
\begin{aligned}
Q_\lambda &= \mathcal{B}_{\lambda_1}(Q_{\lambda'}) \\
&= \sum_{\mu,i} (-1)^i a_{\lambda'\mu}\, q_{\lambda_1+i}\, q_i^\perp(q_\mu) \\
&= \sum_{\substack{\mu \\ \alpha\in\mathbb{N}^{\ell(\mu)}}} (-1)^{|\alpha|}\, 2^{\eta(\alpha)}\, a_{\lambda'\mu}\, q_{\lambda_1+|\alpha|}\, q_{\mu-\alpha}\,.
\end{aligned}
$$

Relation (7.1) is then used if necessary to write this as a linear combination of q_ν for $\nu\in\mathcal{D}$. This decreases (if anything) the lengths of subscripts, so $a_{\lambda\nu}=0$ if $\ell(\nu)>\ell(\lambda)$, using the inductive hypothesis. This proves (a).

Relation (7.1) writes q_n^2 as an element of 2Δ, so the only term not divisible by 2 would arise with all $\alpha_i=0$ and $\mu=\lambda'$, i.e. $(-1)^0\, 2^0\, a_{\lambda'\lambda'}=a_{\lambda\lambda}=1$, completing the proof of (b).

Finally, a term q_ν with

$$\ell(\mu)+1-\ell(\nu)=t>0$$

can arise from $q_{\lambda_1+|\alpha|,\mu_1-\alpha_1,\ldots,\mu_s-\alpha_s}$ only by having $\alpha_i=\mu_i$ for "p" values of i, together with "q" applications of relation (7.1), for some p,q with $p+q\geq t$. But this introduces $p+q$ factors of 2, so the relation

$$a_{\lambda\nu}\in 2^{\ell(\lambda)-\ell(\nu)}\mathbb{Z}$$

follows, since

$$a_{\lambda'\mu}\in 2^{\ell(\lambda)-1-\ell(\mu)}\mathbb{Z}\,.$$

Notes

Our introduction to symmetric functions owes much to Macdonald (1979). In the second edition, Macdonald (1991+), there is, if anything, even more emphasis on characterizing functions by orthogonality and unitriangularity. See also Stanley (1988). The abstract approach emphasizing the coproduct (Hopf algebra), for Schur functions and many other examples, is thoroughly developed in Zelevinsky (1981). See also Geissinger (1977), Hoffman (1979) and Liulevicius (1980). Algebras closer to Zelevinsky's and relevant to the subject of this book are defined and classified in Bean-Hoffman (1988). The Q-functions were first defined by Schur (1911). Many of the formulae in this chapter can be obtained by setting $t = -1$ in formulae within the algebra $\Lambda \otimes \mathbb{Q}(t)$ with inner product for which the Hall–Littlewood functions are an orthogonal basis; Macdonald (1979) III. See also Chapters 9 and 12 and Appendix 9 of this book. Theorem 7.21 appears in Hoffman (1990). It is the analogue of a formula for Schur functions due to Bernstein; Zelevinsky (1981, p. 69). By an argument as in the proof of 7.21, it generalizes immediately to the Hall–Littlewood functions; Hoffman (1990). See Appendix 9 of this book. Bernstein's formula was rediscovered by Frenkel (1986) in the context of the vertex operator algebra approach to representations of affine Kac–Moody algebras. Proceeding analogously, Jing (1989A, B) has independently discovered Theorem 7.21 and the analogue for Hall–Littlewood functions, respectively, by considering "twisted" vertex operators.

8

THE IRREDUCIBLE NEGATIVE REPRESENTATIONS OF $\tilde{S}_n$

The character calculations and Q-functions of the last two chapters can now be combined to obtain the main results on the irreducible projective representations of S_n and A_n. These are Theorems 8.6 and 8.7 below.

Proposition 8.1. *Let λ be a partition of length ℓ, and let μ^* be a refinement of λ, with $\mu^{(1)},\ldots,\mu^{(\ell)}$ being partitions of $\lambda_1,\ldots,\lambda_\ell$, respectively. Let σ be an element of the Young subgroup $S_\lambda = S_{\lambda_1} \times \ldots \times S_{\lambda_\ell} \subset S_{|\lambda|}$. Suppose that σ has cycle type μ^*. Then the order of the centralizer of σ in S_λ is the product*

$$w_{\mu^*}(\lambda) = z_{\mu^{(1)}} \ldots z_{\mu^{(\ell)}} .$$

Sketch of the proof. (See 1.2.15 in James and Kerber (1981) for the direct count of conjugacy classes when $\ell = 1$.) The centralizer of an element (g, h) in a direct product is $C_G(g) \times C_H(h)$, so we need only to prove this when $\ell = 1$. Let the cycles of length k in σ be

$$\begin{array}{llll} (a_{11} & a_{12} & \cdots & a_{1k}) \\ (a_{21} & a_{22} & \cdots & a_{2k}) \\ \cdot & & & \cdot \\ \cdot & & & \cdot \\ \cdot & & & \cdot \\ (a_{f1} & a_{f2} & \cdots & a_{fk}). \end{array}$$

Then the centralizer of σ in the symmetric group is the product $G_1 \times G_2 \times \ldots$, where G_k is the subgroup generated by those permutations which permute the rows above among themselves, together with those which permute the elements in one row and keep the elements in all the other rows fixed. These generating permutations must fix all integers not appearing in the array. The group G_k is known as a monomial group or wreath product. It has order $f!k^f$, the kth factor in the definition of z_μ. Thus $|G_1 \times G_2 \times \ldots| = z_\mu$, as required.

Proposition 8.2. *If* $\lambda \in \mathcal{D}(m)$, *then*

$$q_\lambda = \sum_{\mu \in \mathcal{P}^0(m)} \sum_{\mu^* \in \mathrm{Ref}(\lambda,\mu)} 2^{\ell(\mu)} p_\mu / w_{\mu^*}(\lambda)\,.$$

(Recall that Ref(λ, μ) is the set of refinements $(\mu^{(1)},\ldots,\mu^{(\ell)})$ with $\mu^{(i)} \in \mathcal{P}(\lambda_i)$ such that, when rearranged into weakly decreasing order, the juxtaposed sequence $\mu^{(1)}\mu^{(2)}\ldots$ becomes the partition μ.)

Proof. Multiply together the expressions for $q_{\lambda_1}, q_{\lambda_2}, \ldots$ given by (7.9).

Definition. Let $\Theta^{(m)}$ be the $\mathbb{Z}$-module generated by the characters ξ of those negative representations of $\tilde{S}_m$ such that $\xi(g) = 0$ for all g which project to an element of S_m whose cycle type has an even part. It will transpire that $\Theta^{(m)}$ is the group generated by the characters of self-associate negative representations of $\tilde{S}_m$. Let Θ be the direct sum of the $\Theta^{(m)}$.

By Theorem 3.6, the inverse image of each conjugacy class in S_m is either a conjugacy class $\mathcal{C}$ in $\tilde{S}_m$, or splits as a union of two classes. In the first case, if g is in $\mathcal{C}$, then zg is also in $\mathcal{C}$, and so $\xi(g) = 0$ for any negative character ξ. In the second case, the values of ξ are negatives of each other on the two classes. By Theorem 3.8, this case includes all $\mathcal{C}$ which project to cycle types in $\mathcal{P}^0(m)$.

Definition. For $\lambda \in \mathcal{P}^0(m)$, let g be an element in $\tilde{S}_m$ projecting to cycle type λ and such that $\chi_m(g) > 0$, where χ_m is the basic negative character of $\tilde{S}_m$ defined before Theorem 6.6. Let $f(\lambda) := f(g)$ for any linear combination f of negative characters of $\tilde{S}_m$.

In effect we are fixing one of the two conjugacy classes in $\tilde{S}_m$ which project to cycle type λ in S_m (relative to the fixed choice of χ_m). This will simplify formulae.

Definition. The *odd characteristic* is the group homomorphism

$$\mathrm{och}: \Theta \to \Delta_{\mathbb{C}} = \Delta \otimes \mathbb{C}\,,$$

defined, for f in $\Theta^{(m)}$, by

$$\operatorname{och}(f) = \sum_{\lambda \in \mathcal{P}^0(m)} 2^{\ell(\lambda)/2} f(\lambda) p_\lambda / z_\lambda .$$

Remark. Coefficients in $\mathbb{C}$ are a luxury; the subfield $\mathbb{Q}[\sqrt{2}]$ would suffice, as we shall see.

Example. Let m be an odd integer. By Theorem 6.8 (i), $\chi_m(\lambda)$ is zero unless λ is in $\mathcal{P}^0(m)$, in which case

$$\chi_m(\lambda) = 2^{[\ell(\lambda)-1]/2} .$$

Thus χ_m is in $\Theta^{(m)}$, and

$$\begin{aligned} \operatorname{och}(\chi_m) &= \sum_{\lambda \in \mathcal{P}^0(m)} 2^{\ell(\lambda)/2} 2^{[\ell(\lambda)-1]/2} p_\lambda / z_\lambda \\ &= 2^{-1/2} \sum_{\lambda \in \mathcal{P}^0(m)} 2^{\ell(\lambda)} p_\lambda / z_\lambda \\ &= 2^{-1/2} q_m, \end{aligned}$$

using (7.9). When m is even, χ_m is not in $\Theta^{(m)}$. However, Theorem 6.8 (ii) shows that $\chi_m + \chi_m^a$ is in $\Theta^{(m)}$. A similar calculation to the above shows that $\operatorname{och}(\chi_m + \chi_m^a) = q_m$.

Definition. Let $\lambda \in \mathcal{D}(m)$. Recall, from before Theorem 6.11, the basic negative character χ_λ of the subgroup $\tilde{S}_\lambda$ of $\tilde{S}_m$. Let

$$\xi_\lambda = \chi_\lambda \uparrow \tilde{S}_m ,$$

the induced character.

Proposition 8.3. *The values of the character ξ_λ for $\lambda \in \mathcal{D}(m)$ are as follows. Let $g \in \tilde{S}_m$ project to cycle type μ.*

(i) If $\mu \in \mathcal{P}^0(m)$ and g is conjugate to g_μ (rather than zg_μ), then

$$\xi_\lambda(g) = 2^{[\ell(\mu)-\ell(\lambda)-\varepsilon(\lambda)]/2} z_\mu \sum_{\mu^* \in \operatorname{Ref}(\lambda,\mu)} w_{\mu^*}(\lambda)^{-1} .$$

where $\varepsilon(\lambda)$ is 0 or 1 according as λ is even or odd.

(ii) If $\mu = \lambda$ *is odd, then*

$$\xi_\lambda(g) = \chi_\lambda(g) = \pm[(-1)^{m-\ell(\lambda)-1}\lambda_1\lambda_2 \ldots]^{1/2}.$$

(iii) Otherwise $\xi_\lambda(g) = 0$.

Proof. Formula (D5) of Chapter 4 for induced characters may be written

$$(\theta\uparrow G)(g) = \sum_{\substack{[w]\in G/H \\ wgw^{-1}\in H}} \theta(wgw^{-1}), \tag{**}$$

where θ is a character of H. To calculate, one makes a list $\{h_1,\ldots,h_s\} \subset H$ of elements conjugate in G to g such that every element of H which is conjugate in G to g is conjugate in H to exactly one h_i. Then (**) becomes

$$(\theta\uparrow G)(g) = \sum_{i=1}^{s} \theta(h_i)\, |C_G(h_i)| / |C_H(h_i)| \tag{*}$$

since $|C_G(h_i)| / |C_H(h_i)|$ is the number of cosets $[w]\in G/H$ for which wgw^{-1} is conjugate in H to h_i.

Now let $H = \tilde{S}_\lambda$, $\theta = \chi_\lambda$ and $G = \tilde{S}_m$. The values of χ_λ are given in Theorem 6.11, which reveals the (somewhat unusual) circumstance that, for all $h\in\tilde{S}_\lambda$, the number $\chi_\lambda(h)$ depends only on the conjugacy class in $\tilde{S}_m$ of h.

To prove (iii), when g projects to $\mu \in \mathcal{P}^0(m)$, with $\mu \neq \lambda$ if λ is odd, we have $\chi_\lambda(h) = 0$ for all h in $\tilde{S}_\lambda$ which are conjugate in $\tilde{S}_m$ to g. Thus $\xi_\lambda(g) = 0$, since all terms $\theta(h_i)$ in (*) are zero.

To prove (i), Theorem 6.11 yields

$$\chi_\lambda(h) = 2^{[\ell(\mu)-\ell(\lambda)-\varepsilon(\lambda)]/2}$$

for any h in $\tilde{S}_\lambda$ conjugate to g_μ. The positive sign follows from 6.11 (iii), since g_μ was chosen with $\chi_n(g_\mu) > 0$. The h_i in (*) are the elements

$$h_i = (g_{\mu^{(1)}},\ldots, g_{\mu^{(\ell)}})\in \tilde{S}_{\lambda_1} \tilde{\curlyvee} \ldots \tilde{\curlyvee} \tilde{S}_{\lambda_\ell} = \tilde{S}_\lambda,$$

one for each $\mu^* = (\mu^{(1)},\ldots, \mu^{(\ell)})\in \mathrm{Ref}(\lambda,\mu)$. The element h_i is conjugate to g_μ, not zg_μ. By Proposition 8.1, the factor $|C_G(h_i)| / |C_H(h_i)|$ in (*) is $z_\mu / w_{\mu^*}(\lambda)$, as required.

Finally, to prove (ii), suppose that g projects to cycle type λ, which is odd. By Theorem 6.11 (ii), for any $h\in\tilde{S}_\lambda$ which is conjugate in $\tilde{S}_m$ to g, the value of χ_λ at h is the right-hand side of the formula to be proved. Since $\lambda\in\mathcal{D}$, there is only one refinement $\lambda^* \in \mathrm{Ref}(\lambda, \lambda)$, namely that in which $\lambda^{(i)}$ is the partition (λ_i) with one part. Furthermore, $w_{\lambda^*}(\lambda) = z_\lambda = \lambda_1\lambda_2 \ldots \lambda_\ell$, so (*) yields $\xi_\lambda(g) = \chi_\lambda(g)$.

Now $\mathrm{och}(\xi_\lambda)$ can be calculated, generalizing the previous example.

Proposition 8.4. *If the strict partition λ is even (so that $m-\ell(\lambda)$ is even), then $\xi_\lambda \in \Theta^{(m)}$ and*

$$\mathrm{och}(\xi_\lambda) = 2^{-\ell(\lambda)/2} q_\lambda\,.$$

If λ is odd, then $\xi_\lambda + \xi_\lambda^a \in \Theta^{(m)}$ and

$$\mathrm{och}(\xi_\lambda + \xi_\lambda^a) = 2^{[1-\ell(\lambda)]/2} q_\lambda\,.$$

Proof. Both of these are easy calculations using Propositions 8.2 and 8.3. For example, if λ is even,

$$\begin{aligned}
\mathrm{och}(\xi_\lambda) &= \sum_{\mu \in \mathcal{P}^0(m)} 2^{\ell(\mu)/2}\left(2^{[\ell(\mu)-\ell(\lambda)]/2} z_\mu \sum_{\mu^* \in \mathrm{Ref}(\lambda,\mu)} w_{\mu^*}(\lambda)^{-1}\right) p_\mu / z_\mu \\
&= 2^{-\ell(\lambda)/2} \sum_{\mu \in \mathcal{P}^0(m)} \sum_{\mu^* \in \mathrm{Ref}(\lambda,\mu)} w_{\mu^*}(\lambda)^{-1} 2^{\ell(\mu)} p_\mu \\
&= 2^{-\ell(\lambda)/2} q_\lambda\,.
\end{aligned}$$

The proof for odd λ is left as an exercise.

Remark. Another method for getting this result is to use $\tilde{\otimes}$ to define a multiplication on Θ such that och becomes a ring homomorphism, and such that ξ_λ is a multiple of $\xi_{\lambda_1}\xi_{\lambda_2}\ldots$. But $\xi_{\lambda_i} = \chi_{\lambda_i}$, on which the value of och is given by the previous example. See also Remark (2) at the end of this chapter.

We are about to investigate the connection between the inner product on Δ from 7.7 and the familiar inner product of characters. If ζ and ξ are characters of negative representations of $\tilde{S}_m$, their inner product is

$$<\zeta, \xi> = \sum_{g \in \tilde{S}_m} \zeta(g)\bar{\xi}(g)/(m!2)\,.$$

Since the conjugacy class of elements of cycle type λ in S_m has $m!/z_\lambda$ elements, so have both corresponding conjugacy classes in $\tilde{S}_m$ (when $\lambda \in \mathcal{P}^0(m)$). Also

$$\zeta(zg)\xi(zg) = -\zeta(g)[-\xi(g)] = \zeta(g)\xi(g)\,.$$

If ζ and ξ are in $\Theta^{(m)}$, their values are zero if λ has an even part. Thus, for ζ and ξ in $\Theta^{(m)}$,

$$<\zeta, \xi> = \sum_{\lambda \in \mathcal{P}^0(m)} \zeta(\lambda)\xi(\lambda)/z_\lambda\,.$$

By linearity, this formula holds for any f and g in $\Theta^{(m)}$.

Proposition 8.5. *The map* och *is an isometry.*

Proof. Let f and g be in $\Theta^{(m)}$. Then

$$\begin{aligned}
&< \mathrm{och}(f), \mathrm{och}(g)> \\
&= < \sum_{\lambda \in \mathcal{P}^0(m)} 2^{\ell(\lambda)/2} f(\lambda) p_\lambda / z_\lambda, \sum_{\mu \in \mathcal{P}^0(m)} 2^{\ell(\mu)/2} g(\mu) p_\mu / z_u > \\
&= \sum_{\lambda \in \mathcal{P}^0(m)} 2^{\ell(\lambda)} f(\lambda) g(\lambda) <p_\lambda, p_\lambda> / z_\lambda^2 \\
&= \sum_{\lambda \in \mathcal{P}^0(m)} f(\lambda) g(\lambda) / z_\lambda, \quad \text{using Theorem 7.7 (iii)} \\
&= <f, g>\,,
\end{aligned}$$

as required.

Definition. For a partition λ, recall that

$$\varepsilon(\lambda) = \begin{cases} 0 & \text{if } \lambda \text{ is even;} \\ 1 & \text{if } \lambda \text{ is odd.} \end{cases}$$

Thus $\varepsilon(\lambda)$ is the reduction mod 2 of $|\lambda| - \ell(\lambda)$. Define

$$d(\lambda) = 2^{[\varepsilon(\lambda) - \ell(\lambda)]/2}\,.$$

In the following theorem, the reverse lexicographic order $<$ is used on the set of strict partitions. This theorem is the central result concerning projective representations of S_n and A_n.

Theorem 8.6 (Schur 1911). *Let m be a non-negative integer, which is greater than 1 in the clauses below referring to $\tilde{A}_m$. The irreducible negative representations of $\tilde{S}_m$ and $\tilde{A}_m$ are given as follows. (All partitions below are strict.)*

(i) For each λ *in* $\mathcal{D}(m)$*, there is a negative character* $<\lambda>$ *of* $\tilde{S}_m$ *satisfying:*

(a) $<\lambda>$ *is irreducible;*

(b) $<\lambda> - \xi_\lambda$ *is a linear combination over* $\mathbb{Z}$ *of*

$$\{\xi_\mu : \text{even } \mu<\lambda\} \cup \{\xi_\mu + \xi_\mu^a : \text{odd } \mu<\lambda\} ;$$

$$(c)\ d(\lambda)Q_\lambda = \begin{cases} \text{och}(<\lambda>) & \text{if } \lambda \text{ is even;} \\ \text{och}(<\lambda> + <\lambda>^a) & \text{if } \lambda \text{ is odd.} \end{cases}$$

The characters $<\lambda>$ *are uniquely specified by (b) together with either (a) or (c).*

(ii) The $<\lambda>$*, as* λ *varies over* $\mathcal{D}(m)$*, together with the* $<\lambda>^a$ *when* λ *is odd, are a complete non-redundant list of the irreducible negative characters of* $\tilde{S}_m$*.*

(iii) When λ *is odd (so that* $<\lambda> \neq <\lambda>^a$ *by (ii)), the character* $<\lambda>$ *restricts to an irreducible character for* $\tilde{A}_m$ *(which is also the restriction of* $<\lambda>^a$*). If* λ *is even, then* $<\lambda>^a = <\lambda>$ *and the restriction of* $<\lambda>$ *to* $\tilde{A}_m$ *is a sum of two distinct (conjugate) irreducibles.*

(iv) The restrictions as described in (iii) give a complete non-redundant list of the irreducible negative representations for $\tilde{A}_m$*.*

Remarks. The definition of och together with (i) (c) show that the following are closely related:

(i) the transition matrix between the bases $\{Q_\lambda : \lambda \in \mathcal{D}\}$ and $\{p_\lambda : \lambda \in \mathcal{P}^0\}$ of $\Delta_\mathbb{Q}$; and

(ii) the values of $<\lambda>$ on elements which project to cycle types in $\mathcal{P}^0$.

Although the map och requires coefficients in $\mathbb{Q}(\sqrt{2})$, these values are all integers. The next theorem will show that $<\lambda>$ is zero on almost all the other elements.

The proof of the aspect of (i) (b) that, in the expression for $<\lambda>$, the coefficients of ξ_μ and ξ_μ^a are equal, for $\mu < \lambda$ and both odd, will depend on a character calculation in the next theorem. This one aspect will therefore not be proved until after that calculation.

Theorem 8.7 (Schur 1911). *Let* $\lambda \in \mathcal{D}(m)$ *have length* ℓ*, and let* $g \in \tilde{S}_m$*. First suppose that* λ *is odd.*

(i) If g projects to cycle type λ*, then*

$$<\lambda>(g) = \chi_\lambda(g) = \pm i^{(m-\ell+1)/2} (\lambda_1\lambda_2 \cdots \lambda_\ell/2)^{1/2} .$$

(ii) If g projects to a cycle type which is neither in $\mathcal{P}^0(m)$ *nor equal to* λ*, then* $<\lambda>(g) = 0$*.*

Now suppose instead that λ is even.

(iii) If g does not project to a cycle type in $\mathcal{P}^0(m)$, then $<\lambda>(g) = 0$.

(iv) Let $<\lambda>\downarrow \tilde{A}_m$ be the sum of irreducibles α and α^c. If g is in $\tilde{A}_m$ and projects to cycle type λ, then

$$\alpha(g) - \alpha^c(g) = \pm i^{(m-\ell)/2} (\lambda_1 \lambda_2 \cdots \lambda_\ell)^{1/2}.$$

For all other g, we have $\alpha(g) = \alpha^c(g)$. (Since $\alpha(g) + \alpha^c(g) = <\lambda>(g)$, this calculates α and α^c in terms of $<\lambda>$.)

Remark. All the values of $<\lambda>$ not specified in this theorem are determined in principle by Theorem 8.6 (i) (c), and in practice by a method to be described in Chapter 10.

Proof of Theorem 8.6. Since (i) (a) fixes the $<\lambda>$ up to re-naming, and (i) (c) determines them up to interchanging $<\lambda>$ with $<\lambda>^a$, the uniqueness assertion in (i) is clear.

Given $\lambda \in \mathcal{D}(m)$, write

$$Q_\lambda = \sum_{\mu \in \mathcal{D}(m)} a_{\lambda\mu} q_\mu .$$

By Proposition 7.13 (ii), we find that $d(\lambda) a_{\lambda\mu} / d(\mu) \in \mathbb{Z}$ (usually with powers of 2 to spare), and that $(a_{\lambda\mu})$ is a unitriangular integer matrix:

$$a_{\lambda\mu} = \begin{cases} 1 & \text{if } \lambda = \mu; \\ 0 & \text{if } \lambda < \mu. \end{cases}$$

As a temporary notation, let

$$\eta_\lambda = \begin{cases} \xi_\lambda & \text{if } \lambda \text{ is even;} \\ \xi_\lambda + \xi_\lambda^a & \text{if } \lambda \text{ is odd.} \end{cases}$$

Then Proposition 8.4 states that

$$\operatorname{och}(\eta_\lambda) = d(\lambda) q_\lambda .$$

Thus if $b_{\lambda\mu} := d(\lambda) a_{\lambda\mu} / d(\mu)$ and

$$\phi_\lambda := \sum b_{\lambda\mu} \eta_\mu ,$$

then $(b_{\lambda\mu})$ is also unitriangular with integer entries. Furthermore,

$$\begin{aligned}\operatorname{och}(\phi_\lambda) &= \sum_\mu b_{\lambda\mu} \operatorname{och}(\eta_\mu) \\ &= \sum_\mu d(\lambda) a_{\lambda\mu} d(\mu)^{-1} d(\mu) q_\mu \\ &= d(\lambda) \sum_\mu a_{\lambda\mu} q_\mu \\ &= d(\lambda) Q_\lambda .\end{aligned}$$

Since och is an isometry, we have, using 7.13 (i),

$$<\phi_\lambda, \phi_\lambda> = d(\lambda)^2 <Q_\lambda, Q_\lambda> = d(\lambda)^2 2^{\ell(\lambda)} = 2^{\varepsilon(\lambda)} .$$

Thus, when λ is even, $<\phi_\lambda, \phi_\lambda> = 1$. But ϕ_λ is an integer linear combination of characters, so $\phi_\lambda = \pm<\lambda>$ for some irreducible character $<\lambda>$. Now

$$(c_{\lambda\mu}) := (b_{\lambda\mu})^{-1}$$

is also an integer unitriangular matrix, and

$$\eta_\lambda = \phi_\lambda + \sum_{\mu<\lambda} c_{\lambda\mu} \phi_\mu .$$

But if $\mu \neq \lambda$, then

$$<\phi_\mu, \phi_\lambda> = d(\mu) d(\lambda) <Q_\mu, Q_\lambda> = 0 .$$

Since η_λ is a character, it is a non-negative linear combination of irreducibles. Thus ϕ_λ is irreducible, i.e. $\phi_\lambda = +<\lambda>$, not $-<\lambda>$.

When λ is odd, $<\phi_\lambda,\phi_\lambda> = 2$. Thus $\phi_\lambda = \pm<\lambda> \pm <\lambda>'$ for a pair $<\lambda>$, $<\lambda>'$ of distinct irreducibles. By an argument as above,

$$\phi_\lambda = <\lambda> + <\lambda>' .$$

Since $<\phi_\lambda, \phi_\mu> = 0$ for $\lambda \neq \mu$, the list consisting of all $<\lambda>$, together with $<\lambda>'$ for odd λ, contains no repeats. The number of these irreducibles is $a_m + 2b_m$, in notation from Chapter 3. By Corollary 3.10, this is also the difference between the numbers of conjugacy classes in $\tilde{S}_m$ and in S_m. Thus the list above is a complete non-redundant list of irreducible negative characters for $\tilde{S}_m$, by Proposition 4.4.

At this point, we have proved (i) (a), (i) (c), and (ii), except for showing that $<\lambda>' = <\lambda>^a$.

When λ is even,

$$<\lambda> = \phi_\lambda = \sum b_{\lambda\mu}\eta_\mu = \xi_\lambda + \sum_{\text{ev}\,\mu<\lambda} b_{\lambda\mu}\xi_\mu + \sum_{\text{od}\,\mu<\lambda} b_{\lambda\mu}(\xi_\mu + \xi_\mu^a)\,.$$

The middle summation is over all $\mu \in \mathcal{D}(m)$ for which μ is even and $\mu < \lambda$, denoted "ev $\mu < \lambda$". Similarly, "od $\mu < \lambda$" means μ is odd and $\mu < \lambda$. We shall use this notation on several summations in the rest of this chapter. The equation implies (i) (b) in the case of even λ, as well as that $<\lambda>^a = <\lambda>$.

When λ is odd,

$$\begin{aligned}\xi_\lambda + \xi_\lambda^a &= \sum c_{\lambda\mu}\phi_\mu \\ &= <\lambda> + <\lambda>' + \sum_{\text{ev}\,\mu<\lambda} c_{\lambda\mu}<\mu> + \sum_{\text{od}\,\mu<\lambda} c_{\lambda\mu}(<\mu> + <\mu>').\end{aligned}$$

Suppose, inductively on the reverse lexicographic order, that $<\mu>' = <\mu>^a$ for all odd $\mu < \lambda$. Writing, for non-negative integers b, b', $d_{\lambda\mu}$, $e_{\lambda\mu}$ and $f_{\lambda\mu}$,

$$\xi_\lambda = b<\lambda> + b'<\lambda>' + \sum_{\text{ev}\,\mu<\lambda} d_{\lambda\mu}<\mu> + \sum_{\text{od}\,\mu<\lambda} (e_{\lambda\mu}<\mu> + f_{\lambda\mu}<\mu>')\,,$$

we have

$$\xi_\lambda^a = b<\lambda>^a + b'<\lambda>'^a + \sum_{\text{ev}\,\mu<\lambda} d_{\lambda\mu}<\mu> + \sum_{\text{od}\,\mu<\lambda} (f_{\lambda\mu}<\mu> + e_{\lambda\mu}<\mu>')\,.$$

Add these two equations and compare the result to the previous equation for $\xi_\lambda + \xi_\lambda^a$. Since $<\lambda>$ and $<\lambda>'$ are distinct from all $<\mu>$ and $<\mu>^a$ for $\mu < \lambda$, we must have $<\lambda>^a$ equal to either $<\lambda>$ or $<\lambda>'$. But $<\lambda>^a = <\lambda>$ would yield the contradiction $2b = 1$, so $<\lambda>^a = <\lambda>'$, completing the induction. Furthermore, one of b or b' equals 1, and the other is zero. By switching the names of $<\lambda>$ and $<\lambda>'$ if necessary, we have $b = 1$ and $b' = 0$. Inverting the above equation for ξ_λ yields

$$<\lambda> = \xi_\lambda + \sum_{\text{ev}\,\mu<\lambda} k_{\lambda\mu}\xi_\mu + \sum_{\text{od}\,\mu<\lambda} (\ell_{\lambda\mu}\xi_\mu + m_{\lambda\mu}\xi_\mu^a)$$

for integers $k_{\lambda\mu}$, $\ell_{\lambda\mu}$ and $m_{\lambda\mu}$. For (ii) (b), it remains only to prove that $\ell_{\lambda\mu} = m_{\lambda\mu}$ for all odd λ and μ. As mentioned in the remarks after the statement of Theorem 8.6, we shall leave this until after part of Theorem 8.7 has been proved.

The only remaining statements in Theorem 8.6 are those referring to the restrictions to $\tilde{A}_m$. These are all immediate consequences of what has been proved, using Theorem 4.2 and Corollary 4.3, which describe the irreducibles for an index two subgroup quite generally.

Proof of Theorem 8.7 (i), (ii) and (iii). Recall from Theorem 3.8 that zg is conjugate to g if and only if g projects to a cycle type which is neither in $\mathcal{P}^0$ nor an odd partition in $\mathcal{D}$. Recall from Proposition 8.3 that $\xi_\mu(g) \neq 0$ implies that either (i) μ is even and g projects to a cycle type in $\mathcal{P}^0$, or (ii) μ is odd and g projects to a cycle type which is either in $\mathcal{P}^0$ or equal to μ. Write

$$<\lambda> = \xi_\lambda + \sum_{\text{ev}\,\mu<\lambda} k_{\lambda\mu}\xi_\mu + \sum_{\text{od}\,\mu<\lambda} (\ell_{\lambda\mu}\xi_\mu + m_{\lambda\mu}\xi_\mu^a)$$

for integers $k_{\lambda\mu}$, $\ell_{\lambda\mu}$ and $m_{\lambda\mu}$. (So far we have only proved $\ell_{\lambda\mu} = m_{\lambda\mu}$ when λ is even.)

To prove (iii), suppose that g projects to a cycle type not in $\mathcal{P}^0(m)$ and that λ is even. By the paragraph above, $\xi_\lambda(g) = 0$, and $\xi_\mu(g) = 0$ for all even μ. Now $\ell_{\lambda\mu} = m_{\lambda\mu}$, so it suffices to prove $\xi_\mu^a(g) = -\xi_\mu(g)$ for μ odd. Denote the relation of conjugacy by $\sim$. If $g \not\sim zg$, then the paragraph above states that g projects to a cycle type which is odd, so

$$\xi_\mu^a(g) = \sigma(g)\xi_\mu(g) = -\xi_\mu(g) .$$

On the other hand, if $g \sim zg$, then

$$\xi_\mu^a(g) = 0 = -\xi_\mu(g) ,$$

since ξ_μ^a and ξ_μ are negative characters. (Of course, $<\lambda>(g) = 0$ directly for the same reason, when $g \sim zg$.)

To prove (i) and (ii) requires a slightly more delicate argument. Here λ is odd, so $\lambda \in \mathcal{P}^0$. Thus, if g projects to cycle type λ and $\mu \neq \lambda$, then $\xi_\mu(g) = 0$ by the second previous paragraph. Hence

$$<\lambda>(g) = \xi_\lambda(g) + \sum_{\text{ev}\,\mu<\lambda} k_{\lambda\mu}0 + \sum_{\text{od}\,\mu<\lambda} 0 = \xi_\lambda(g) ,$$

which proves (i), in view of Proposition 8.3 (ii). Now

$$\sum_{h\in \tilde{S}_m} |<\lambda>(h)|^2 = m!2 ,$$

and

$$\sum_{h\in \tilde{A}_m} |<\lambda>(h)|^2 = m! ,$$

since $<\lambda>\downarrow \tilde{A}_m$ is irreducible. Subtracting,

$$\sum_{h\in \tilde{S}_m\setminus\tilde{A}_m} |<\lambda>(h)|^2 = m!\,.$$

But

$$|<\lambda>(g)|^2 - \lambda_1\lambda_2\ldots\lambda_\ell/2 = z_\lambda/2$$

by part (i). The conjugacy classes of g and zg in $\tilde{S}_m$ contain, in total, $m!2/z_\lambda$ elements by Proposition 8.1. Thus the elements conjugate to g and zg in the last summation gobble up the whole $m!$, and so $<\lambda>(h) = 0$ if h projects to an odd cycle type other than λ itself. If h projects to an even cycle type not in $\mathcal{P}^0(m)$, then $h \sim zh$, so $<\lambda>(h) = 0$. This completes the proof of (ii), and so, of all except (iv), in Theorem 8.7.

Completion of the proof of Theorem 8.6 (i) (b). For odd λ in $\mathcal{D}$, we have

$$<\lambda> = \xi_\lambda + \sum_{\text{ev}\,\mu<\lambda} k_{\lambda\mu}\xi_\mu + \sum_{\text{od}\,\mu<\lambda} (\ell_{\lambda\mu}\xi_\mu + m_{\lambda\mu}\xi^a_\mu) \qquad (*)$$

and we must prove $\ell_{\lambda\mu} = m_{\lambda\mu}$ for all μ in the second sum. Fix a strict odd $\nu < \lambda$. By Theorem 8.7 (ii), $<\lambda>(g_\nu) = 0$. By Proposition 8.3, $\xi_\lambda(g_\nu) = 0 = \xi_\mu(g_\nu)$ for all $\mu \neq \nu$ in $\mathcal{D}$, and $\xi^a_\nu(g_\nu) = -\xi_\nu(g_\nu) \neq 0$. Evaluating (*) at g_ν therefore yields

$$0 = (\ell_{\lambda\nu} - m_{\lambda\nu})\xi_\nu(g_\nu)\,,$$

and so $\ell_{\lambda\nu} = m_{\lambda\nu}$ as required.

Proof of Theorem 8.7 (iv). Proceed by induction on lexicographic order. Assume that $\lambda \in \mathcal{D}(m)$ is even and for all even strict $\mu < \lambda$, the statement holds. For each $\mu \in \mathcal{D}(m)$ let

$$<\mu>\downarrow \tilde{A}_m = \begin{cases} \alpha_\mu + \alpha^c_\mu & \text{if } \mu \text{ is even;} \\ \alpha_\mu & \text{if } \mu \text{ is odd.} \end{cases}$$

Let $g_\lambda \in \tilde{A}_m$ be an element which projects to cycle type λ. Let

$$\chi_\lambda \downarrow \tilde{A}_\lambda = \beta_\lambda + \beta^c_\lambda\,.$$

Let $\gamma_\lambda = \beta_\lambda \uparrow \tilde{A}_m$. Then $\gamma^c_\lambda = \beta^c_\lambda \uparrow \tilde{A}_m$, since, whenever one has a diagram such as

$$\begin{array}{ccc} \tilde{A}_\lambda & \subset & \tilde{A}_m \\ \text{index } 2 \cap & & \cap \text{ index } 2\,, \\ \tilde{S}_\lambda & \subset & \tilde{S}_m \end{array}$$

conjugation commutes with inducing, as may be checked, for example, by calculating characters. Thus

$$\begin{aligned} \gamma_\lambda + \gamma_\lambda^c &= (\beta_\lambda + \beta_\lambda^c) \uparrow \tilde{A}_m = (\chi_\lambda \downarrow \tilde{A}_\lambda) \uparrow \tilde{A}_m \\ &= (\chi_\lambda \uparrow \tilde{S}_m) \downarrow \tilde{A}_m = \xi_\lambda \downarrow \tilde{A}_m\,. \end{aligned}$$

The second last equality is another property of such a diagram which follows from character theory; see also Serre (1977) (7.3). Writing

$$\xi_\lambda = <\lambda> + \sum_{\text{ev}\,\mu<\lambda} a_{\lambda\mu} <\mu> + \sum_{\text{od}\,\mu<\lambda} a_{\lambda\mu}(<\mu> + <\mu>^a)$$

and restricting to $\tilde{A}_m$, we obtain

$$\gamma_\lambda + \gamma_\lambda^c = \alpha_\lambda + \alpha_\lambda^c + \sum_{\text{ev}\,\mu<\lambda} a_{\lambda\mu}(\alpha_\mu + \alpha_\mu^c) + \sum_{\text{od}\,\mu<\lambda} 2a_{\lambda\mu}\alpha_\mu\,.$$

Thus, for some $b, b', b_{\lambda\mu}, b'_{\lambda\mu}$ in $\mathbb{N}$,

$$\gamma_\lambda = b\alpha_\lambda + b'\alpha_\lambda^c + \sum_{\text{ev}\,\mu<\lambda} (b_{\lambda\mu}\alpha_\mu + b'_{\lambda\mu}\alpha_\mu^c) + \sum_{\text{od}\,\mu<\lambda} b_{\lambda\mu}\alpha_\mu\,.$$

Conjugating, adding, and comparing to the previous equation implies that

$$\gamma_\lambda = \alpha_\lambda + \sum_{\text{ev}\,\mu<\lambda} (b_{\lambda\mu}\alpha_\mu + b'_{\lambda\mu}\alpha_\mu^c) + \sum_{\text{od}\,\mu<\lambda} a_{\lambda\mu}\alpha_\mu\,,$$

where we have switched the names of α_λ and α_λ^c if necessary. Conjugating and subtracting,

$$\gamma_\lambda - \gamma_\lambda^c = \alpha_\lambda - \alpha_\lambda^c + \sum_{\text{ev}\,\mu<\lambda} (b_{\lambda\mu} - b'_{\lambda\mu})(\alpha_\mu - \alpha_\mu^c)\,.$$

Evaluating at g_λ,

$$(\gamma_\lambda - \gamma_\lambda^c)(g_\lambda) = (\alpha_\lambda - \alpha_\lambda^c)(g_\lambda)\,,$$

since the inductive hypothesis includes the assumption that $\alpha_\mu(g_\lambda) = \alpha_\mu^c(g_\lambda)$ for all μ in the summation. By the same argument as in the proof of Proposition 8.3 (ii) which showed $\xi_\lambda(g') = \chi_\lambda(g')$ when λ is odd and g' projects to cycle type λ, we obtain

$$(\gamma_\lambda - \gamma_\lambda^c)(g_\lambda) = (\beta_\lambda - \beta_\lambda^c)(g_\lambda)\,.$$

But by Theorem 6.8

$$(\beta_\lambda - \beta_\lambda^c)(g_\lambda) = \pm i^{[m-\ell(\lambda)]/2}(\lambda_1\lambda_2 \ldots)^{1/2}\,,$$

yielding the first assertion in 8.7 (iv). Now

$$\begin{aligned}\sum_{g\in\tilde{A}_m} &|(\alpha_\lambda - \alpha_\lambda^c)(g)|^2 \\ &= \sum_{g\in\tilde{A}_m}\{|\alpha_\lambda(g)|^2 + |\alpha_\lambda^c(g)|^2 - \alpha_\lambda(g)\overline{\alpha_\lambda^c(g)} - \overline{\alpha_\lambda(g)}\alpha_\lambda^c(g)\} \\ &= m!2\,,\end{aligned}$$

since α_λ and α_λ^c are distinct irreducibles for $\tilde{A}_m$. The set of elements which are conjugate to either g_λ or zg_λ in $\tilde{A}_m$ has size $m!2/\lambda_1\lambda_2\ldots\lambda_\ell$, by Propositions 8.1 and 3.9, so these elements "use up" the entire $m!2$ in the above sum of squares. Thus $(\alpha_\lambda - \alpha_\lambda^c)(g) = 0$ for all $g \in \tilde{A}_m$ which do not project to cycle type λ, completing the proof (which also shows that $b_{\lambda\mu} = b'_{\lambda\mu} = a_{\lambda\mu}/2$ from the earlier part of the proof).

Remarks. (1) By Theorem 8.7, the free abelian group with basis

$$\{<\lambda>: \text{ev}\,\lambda \in \mathcal{D}\} \cup \{<\lambda> + <\lambda>^a : \text{od}\,\lambda \in \mathcal{D}\}$$

is exactly the group of those virtual characters β such that $\beta(g) = 0$ unless g projects to a cycle type in $\mathcal{P}^0$, i.e. the group Θ. But by Theorem 8.6 (ii), it is also the group of $\mathbb{Z}$-linear combinations of self-associate characters. This establishes the statement made about Θ in its definition.

(2) It is clear from Theorem 8.6 that och takes values in the subgroup $\Delta\otimes\mathbb{Q}[\sqrt{2}]$ of $\Delta\otimes\mathbb{C}$. One can extend och to a $\mathbb{Q}[\sqrt{2}]$-linear map

$$\Theta\otimes\mathbb{Q}[\sqrt{2}] \to \Delta\otimes\mathbb{Q}[\sqrt{2}]\,.$$

By Propositions 8.4 and 8.5, this map is an isometric isomorphism. Using the operation $\tilde{\otimes}$, the group Θ can be made into a ring (see Appendix 8 for an algebra which carries more information than this ring). Then och is an isometric isomorphism of $\mathbb{Q}[\sqrt{2}]$-algebras.

(3) It should be pointed out that, despite appearances, the material of this chapter has no genuine dependence on symmetric function theory. All that is used is the graded $\mathbb{Z}$-algebra generated by symbols $\{q_1, q_2, \ldots\}$ satisfying

relations (7.1), together with the unique coproduct and inner product satisfying (7.3) and (7.7) (ii), plus the fact that Gram–Schmidt produces a basis $\{Q_\lambda\}$ satisfying (7.13). Our proofs of these facts are actually independent of the interpretation of q_n as a symmetric function. One could make $\Theta \otimes \mathbb{R}$ itself into such an algebraic object. Appendix 8 pursues this point of view, working with a slightly more refined Hopf algebra with inner product.

(4) In Chapter 6, the representation χ_α of $\tilde{S}_{\alpha_1} \tilde{\curlyvee} \tilde{S}_{\alpha_2} \tilde{\curlyvee} \ldots$ was defined for *any* sequence α of positive integers. Inducing this yields the character ξ_α of $\tilde{S}_{|\alpha|}$. Any question concerning these, such as expressing them in terms of a basis $\{\xi_\lambda, \xi_\lambda^a : \lambda \in \mathcal{D}\}$ or $\{<\lambda>, <\lambda>^a : \lambda \in \mathcal{D}\}$, translates immediately into the "same" question for the functions q_α, via the map och. Several such questions (for example, writing Q_λ as a polynomial in the q_i) are answered in Chapters 9 and 12. Of course, this correspondence works modulo the indeterminacy of ξ_α up to replacement by its associate.

Notes

In Macdonald (1979) I. 7, the Schur functions (defined in Appendix 9 of this book) and a map ch are used to describe the basic facts concerning linear representations of S_n. An important role is played by the inner product on Λ for which the Schur functions form an orthonormal basis. The proof of Theorem 8.6, using Q-functions and the map och, is intended to appear analogous to that description. A missing analogue is the ring structure on Θ; see Appendix 8 and its notes. All the results in this chapter are due to Schur (1911). Maps which are multiples of och are used in recent treatments of these results in Stembridge (1989), where the required properties of Q_λ are derived from a combinatorial definition, and in Jozefiak (1989), which uses Schur's (Pfaffian) definition of Q_λ as well as representations of $\mathbb{Z}/2$-graded algebras. Earlier, Sergeev (1984) had defined a certain characteristic map, in analogy to Macdonald's, taking values in the algebra of Q-functions, and determining the characters of one of the seven sequences of essential double covers of the hyperoctahedral groups.

Appendix 8

An algebra of $\tilde{S}_n$-representations

Here we indicate an alternative route to Theorem 8.6, that is, to a description of the irreducible representations of $\tilde{S}_n$ as linear combinations of other representations. We use no computations of characters nor manipulations with symmetric functions. The prerequisites for this are most of the material in the first three chapters together with the part of Appendix 6 before Proposition A6.4. The trace calculation in Appendix 6 will not be used till Appendix 12, where we recalculate all the characters, based on the approach here.

Amongst the advantages of this method are that it treats $\tilde{A}_n$ and $\tilde{S}_n$ simultaneously and symmetrically, and describes some additional structure, an algebra built from projective representations.

Recall the category $\mathscr{G}$ of objects (G, z, σ) introduced before Proposition 3.1. In analogy with $\mathbb{Z}/2$-graded modules, we define a $\mathbb{Z}/2$-*graded representation* of (G, z, σ) to be an ordered pair (V_0, V_1), of complex vector spaces, together with a linear action of G on $V_0 \oplus V_1$ such that $gV_i = V_{i+\sigma(g)}$ for $i=0$ or 1 and $g \in G$, where subscripts are read (mod 2). We shall consider only *negative* representations; that is, those for which $zv = -v$ for all v. A *morphism* of two $\mathbb{Z}/2$-graded representations of (G, z, σ) is a linear transformation which commutes with the action of G and preserves the grading.

The motivation is the following. For $n \geq 2$, the $\mathbb{Z}/2$-graded representations of $\tilde{S}_n$ are in $1-1$ correspondence with representations of $\tilde{A}_n$ (see Proposition A8.1 below). On the other hand, for $n = 0$ and 1, it is the $\mathbb{Z}/2$-graded representations of $\tilde{S}_n$, rather than the ungraded representations of $\tilde{A}_n$, which are needed in forming the algebra (see the paragraph before Theorem A8.8). Furthermore, the algebra multiplication appears more natural using $\mathbb{Z}/2$-graded representations, even in cases where they could be replaced by representations of $\ker \sigma$.

If $\sigma = 0$, it is clear that the set of isomorphism classes of $\mathbb{Z}/2$-graded negative representations of (G, z, σ) is in $1-1$ correspondence with the set of isomorphism classes of ordered pairs of negative representations of (G, z).

Proposition A8.1. *If $\sigma \neq 0$, there is a bijection from the set of isomorphism classes of $\mathbb{Z}/2$-graded negative representations of* (G, z, σ) *to the set of isomorphism classes of negative representations of* $(\ker \sigma, z)$.

Proof. Map $[(V_0, V_1)]$ to $[V_0]$. To define an inverse, choose g_1 with $\sigma(g_1) = 1$. Let $\ker \sigma$ act on W, and map $[W]$ to $[(W, W)]$, where the action of G on $W \oplus W$ is

$$g(w, w') = \begin{cases} (gw, g_1^{-1} g g_1 w') \text{ if } \sigma(g) = 0; \\ (g g_1 w', g_1^{-1} g w) \text{ if } \sigma(g) = 1. \end{cases}$$

It is straightforward to check that this is well-defined and provides an inverse.

The direct sum of (V_0, V_1) and (W_0, W_1) is $(V_0 \oplus W_0, V_1 \oplus W_1)$ with the obvious action. This is invariant under isomorphism and makes the set of isomorphism classes into an abelian monoid.

For each object G in $\mathcal{G}$, define T^*G to be the $\mathbb{Z}/2$-graded abelian group $T^0G \oplus T^1G$, where T^0G and T^1G are defined as follows. Let T^1G be the group generated by the abelian monoid of isomorphism classes of negative representations of G. Thus T^1G is free abelian, with a canonical $\mathbb{Z}$-basis consisting of the irreducibles. Let T^0G be the group generated by the abelian monoid of isomorphism classes of $\mathbb{Z}/2$-graded negative representations of G.

It might seem at least as natural to reverse the grading of T^*. The reason for the present convention is contained in the definition of $\boxtimes^-$ below.

A non-zero $\mathbb{Z}/2$-graded representation is *irreducible* if it cannot be written as a direct sum of non-zero summands. (By the usual argument, this is equivalent to having no $\mathbb{Z}/2$-graded invariant subspaces except the trivial ones.)

Proposition A8.2. *Every $\mathbb{Z}/2$-graded representation is a direct sum of irreducibles, which are unique up to order and isomorphism. Thus T^0G is a free abelian group, with the irreducibles forming a basis.*

Proof. With a few verifications concerning grading, this follows by the usual proof of the analogous result in the ungraded case. In particular, uniqueness follows because the only morphisms between irreducibles are zero and isomorphisms, and all endomorphisms are scalar multiples of the identity. An alternative proof uses Proposition A8.1 (which shows that $T^0G \cong T^1 \ker\sigma$ if $\sigma \neq 0$) and $T^0G \cong T^1G \oplus T^1G$ if $\sigma = 0$.

We shall make T^*G into a module over the commutative ring K defined by

$$K := \mathbb{Z}[\kappa]/(\kappa^3 - 2\kappa).$$

This ring is given a $\mathbb{Z}/2$-grading by requiring $\kappa \in K^{(1)}$. Thus

$$K^{(0)} \cong \mathbb{Z} \oplus \mathbb{Z} \text{ with basis } \{1, \rho\} \text{ where } \rho := \kappa^2 - 1,$$

$$\text{and } K^{(1)} \cong \mathbb{Z} \qquad \text{with basis } \{\kappa\}.$$

Construct an action of K on T^*G as follows. For $[(V_0, V_1)] \in T^0G$, define

$$\kappa[(V_0, V_1)] = [V_0 \oplus V_1] \in T^1G.$$

For $[W] \in T^1G$, first let $\rho(W)$ be W as a vector space, but with action $*$, where $g*w = (-1)^{\sigma(g)}gw$. Thus $\rho(W)$ is the *associate* of W, as in Chapter 3. The notation is chosen so that $[\rho(W)] = \rho[W]$. Now define $\kappa(W)$ to be $W \oplus \rho(W)$ as a module, and give it the following $\mathbb{Z}/2$-grading:

$$\kappa(W)^{(j)} = \{(w, (-1)^j w) : w \in W\}.$$

Compare this to the definition of M^*_{2n-1} in Appendix 6.

It is relatively straightforward to check the following:

(i) these actions and gradings are well-defined and invariant under direct sum and isomorphism;

(ii) the definition of ρ on an ungraded W is compatible with the relation $\kappa^2 = 1+\rho$;

(iii) the relation $\kappa^3 = 2\kappa$ holds for the action of κ, using (ii) and the relation $\kappa\rho = \kappa$;

(iv) the action of ρ on T^0G is determined by $\rho(V_0, V_1) = (V_1, V_0)$.

Note that $\rho^2 = 1$. These statements establish the following theorem.

Theorem A8.3. *The above definitions make T^*G into a $\mathbb{Z}/2$-graded module over K.*

Definition. An element x of $T^0G \cup T^1G$ is a *special irreducible* if and only if x is the isomorphism class of an irreducible representation and $\rho x \neq x$.

Theorem A8.4. *For each object G, determine integers r_0 and r_1, as follows:*

If $\sigma \neq 0$, let

$r_0 + 2r_1 =$ *the number of conjugacy classes in $G/\{1, z\}$ whose inverse images consist of two conjugacy classes in G;*

$2r_0 + r_1 =$ *the number of classes in* $\ker\sigma/\{1, z\}$ *which similarly "split in* $\ker\sigma$*".*

If $\sigma = 0$, let $r_1 = 0$ and

$r_0 =$ *the number of conjugacy classes in $G/\{1, z\}$ which "split in G".*

Then:

(i) the module T^*G is free over K of rank r_0+r_1, with r_i generators in T^iG;

(ii) a K-basis for T^*G consists of special irreducibles, with exactly one chosen from each pair x, ρx. The set of all irreducibles is the union of the triples $\{x, \rho x, \kappa x\}$.

Proof. When $\sigma = 0$, this is obvious. When $\sigma \neq 0$, it is essentially the same as Theorem 4.2 and Corollary 4.3. Note that the action of κ corresponds to inducing $T^0 \to T^1$ and to restricting $T^1 \to T^0$, when we identify T^0G with $T^1\ker\sigma$. The action of ρ on T^0 corresponds to taking conjugates, and on T^1 to taking associates. For a direct proof of this theorem, which does not use Proposition A8.1, see Hoffman and Humphreys (1986) (2.10).

Now we define some more structure on T^*G, namely, a K-valued inner product $\langle\, , \rangle$. This will be uniquely specified by requiring some K-basis of special irreducibles to be orthonormal (using Theorem A8.4). Of course, $\langle\, , \rangle$ is also required to be bilinear over K. Note that $\langle\, , \rangle$ is symmetric, and is independent of the choices from the pairs x, ρx. The only other orthonormal bases are obtained by taking the negatives of some or all of the chosen special irreducibles.

Note that G' does not mean commutator subgroup in the discussion below.

For two objects G and G', recall the construction $G\tilde{\curlyvee}G'$ of Chapter 3. We shall now define an operation

$$\boxtimes^- : T^jG \times T^kG' \to T^{j+k}(G\tilde{\curlyvee}G').$$

Its effect on $([V], [V'])$ will be denoted $[V\boxtimes^- V']$. There will be four cases depending on which ingredients are graded. Components of $\boxtimes^-$ are closely related to, and may be used to redefine, the operation $\widetilde{\otimes}$ of Chapter 5; see the notes after this appendix.

In the two cases when V is $\mathbb{Z}/2$-graded, define $V\boxtimes^- V'$ as a module to be $(V_0 \oplus V_1) \otimes V'$ with action determined as follows. Let $(g, g') \in G\tilde{\curlyvee}G'$, $v \in V_j$ and $v' \in V'$. Define

$$(g, g')(v \otimes v') = (-1)^{j\sigma'(g')}(gv) \otimes (g'v').$$

If V' is also $\mathbb{Z}/2$-graded, define the grading on $V\boxtimes^- V'$ by

$$(V\boxtimes^- V')_0 = V_0 \otimes V'_0 + V_1 \otimes V'_1;$$

$$(V\boxtimes^- V')_1 = V_1 \otimes V'_0 + V_0 \otimes V'_1.$$

In the third case, when V is ungraded but V' is graded, use $V \otimes (V'_0 \oplus V'_1)$ with action determined by

$$(g, g')(v \otimes v') = (-1)^{\sigma(g)[j+\sigma'(g')]}(gv) \otimes (g'v')$$

if $v' \in V'_j$. Straightforward calculations similar to those in Proposition 5.6 show that what has so far been specified is a well-defined operation, invariant under isomorphism and biadditive with respect to direct sum.

In the final case, both V and V' are ungraded. Define (with $i^2 = -1$)

$$(V \boxtimes^- V')_0 := \text{Span}\{(v, v) \otimes (v', v') + i(v, -v) \otimes (v', -v') : v \in V, v' \in V'\}$$

$$(V \boxtimes^- V')_1 := \text{Span}\{(v, v) \otimes (v', -v') - i(v, -v) \otimes (v', v') : v \in V, v' \in V'\}.$$

These are to be regarded as subspaces of $(V \oplus \rho V) \boxtimes^- (V' \oplus \rho V')$. Here we are using $\boxtimes^-$ as in the first two cases defined above, with $V \oplus \rho V$ regarded as the $\mathbb{Z}/2$-graded representation κV, and similarly for V'. The sum of the subspaces $(V \boxtimes^- V')_i$ immediately above is easily seen to be direct, and invariant under the action of $G \tilde{\curlyvee} G'$. Thus we have defined a graded representation $V \boxtimes^- V'$. Since this construction is somewhat complicated and unnatural, we give it below in two other forms.

Firstly, let $\nabla^k(v, v')$ be an abbreviation for the element in the definition of $(V \boxtimes^- V')_k$ in the paragraph above. Then the following is the formula for the action:

$$(g, g')\nabla^k(v, v') = \left((-1)^{k+\sigma'(g')}i\right)^{\sigma(g)} \nabla^{k+\sigma(g)+\sigma'(g')}(gv, g'v').$$

Note that, for $k \in \mathbb{Z}/2$, the power i^k is defined by $i^0 = 1$ and $i^1 = i$. In particular,

$$i^{k+\ell} = (-1)^{k\ell} i^k i^\ell, \text{ not } i^k i^\ell.$$

The function ∇^k is bilinear. Using it, one can verify associativity of $\boxtimes^-$ (modulo the identification $[(g, h), \ell] \leftrightarrow [g, (h, \ell)]$) by considering the eight cases according to which of U, V or W is graded. For example, in the case $(U, V, W) \in T^0 \times T^1 \times T^0$, if $u \in U_i$ and $w \in W_k$, the actions of $[(g, h), \ell]$ on $(u \otimes v) \otimes w$, and of $[g, (h, \ell)]$ on $u \otimes (v \otimes w)$, are not the "same"; but the map sending $(u \otimes v) \otimes w$ to $(-1)^{ik} u \otimes (v \otimes w)$ gives an isomorphism from $(U \boxtimes^- V) \boxtimes^- W$ to $U \boxtimes^- (V \boxtimes^- W)$. When $(U, V, W) \in T^1 \times T^1 \times T^1$, an isomorphism is given by

$$\nabla^k(u, v) \otimes w \mapsto i^k u \otimes (\nabla^1(v, w) + (-1)^k \nabla^0(v, w)).$$

Even with associativity, the naturalness of using these four components of $\boxtimes^-$ together is not compelling. However there is a related, more natural, map

$$\mathrm{HOM}^- : T^*H \times T^*(G\tilde{\curlyvee}H) \to T^*G.$$

To define it, first define a subset $\mathrm{HOM}_H^-(U, W)$ of linear maps $\phi: U \to W$, where $[U] \in T^*H$ and $[W] \in T^*(G\tilde{\curlyvee}H)$.

(i) If both U and W are graded, $\phi \in \mathrm{HOM}_H^-(U, W)_j$ if and only if $\phi(hu) = h\phi(u)$ and $\phi(U_i) \subset W_{i+j}$.

(ii) If exactly one is graded, $\phi \in \mathrm{HOM}_H^-(U, W)$ if and only if $\phi(hu) = h\phi(u)$.

(iii) If neither is graded, $\phi \in \mathrm{HOM}_H^-(U, W)_j$ if and only if $\phi(hu) = (-1)^{j\sigma(h)} h\phi(u)$.

In all these formulae, the action of H on W is $hw = (1, h)w$.

Define the action of G on $\mathrm{HOM}_H^-(U, W)$ as follows:

$(g\phi)(u) = (-1)^{j\sigma(g)}(g, 1)\phi(u)$ if U is graded, $u \in U_j$;

$(g\phi)(u) = (g, 1)\phi(u)$ if both U and W are ungraded;

$(g\phi)_j(u) = i^{\sigma(g)}(-1)^{j\sigma(g)}(g,1)\phi_{j+\sigma(g)}(u)$ if U is ungraded, W is graded and ϕ_j is the component of ϕ in W_j.

The promised map $T^*H \times T^*(G\tilde{\curlyvee}H) \to T^*G$ then takes (U, W) to $\mathrm{HOM}_H^-(U, W)$.

The connection between HOM^- and $\boxtimes^-$ is as follows. Firstly, there is an evaluation map of $G \tilde{\curlyvee} H$-modules

$$\mathrm{HOM}^-(U, W) \boxtimes^- U \to W$$

given as follows:

$$\begin{array}{ll} \phi \otimes u \mapsto \phi(u) & \text{if } W \in T^1; \\ \phi \otimes u \mapsto (-1)^{ik}\phi(u) & \text{if } (U, W) \in T^0 \times T^0,\ \phi \in \mathrm{HOM}_i^- \text{ and } u \in U_k; \\ \nabla^j(\phi, u) \mapsto \phi_j(u) & \text{if } (U, W) \in T^1 \times T^0. \end{array}$$

Taking $U = V_2 \boxtimes^- V_3$, this yields an adjointness isomorphism of G_1-modules:

$$\mathrm{HOM}^-_{G_2\tilde{\curlyvee}G_3}(V_2 \boxtimes^- V_3, W) \to \mathrm{HOM}^-_{G_2}\left[V_2, \mathrm{HOM}^-_{G_3}(V_3, W)\right],$$

where V_2, V_3 and W are acted on by G_2, G_3 and $G_1\tilde{\curlyvee}G_2\tilde{\curlyvee}G_3$ respectively.

This adjointness has the bonus of giving a non-computational proof of the associativity of $\boxtimes^-$: taking U to be either $(V_2 \boxtimes^- V_3) \boxtimes^- V_4$ or $V_2 \boxtimes^- (V_3 \boxtimes^- V_4)$, the representation

$$\mathrm{HOM}^{-}_{G_2 \curlyvee G_3 \curlyvee G_4}(U, W)$$

reduces in two steps to

$$\mathrm{HOM}^{-}_{G_2}\left\{ V_2, \mathrm{HOM}^{-}_{G_3}\left[V_3, \mathrm{HOM}^{-}_{G_4}(V_4, W) \right] \right\} .$$

A case by case proof that

$$\kappa(V \boxtimes^{-} V') \cong \kappa(V) \boxtimes^{-} V' \cong V \boxtimes^{-} \kappa(V')$$

is not hard to find (see Hoffman and Humphreys, 1986 (2.22)). However, it is rather easier to see that

$$\mathrm{HOM}^{-}(\kappa U, W) \cong \mathrm{HOM}^{-}(U, \kappa W), \cong \kappa \mathrm{HOM}^{-}(U, W),$$

from which the K-bilinearity of $\boxtimes^{-}$ follows by adjointness. In summary we have

Theorem A8.5. *The operation $\boxtimes^{-}$ is well-defined, bilinear over K, and associative.*

Before considering $\boxtimes^{-}$ further, let us digress to relate HOM^{-} to the inner product. The $\mathbb{Z}$-valued inner product of two representations in the standard theory agrees with the dimension of the space of invariant linear transformations between the representations. The analogous formula in our case is obtained by letting $G = \tilde{S}_0 = \{1, z\}$ in the definition of HOM^{-}, yielding a map

$$T^{*}H \times T^{*}H \cong T^{*}H \times T^{*}(\tilde{S}_0 \curlyvee H) \to T^{*}\tilde{S}_0 \cong K.$$

It is elementary to show that this agrees with our inner product $< , >$. More explicitly, $<U, W>$ is:

$$\begin{array}{ll} \dim \mathrm{HOM}^{-}(U, W)\,\kappa & \text{if exactly one is graded;} \\ \dim \mathrm{HOM}^{-}(U, W)_0\; 1 + \dim \mathrm{HOM}^{-}(U, W)_1\,\rho & \text{otherwise.} \end{array}$$

So far, the constructions $\boxtimes^{-}$ and HOM^{-} play the same rôle for the functor T^{*} on $\mathcal{G}$ as the role played by the usual tensor product and HOM for the representation ring functor on the category of just plain groups. The following is therefore not surprising.

Theorem A8.6. *If V and V′ are special irreducibles (that is, irreducibles W for which $W \not\cong \rho(W)$), then $V \boxtimes^{-} V'$ is also a special irreducible. The operation $\boxtimes^{-}$ determines an isomorphism of K-modules*

$$T^*G \otimes_K T^*G' \to T^*(G\tilde{\curlyvee}G').$$

In particular, all special irreducibles for $G\tilde{\curlyvee}G'$ *are obtained by applying* $\boxtimes^-$ *to pairs of irreducibles for* G *and* G'.

Proof. Let V, V', W and W' be special irreducibles. Identifying $T^*(\{1, z\})$ with K,

$$\mathrm{HOM}^-(V, W) = \begin{cases} 1 & \text{if } V \cong W; \\ \rho & \text{if } V \cong \rho W; \\ 0 & \text{otherwise.} \end{cases}$$

More generally, it is straightforward to see that

$$\mathrm{HOM}^-(V', W\boxtimes^- W') \cong \begin{cases} W & \text{if } V' \cong W'; \\ \rho W & \text{if } V' \cong \rho W'; \\ 0 & \text{otherwise.} \end{cases}$$

See Adams (1969), p. 73, for analogous manipulations with HOM and $\otimes$. Also, by adjointness,

$$\mathrm{HOM}^-(V\boxtimes^- V', W\boxtimes^- W') \cong \mathrm{HOM}^-(V, \mathrm{HOM}^-(V', W\boxtimes^- W')).$$

Combining these shows that $\mathrm{HOM}^-(V\boxtimes^- V', V\boxtimes^- V') = 1$, so $V\boxtimes^- V'$ is a special irreducible. It also shows that $V\boxtimes^- V'$ is distinct from $W\boxtimes^- W'$ except when $([V], [V'])$ is either $([W], [W'])$ or $([\rho W], [\rho W'])$.

It remains to show that all the special irreducibles for $G\tilde{\curlyvee}G'$ have the form $V\boxtimes^- V'$. This follows from an argument, using the sum of squares of dimensions, which is almost identical to that in the proof of Theorem 5.9. Alternatively, and more in the style of the above manipulations, it suffices to show that each irreducible W for $G\tilde{\curlyvee}G'$ is the codomain for a non-trivial $G\tilde{\curlyvee}G'$-invariant map from a representation of the form $V\boxtimes^- V'$. For this, take the evaluation map with $V = \mathrm{HOM}^-(V', W)$ for an irreducible constituent V' of the restriction of W to G'.

Let $\phi: G \to G'$ be a $\mathcal{G}$-map; that is, a group homomorphism such that $\phi(z) = z'$ and $\sigma'\phi(g) = \sigma(g)$ for all g. If V' is a representation of G', we denote by ϕ^*V' the representation of G obtained by "restricting V' along ϕ"; that is, ϕ^*V' is V' as a vector space, with G acting by $gv' := \phi(g)v'$. Similarly, in the graded case, $\phi^*(V'_0, V'_1)$ is just (V'_0, V'_1) as a pair of vector spaces, with action $gv' := \phi(g)v'$. Obviously ϕ^* is invariant under isomorphism, and preserves direct sums and the action of elements of K (that is, $\phi^*\kappa V' \cong \kappa\phi^* V'$). Thus it induces a map of K-modules, also denoted

$$\phi^* : T^*G' \to T^*G.$$

Let $\tau : G \tilde{\mathbf{Y}} G' \to G' \tilde{\mathbf{Y}} G$ be the $\mathcal{G}$-isomorphism in the proof of Theorem 3.3 (b).

Proposition A8.7. *If* $V \in T^iG$ *and* $V' \in T^jG'$, *then*

$$V \boxtimes^- V' \cong \rho^{ij}\tau^*(V' \boxtimes^- V).$$

Proof. When exactly one of V or V' is graded, the map $v \otimes v' \mapsto v' \otimes v$ is the required isomorphism. When both are graded, one uses the map $v \otimes v' \mapsto (-1)^{k\ell} v' \otimes v$ where $v \in V_k$ and $v' \in V'_\ell$. When neither is graded we have $ij = 1$. It can be checked that the map $\nabla^k(v, v') \mapsto (-i)^k \nabla^{k+1}(v', v)$ is an isomorphism from $V \boxtimes^- V'$ to $\rho\tau^*(V' \boxtimes^- V)$. One uses the "same $\sqrt{-1}$" for i as in the formula for the action of (g, g') on $\nabla^k(v, v')$, and the identity $i^{a+b} = (-1)^{ab} i^a i^b$.

Now suppose $\phi : G \to G'$ is an injective $\mathcal{G}$-map and that V is a graded representation of G. Then one can "induce along ϕ" to produce a graded representation of G', called $\phi_* V$, such that, using the usual inducing in T^1G, one obtains a morphism of K-modules, also denoted

$$\phi_* : T^*G \to T^*G'.$$

This is done in exact analogy with the ungraded case. See Serre (1977) p. 28 for a characterless definition of induced representations.

Working directly from the definition of HOM^-, if $\psi : H \to H'$ is an injective $\mathcal{G}$-map, $U \in T^*H$, and $W' \in T^*(G\tilde{\mathbf{Y}}H')$, then we have the reciprocity property:

$$\mathrm{HOM}^-(\psi_* U, W') \cong \mathrm{HOM}^-(U, (1\tilde{\mathbf{Y}}\psi)^*(W')).$$

This specializes to

$$<\psi_* x, y>_{H'} = <x, \psi^* y>_H \text{ for } x \in T^*H, y \in T^*H'.$$

A relevant fact here is that if $\phi : G \to G$ is the inner automorphism corresponding to $g \in G$, then, of course, $\phi_* = \phi^*$ is the identity on T^1G, but $\phi_* = \phi^* = \rho^{\sigma(g)}$ on T^0G.

We are now in a position to introduce the promised algebra structure. Given non-negative integers k and ℓ, let

$$\phi = \phi_{k,\ell} : \tilde{S}_k \,\tilde{\mathbf{Y}}\, \tilde{S}_\ell \to \tilde{S}_{k+\ell}$$

be the $\mathcal{G}$-embedding onto the subgroup $<\tilde{S}_k, \tilde{S}_\ell>$ given in the example before Theorem 3.4. Then the algebra multiplication is defined to be the following composite:

$$T^*\tilde{S}_k \times T^*\tilde{S}_\ell \xrightarrow{\boxtimes^-} T^*(\tilde{S}_k \tilde{\mathbf{Y}} \tilde{S}_\ell) \xrightarrow{\phi_*} T^*\tilde{S}_{k+\ell}.$$

This makes $\bigoplus\limits_{n=0}^{\infty} T^*\tilde{S}_n$ into a K-algebra; that is, the multiplication is K-bilinear.

Theorem A8.8. *The algebra* $\oplus T^*\tilde{S}_n$ *is associative, and is pseudo-commutative in that* $xy = \rho^{ij+k\ell} yx$ *for* $x \in T^i\tilde{S}_k$ *and* $y \in T^j\tilde{S}_\ell$.

Proof. The associativity follows immediately from the associative property of $\boxtimes^-$, functoriality of inducing, and the fact that the following diagram commutes:

$$\begin{array}{ccc}
\tilde{S}_a \tilde{\mathbf{Y}} \tilde{S}_b \tilde{\mathbf{Y}} \tilde{S}_c & \xrightarrow{\phi_{a,b} \tilde{\mathbf{Y}} 1} & \tilde{S}_{a+b} \tilde{\mathbf{Y}} \tilde{S}_c \\
{\scriptstyle 1\tilde{\mathbf{Y}}\phi_{b,c}} \downarrow & & \downarrow {\scriptstyle \phi_{a+b,c}} \\
\tilde{S}_a \tilde{\mathbf{Y}} \tilde{S}_{b+c} & \xrightarrow{\phi_{a,b+c}} & \tilde{S}_{a+b+c}
\end{array}$$

We have suppressed the isomorphism in Theorem 3.3 (c) which expresses the associativity of the operation $\tilde{\mathbf{Y}}$.

Pseudo-commutativity is the commutativity of the following diagram:

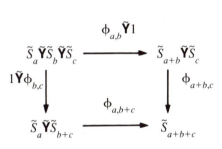

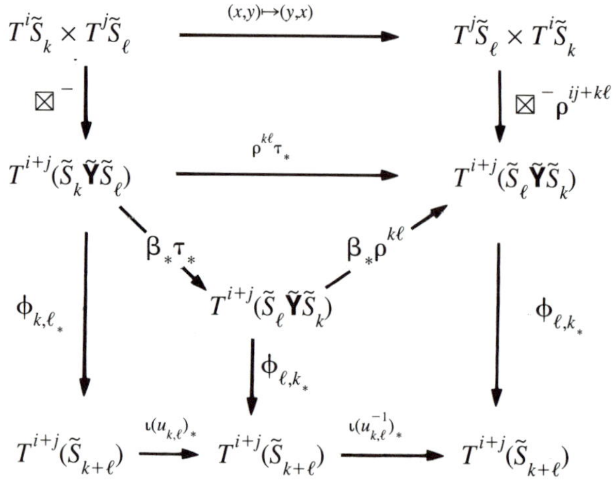

Here β is $(x \mapsto z^{k\ell\sigma(x)}x)$; $u_{k,\ell}$ is $(t_1t_2...t_{k+\ell-1})^{\ell}$; and $\iota(x)$ is $(y \mapsto xyx^{-1})$. The upper diagram commutes by Proposition A8.7. The middle one does since β^2 is the identity. As for the lower left, one can check that $\phi_{\ell,k}\beta\tau = \iota(u_{k,\ell})\phi_{k,\ell}$ by checking on generators $(t_i, 1)$ for $i < k$ and $(1, t_j)$ for $j < \ell$. (Recall that $\tau(g, h) = z^{\sigma(g)\sigma(h)}(h, g)$.) Finally, the lower right diagram commutes when $i+j = 0$ because then

$$\beta_* = 1, \ \iota(u)_* = \rho^{\sigma(u)} \text{ and } \sigma(u_{k,\ell}) = k\ell.$$

It commutes when $i+j = 1$ because then

$$\beta_* = \rho^{k\ell} \text{ and } \iota(u)_* = 1.$$

As well as the properties given in the above theorem, the algebra $\oplus T^*\tilde{S}_n$ has $1 \in T^0\tilde{S}_0$. The identity element is the isomorphism class of $(\mathbb{C}, 0)$, where the trivial object $\tilde{S}_0 = \{1, z\}$ acts in the only way it can.

One can reverse the arrows defining the multiplication, using Theorem A8.6, and produce a *comultiplication*

$$T^*\tilde{S}_{k+\ell} \xrightarrow{\phi^*} T^*(\tilde{S}_k \tilde{\curlyvee} \tilde{S}_\ell) \xrightarrow{(\boxtimes^-)^{-1}} T^*\tilde{S}_k \otimes_K T^*\tilde{S}_\ell.$$

This was crucial in earlier analysis of the structure of $\oplus T^*\tilde{S}_n$; Hoffman and Humphreys (1986). It is not needed in the proofs here, which are shorter and yield more information about the irreducibles. The multiplication and comultiplication are mutually well behaved, in that one has a Hopf algebra over K. It will be convenient for some calculations to have a formula which is equivalent to the Hopf algebra property. We state and prove this below without reference to Hopf algebras. It should be emphasized that no results from the theory of Hopf algebras are needed for the calculation of $T^*\tilde{S}_n$ in this appendix.

Lemma A8.9. *Let a, b, c_1 and c_2 be non-negative integers with $a+b = c_1+c_2$. Let $x_i \in T^*\tilde{S}_{c_i}$ for $i=1,2$. For each matrix $M = \begin{pmatrix} a_1 & b_1 \\ a_2 & b_2 \end{pmatrix}$ with $a_1+a_2 = a$, $b_1+b_2 = b$, $a_i+b_i = c_i$, use Theorem A8.6 to choose elements y_α, y_β, u_α, u_β, and indexing sets $\mathcal{A}(M)$ and $\mathcal{B}(M)$, so that*

$$\phi^*_{a_1,b_1}(x_1) = \sum_{\alpha \in \mathcal{A}(M)} y_\alpha \boxtimes^- u_\alpha$$

and

$$\phi^*_{a_2,b_2}(x_2) = \sum_{\beta \in \mathcal{B}(M)} y_\beta \boxtimes^- u_\beta.$$

Then

$$\phi^*_{a,b}(x_1x_2) = \sum \rho^{n(\alpha,\,\beta)}(y_\alpha y_\beta) \boxtimes^- (u_\alpha u_\beta),$$

where the summation is over all (α, β) in $\mathcal{A}(M) \times \mathcal{B}(M)$ for all M, and where $n(\alpha,\beta)$ is defined by

$$u_\alpha y_\beta = \rho^{n(\alpha,\beta)} y_\beta u_\alpha.$$

This is an application of the Mackey double coset formula which we shall now state. One is given injective $\mathcal{G}$-maps

$$\phi: G' \to G \quad \text{and} \quad \psi: G'' \to G,$$

and, for Δ ranging over some indexing set, a complete set $\{g_\Delta\}$ of $(\phi G', \psi G'')$-double coset representatives in G. Finally, for each Δ, choose a $\mathcal{G}$-object G_Δ and injective $\mathcal{G}$-maps

$$\phi_\Delta: G_\Delta \to G'' \quad \text{and} \quad \psi_\Delta: G_\Delta \to G'$$

such that

$$\psi\phi_\Delta(G_\Delta) = (g_\Delta(\phi G')g_\Delta^{-1}) \cap (\psi G'')$$

and such that the following commutes:

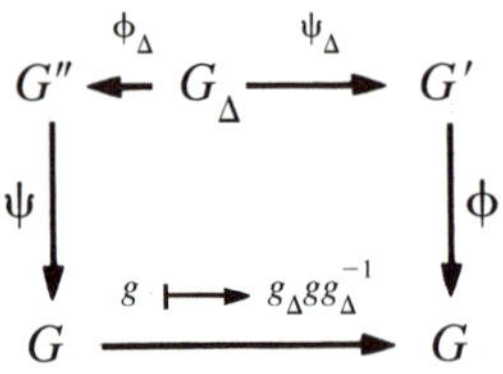

Then the Mackey formula for T^* is

Theorem A8.10. $\phi^* \circ \psi_* = \sum_\Delta \rho^{\sigma(g_\Delta)} \psi_{\Delta_*} \circ \phi_\Delta^*$

It is proved exactly as in the usual representation theory; Serre (1977) (7.3). See also Hoffman and Humphreys (1986) (2.16).

To prove Lemma A8.9, apply this theorem, taking $\phi = \phi_{a,b}$ and $\psi = \phi_{c_1,c_2}$. We need only observe that in our example we can let Δ range over the matrices in the lemma, taking

$$G_\Delta = \tilde{S}_{a_1} \mathbf{\tilde{Y}} \tilde{S}_{b_1} \mathbf{\tilde{Y}} S_{a_2} \mathbf{\tilde{Y}} \tilde{S}_{b_2},$$

$$\phi_\Delta = \phi_{a_1,b_1} \mathbf{\tilde{Y}} \phi_{a_2,b_2}, \ \psi_\Delta = (\phi_{a_1,a_2} \mathbf{\tilde{Y}} \phi_{b_1,b_2}) \circ (1 \mathbf{\tilde{Y}} \tau \mathbf{\tilde{Y}} 1),$$

and

$$g_\Delta = (t_{a_1} t_{a_1+1} \ldots t_{a_1+a_2+b_1-1})^{b_1}.$$

This double coset decomposition of $\tilde{S}_n$ is the double cover of the well known decomposition of S_n; see James and Kerber (1981) (1.3.10).

Now let us proceed to the calculation of the algebra $\oplus T^* \tilde{S}_n$ and its irreducibles. We shall give a K-basis consisting of products of certain basic special irreducibles, plus a simple inductive formula for all the irreducibles as K-linear combinations of this basis.

For each $x \in T^i \tilde{S}_k$, define an operator $x^\perp$ on $\oplus T^* \tilde{S}_n$, which reduces $\mathbb{N}$-grading by k and $\mathbb{Z}/2$-grading by i as follows:

$$\langle x^\perp(y), u \rangle = \langle y, xu \rangle.$$

Equivalently, if $\{a_\lambda\}$ is a K-basis of special irreducibles for $\oplus T^* \tilde{S}_n$, we have

$$x^\perp(y) = \sum_\lambda \langle y, x a_\lambda \rangle a_\lambda.$$

Thus $x^\perp$ is a morphism of K-modules. We have

$$(xy)^\perp = y^\perp x^\perp = \rho^{ij+k\ell} x^\perp y^\perp$$

for $x \in T^i\tilde{S}_k$ and $y \in T^j\tilde{S}_\ell$.

The basic special irreducibles c_n are defined to be the isomorphism classes of irreducible Clifford modules from Appendix 6. Recall the inclusions $\tilde{S}_n \subset B_{n-1} \subset C_n$, and the modules N_k defined before Lemma A6.4. Let

$$c_{2k+1} := [N_{2k}] \in T^0\tilde{S}_{2k+1}$$

and

$$c_{2k} := [N_{2k-1}] \in T^1\tilde{S}_{2k}.$$

It will not be necessary to make an explicit choice between N_j and N'_j. Note that $\rho c_i = [N'_{i-1}]$. As remarked before Proposition A6.1, the element c_n is irreducible because $\tilde{S}_n$ generates the algebra B_{n-1}. Thus we have indeed defined a sequence of special irreducibles.

Proposition A8.11. *The above elements satisfy the following relations*:

(*i*) $c_k c_\ell = \rho^{k+\ell+1} c_\ell c_k$.

(*ii*) $c_n^2 = (-1)^{n+1}\kappa(c_{2n} + \kappa \sum_{i=1}^{n-1} (-1)^i c_{2n-i} c_i)$.

(*iii*) *For all y and all* $x \in T^i\tilde{S}_k$,

$$c_n^\perp(xy) = \rho^{in+kn+i} x c_n^\perp(y) + c_n^\perp(x) y + \kappa \sum_{j=1}^{n-1} c_j^\perp(x) c_{n-j}^\perp(y).$$

(*iv*) *For all* $k > 0$,

$$\kappa(c_k^\perp + (-1)^k c_k^\perp + \kappa \sum_{j=1}^{k-1} (-1)^j c_j^\perp c_{k-j}^\perp) = 0\,.$$

(*v*) *For all* $i < n$, $c_i^\perp(c_n) = \kappa c_{n-i}$; *whereas* $c_n^\perp(c_n) = 1$.

To prove this we need the following calculation.

Lemma A8.12. *For all positive a and b*,

$$\phi^*_{a,b}(c_{a+b}) = \kappa c_a \boxtimes^- c_b.$$

Proof. Consider the diagram:

$$
\begin{array}{ccccc}
(g, h) & & \tilde{S}_a \,\tilde{\Upsilon}\, \tilde{S}_b & \xrightarrow{\ \phi\ } & \tilde{S}_{a+b} \\
\big\downarrow & & \big\downarrow & & \big\downarrow \\
g \otimes h & & C(W^{a-1}) \tilde{\boxtimes} C(W^{b-1}) & & C(W^{a+b-1}) = B_{a+b-1} \\
 & & \big\downarrow & & \big\downarrow \\
 & & C(\mathbb{R}^a) \tilde{\boxtimes} C(\mathbb{R}^b) & \xrightarrow{\ \cong\ } & C(\mathbb{R}^{a+b}) = C_{a+b}
\end{array}
$$

The symbol $\tilde{\boxtimes}$ denotes the $\mathbb{Z}/2$-graded tensor product of algebras, so that, for two $\mathbb{Z}/2$-graded algebras D and E, the module $D_i \otimes E_j$ is one of the two summands of $(D \tilde{\boxtimes} E)_{i+j}$, and

$$(d \otimes e)(d' \otimes e') = (-1)^{jk} dd' \otimes ee'$$

if $e \in E_j$ and $d' \in D_k$. The operator κ acts on Clifford modules, so that we have, for example, $\kappa(M_n) = \kappa(M'_n) = M_n^*$. Define $p(n)$ to be 0 if n is even and 1 if n is odd. Let $\tilde{\boxtimes}$ also denote the $\mathbb{Z}/2$-graded tensor product of $\mathbb{Z}/2$-graded Clifford modules: if M and M' are modules for Clifford algebras C and C' respectively, then $M \tilde{\boxtimes} M'$ is a module for the algebra $C \tilde{\boxtimes} C'$. In each $\mathbb{Z}/2$-grading it is the direct sum of the usual two components, and the action of the algebra satisfies the expected sign conventions, namely

$$(c \otimes c')(m \otimes m') = (-1)^{jk}(cm) \otimes (c'm')$$

for $c' \in C'_j$, $m \in M_k$. Then the following diagram shows the effect of restricting modules along the embeddings in the previous diagram:

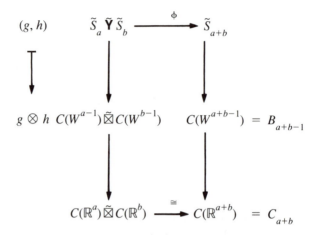

$$\begin{array}{ccc}
\kappa^{p(a)+1-p(ab)} C_a \boxtimes^- \kappa^{p(b)+1-p(ab)} C_b & \overset{\phi^*}{\longleftarrow\!\dashv} & \kappa^{p(a+b)+1} C_{a+b} \\
\uparrow & & \uparrow \\
\kappa^{p(a)+1-p(ab)} N_{a-1} \tilde{\boxtimes} \kappa^{p(b)+1-p(ab)} N_{b-1} & & \kappa^{p(a+b)+1} N_{a+b-1} \\
\uparrow & & \uparrow \\
\kappa^{p(a)-p(ab)} M_a \tilde{\boxtimes} \kappa^{p(b)-p(ab)} M_b & \longleftarrow\!\dashv & \kappa^{p(a+b)} M_{a+b}
\end{array}$$

The powers of κ have been inserted to make all the modules $\mathbb{Z}/2$-graded. In that case, since $\tilde{\boxtimes}$ and $\boxtimes^-$ are defined using the same formulae, we find that, for Clifford modules U, V and representations $\overline{U}$, $\overline{V}$, if $U \mapsto \overline{U}$ and $V \mapsto \overline{V}$, then $U \tilde{\boxtimes} V \mapsto \overline{U} \boxtimes^- \overline{V}$. The justification for the given restrictions between Clifford modules is equality of dimensions. For the other two vertical arrows, it is the definition of c_n.

We can now conclude that ϕ^* maps c_{a+b} to $\kappa c_a \boxtimes^- c_b$ as required, using the following:

(a) the diagram is commutative;

(b) the exponent of κ on the top right is one less than the sum of the exponents on the top left;

(c) in a free K-module, $\kappa^\ell x = 0$ implies $\kappa x = 0$.

Proof of Proposition A8.11. Part (i) is clear from pseudo-commutativity.

To prove (v), let u be any element. Then, using Lemma A8.12 for the fourth equality,

$$\begin{aligned}
<c_i^\perp(c_n), u> &= <c_n, c_i u> = <c_n, (\phi_{i,n-i})_*(c_i \boxtimes^- u)> \\
&= <\phi^*_{i,n-i} c_n, c_i \boxtimes^- u> = <\kappa c_i \boxtimes^- c_{n-i}, c_i \boxtimes^- u> \\
&= \kappa <c_i, c_i> <c_{n-i}, u> = <\kappa c_{n-i}, u>,
\end{aligned}$$

as required. Also $<c_n^\perp(c_n), 1> = <c_n, c_n 1> = 1$, so $c_n^\perp(c_n) = 1$.

To prove (iii), let $u \in T^*\tilde{S}_k$ be a special irreducible and write $\sum_{s+t=k} \phi^*_{s,t}(u)$ as $\sum_\beta u_\beta' \boxtimes^- u_\beta''$. Let $m(j)$ be 0 if $j = 0$ or n, and be 1 if $0 < j < n$. Then, for all x, y

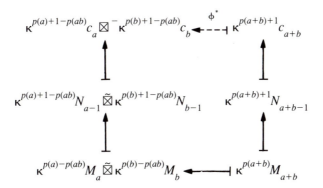

and u,

$$
\begin{aligned}
\langle c_n^{\perp}(xy), u\rangle &= \langle xy, c_n u\rangle \\
&= \langle \phi_*(x \boxtimes^{-} y), c_n u\rangle = \langle x \boxtimes^{-} y, \phi^*(c_n u)\rangle \\
&= \sum_{j,\beta} \langle x \boxtimes^{-} y, \rho^{n(j,\beta)} \kappa^{m(j)} (c_j u_\beta{}') \boxtimes^{-} (c_{n-j} u_\beta{}'')\rangle
\end{aligned}
$$

using Lemmas A8.9 and A8.12 (and where c_0 is interpreted to be 1). Continuing, we obtain

$$
\begin{aligned}
&\langle x \boxtimes^{-} y, \\
&\quad \sum_{\beta} (\rho^{in+kn+i} u_\beta{}' \boxtimes^{-} c_n u_\beta{}'' + c_n u_\beta{}' \boxtimes^{-} u_\beta{}'' + \kappa \sum_{j=1}^{n-1} (c_j u_\beta{}') \boxtimes^{-} (c_{n-j} u_\beta{}''))\rangle \\
&= \langle \rho^{in+kn+i} x \boxtimes^{-} c_n^{\perp}(y) + c_n^{\perp}(x) \boxtimes^{-} y + \kappa \sum_{j=1}^{n-1} c_j^{\perp}(x) \boxtimes^{-} c_{n-j}^{\perp}(y), \\
&\qquad \sum_{\beta} u_\beta{}' \boxtimes^{-} u_\beta{}''\rangle.
\end{aligned}
$$

Replacing $\sum_{\beta} u_\beta{}' \boxtimes^{-} u_\beta{}''$ by $\sum \phi_{s,t}^*(u)$ and using adjointness of ϕ^* and ϕ_* then gives

$$
\langle \rho^{in+kn+i} x c_n^{\perp}(y) + c_n^{\perp}(x) y + \kappa \sum_{j=1}^{n-1} c_j^{\perp}(x) c_{n-j}^{\perp}(y), u\rangle
$$

as required.

Note that (iv) is trivial when k is odd, and is immediate from (ii) when $k = 2n$.

It remains to prove (ii). This follows by taking $m = 2n$ in the following lemma. One substitutes the right-hand sides below into the equation to be proved. This lemma gives a preview of the decompositions of products of c_j into irreducibles.

Lemma A8.13. *For each $m > 2$, there is a sequence $c_m, y_1, y_2, \ldots$ of distinct special irreducibles such that, for $1 \leq s < m/2$,*

$$c_s c_{m-s} = y_s + \kappa c_m + \kappa^2 \sum_{t=1}^{s-1} y_t .$$

When $m = 2n \geq 2$,

$$c_n^2 = \kappa c_{2n} + \kappa^2 \sum_{t=1}^{n-1} y_t .$$

Proof. For $1 \leq s < m$,

$$<c_s c_{m-s}, c_m> = <c_{m-s}, c_s^{\perp}(c_m)> = <c_{m-s}, \kappa c_{m-s}> = \kappa .$$

Similar arguments, using (iii) of Proposition A8.11, yield the following in mechanical fashion:

$$<c_n^2, c_{2n}> = \kappa; \quad <c_n^2, c_n^2> = (2n-1)\kappa^2;$$

$$<c_s c_{m-s}, c_s c_{m-s}> = 1+(2s-1)\kappa^2 \text{ for } 1 \leq s < m/2; \qquad \text{(I)}$$

$$<c_s c_{m-s}, c_t c_{m-t}> = 2s\kappa^2 \text{ for } 1 \leq s < t < m/2;$$

$$<c_n^2, c_s c_{2n-s}> = 2s\kappa^2 \text{ for } 1 \leq s < n. \qquad \text{(II)}$$

When $m = 2$, $<c_1^2, c_2> = \kappa$, so κc_2 occurs in the decomposition of c_1^2. But $<c_1^2, c_1^2> = \kappa^2$, so we have $c_1^2 = \kappa c_2$, as required.

Now assume $m > 2$. We prove the first equation by induction on s. For $s = 1$, we have $<c_1 c_{m-1}, c_m> = \kappa$, so $c_1 c_{m-1} = \kappa c_m + y_1$ for some representation y_1 with $<y_1, \kappa c_m> = 0$. But $<c_1 c_{m-1}, c_1 c_{m-1}> = 1+\kappa^2$, so $<y_1, y_1> = 1$ and therefore y_1 is a special irreducible, as required. For the inductive step, suppose given distinct special irreducibles $c_m, y_1, ..., y_{s-1}$ such that

$$c_r c_{m-r} = y_r + \kappa c_m + \kappa^2 \sum_{t=1}^{r-1} y_t \text{ for } 1 \leq r < s.$$

Repeated application of (I) yields

$$<c_s c_{m-s}, y_r> = \kappa^2 \text{ for } 1 \leq r < s.$$

Since $<c_s c_{m-s}, c_m> = \kappa$, we must have

$$c_s c_{m-s} = y_s + \kappa c_m + \kappa^2 \sum_{t=1}^{s-1} y_t,$$

for some representation y_s with

$$<y_s, c_m> = 0 = <y_s, y_r> \text{ for } r < s.$$

But then

$$\begin{aligned} 1 + (2s-1)\kappa^2 &= <c_s c_{m-s}, c_s c_{m-s}> \\ &= <y_s, y_s> + (s-1)\kappa^4 + \kappa^2. \end{aligned}$$

Thus $<y_s, y_s> = 1$, completing the induction.

To prove the second equation when $m = 2n > 2$, repeated application of (II) yields $<c_n^2, y_t> = \kappa^2$ for $1 \leq t < n$, so

$$c_n^2 = v + \kappa c_{2n} + \kappa^2 \sum_{t=1}^{n-1} y_t$$

for some representation v. But then

$$(2n-1)\kappa^2 = <c_n^2, c_n^2> = <v, v> + (n-1)\kappa^4 + \kappa^2.$$

Thus $<v, v> = 0$, so $v = 0$, as required.

This last proof gives an inkling of how, in principle, one can build up a "repertoire" of irreducibles by multiplying c_n's together and calculating inner products. We now define some operators $\mathcal{A}_\ell$ which turn this into a systematic science.

For each $\ell > 0$, define an operator $\mathcal{A}_\ell$ on $\oplus T^*\tilde{S}_n$ by

$$\mathcal{A}_\ell(x) = c_\ell x + \kappa \sum_{i>0} (-1)^i c_{\ell+i} c_i^\perp(x).$$

It is clear that the summation is finite. The operator $\mathcal{A}_\ell$ is a K-endomorphism which increases $\mathbb{N}$-grading by ℓ and the $\mathbb{Z}/2$-grading by $\ell+1$.

Next, for each strict partition $\lambda = (\lambda_1 > \lambda_2 > ... > \lambda_r > 0)$, define elements of $T^*\tilde{S}_{|\lambda|}$ as follows:

$$a_\lambda := \mathcal{A}_{\lambda_1} \mathcal{A}_{\lambda_2} \cdots \mathcal{A}_{\lambda_r}(1)$$

and

$$c_\lambda := c_{\lambda_1} c_{\lambda_2} \ldots c_{\lambda_r} .$$

By Proposition A8.11 (i) and (ii), the subalgebra generated by all the c_n has a set of K-module generators consisting of all the c_λ.

Theorem A8.14. (*i*) *The set* $\{a_\lambda : \lambda \in \mathcal{D}\}$ *is "the" K-basis of special irreducibles for* $\oplus T^* \tilde{S}_n$.

(*ii*) *If* $\mathcal{D}$ *has the reverse lexicographic order, then the matrix L which gives the* c_λ *as linear combinations of the* a_λ *is lower unitriangular.*

Note that (i) tells us that all the irreducible projective representations of S_n and A_n for $n \geq 4$ are the a_λ, ρa_λ and κa_λ, modulo the identifications of $T^1\tilde{A}_n$ with $T^0\tilde{S}_n$ for $n > 1$, and modulo the exceptional cocyles which A_n possesses for $n = 6$ and 7.

Corollary A8.15. (*i*) *The set* $\{c_\lambda : \lambda \in \mathcal{D}\}$ *is a K-basis for* $\oplus T^* \tilde{S}_n$.

(*ii*) *The algebra* $\oplus T^* \tilde{S}_n$ *is isomorphic to the quotient of the free algebra with generators* $c_1, c_2, \ldots$ *by the relations in Proposition A8.11* (*i*) *and* (*ii*).

Proof of A8.15. Part (i) is immediate from the invertibility of L in Theorem A8.14 (ii). The quotient algebra maps canonically into $\oplus T^* \tilde{S}_n$. But its defining relations obviously imply that $\{c_\lambda : \lambda \in \mathcal{D}\}$ is a set of K-module generators for that quotient algebra, so the map is an isomorphism.

Proof of A8.14. Using reverse lexicographic order, the definitions of $\mathcal{A}_\ell$ and of a_λ immediately lead to the existence of $\beta_{\lambda\mu} \in K$ such that

$$a_\lambda = c_\lambda + \kappa \sum_{\mu < \lambda} \beta_{\lambda\mu} c_\mu . \tag{A}$$

Furthermore, we shall see that

$$<a_\nu, c_\lambda> = \begin{cases} 1 & \text{if } \nu = \lambda; \\ 0 & \text{if } \nu > \lambda. \end{cases} \tag{B}$$

Postponing the proof of (B), if $\lambda \leq \nu$,

$$<a_\nu, a_\lambda> \;=\; <a_\nu, c_\lambda> + \kappa \sum_{\mu<\lambda} \beta_{\lambda\mu} <a_\nu, c_\mu>$$

$$= <a_\nu, c_\lambda> = \begin{cases} 0 \ \text{ if } \lambda < \nu; \\ 1 \ \text{ if } \lambda = \nu. \end{cases}$$

By Theorem A8.4 and Corollary 4.3, the number of a_λ in $T^*\tilde{S}_n$ is its rank as a free K-module. Hence $\{a_\lambda\}$ is a basis of special irreducibles, yielding (i); and (ii) is clear from (A).

To prove (B), since, when $|\nu| = \sum \lambda_i$,

$$<a_\nu, c_{\lambda_1} \ldots c_{\lambda_s}> = <c^\perp_{\lambda_s} \ldots c^\perp_{\lambda_1}(a_\nu), 1> ,$$

it suffices to prove (C) and (D) below, where $\ell > m > \ldots > 0$:

$$c^\perp_\ell(a_{\ell,m,\ldots}) = a_{m,\ldots}; \tag{C}$$

$$c^\perp_{\ell+n}(a_{\ell,m,\ldots}) = 0, \ \text{if } n > 0. \tag{D}$$

These clearly hold when the partition $(\ell, m, \ldots)$ has length one. (We take $a_\varnothing$ to be 1 for the empty sequence $\varnothing$.) Proceed by induction on the length. Denote the partition $(m,\ldots)$ as μ.

To prove (C)

$$c^\perp_\ell(a_{\ell,m,\ldots}) = c^\perp_\ell(c_\ell a_\mu) + \kappa \sum_{i>0} (-1)^i c^\perp_\ell\left(c_{\ell+i} c^\perp_i(a_\mu)\right) \quad \text{(by definition)}$$

$$= c^\perp_\ell(c_\ell) a_\mu + \rho c_\ell c^\perp_\ell(a_\mu) + \kappa \sum_{j=1}^{\ell-1} c^\perp_{\ell-j}(c_\ell) c^\perp_j(a_\mu)$$

$$+ \kappa \sum_{i>0} (-1)^i \left(c^\perp_\ell(c_{\ell+i}) c^\perp_i(a_\mu) + c_{\ell+i} c^\perp_\ell c^\perp_i(a_\mu) + \kappa \sum_{j=1}^{\ell-1} c^\perp_{\ell-j}(c_{\ell+i}) c^\perp_j c^\perp_i(a_\mu) \right)$$

by A8.11 (iii)

$$= a_\mu + \kappa^2 \sum_{j=1}^{m} c_j c_j^\perp(a_\mu) + \kappa^2 \sum_{i>0} (-1)^i c_i c_i^\perp(a_\mu) +$$

$$\kappa^3 \sum_{i>0} \sum_{j=1}^{m} (-1)^i c_{i+j} c_i^\perp c_j^\perp(a_\mu)$$

(since $c_j^\perp(a_\mu)=0$ for $j > m$ by the inductive hypothesis)

$$= a_\mu + \kappa^2 \sum_{k=1}^{m} c_k \left(c_k^\perp + (-1)^k c_k^\perp + \kappa \sum_{i=1}^{k-1} (-1)^i c_i^\perp c_{k-i}^\perp \right)(a_\mu)$$

$$= a_\mu,$$

as required, using Proposition A8.11 (iv).

To prove (D),

$$c_{\ell+n}^\perp(a_{\ell,m,\ldots}) = c_{\ell+n}^\perp(c_\ell a_\mu) + \kappa \sum_{i>0} (-1)^i c_{\ell+n}^\perp\left(c_{\ell+i} c_i^\perp(a_\mu)\right)$$

$$= \kappa c_\ell^\perp(c_\ell) c_n^\perp(a_\mu) + \sum_{j=n+1}^{m} \kappa c_{\ell+n-j}^\perp(c_\ell) c_j^\perp(a_\mu)$$

$$+ \kappa \sum_{i>0} (-1)^i \left(c_{\ell+n}^\perp(c_{\ell+i}) c_i^\perp(a_\mu) + \kappa \sum_{j=1}^{m} c_{\ell+n-j}^\perp(c_{\ell+i}) c_i^\perp c_j^\perp(a_\mu) \right)$$

by A8.11 (iii)

$$= \kappa c_n^\perp(a_\mu) + \kappa^2 \sum_{j=n+1}^{m} c_{j-n} c_j^\perp(a_\mu) + \kappa(-1)^n c_n^\perp(a_\mu)$$

$$+ \kappa^2 \sum_{i=n+1}^{m} (-1)^i c_{i-n} c_i^\perp(a_\mu) + \kappa^2 \sum_{i=1}^{m} \sum_{j=1}^{m} (-1)^i c_{\ell+n-j}^\perp(c_{\ell+i}) c_i^\perp c_j^\perp(a_\mu)$$

(since $c_i^\perp(a_\mu) = 0$ for $i > m$ by the inductive hypothesis)

$$= \left(1+(-1)^n\right) \kappa c_n^\perp(a_\mu) + \kappa^2 \sum_{k=n+1}^{m} \left(1 + (-1)^k\right) c_{k-n} c_k^\perp(a_\mu)$$

$$+ \kappa^2 \sum_{\substack{i+j=n\\ i\geq 1, j\geq 1}} (-1)^i c_i^\perp c_j^\perp (a_\mu) + \kappa^3 \sum_{\substack{i+j>n\\ i\geq 1, j\geq 1}} (-1)^i c_{i+j-n} c_i^\perp c_j^\perp (a_\mu)$$

$$= \kappa \left(c_n^\perp + (-1)^n c_n^\perp + \kappa \sum_{i=1}^{n-1} (-1)^i c_i^\perp c_{n-i}^\perp \right) (a_\mu)$$

$$+ \kappa \sum_{k>n} c_{k-n} \kappa \left(c_k^\perp + (-1)^k c_k^\perp + \kappa \sum_{i=1}^{k-1} (-1)^i c_i^\perp c_{k-i}^\perp \right) (a_\mu)$$

$$= 0 + \kappa \sum_{k>n} c_{k-n} 0 = 0,$$

again using Proposition A8.11 (iv).

This completes the proof of Theorem A8.14.

The notation in this appendix is related to that in Theorem 8.6 as follows. When λ is an even partition, we have $<\lambda> = \kappa a_\lambda$. When λ is an odd partition,

$$<\lambda> + <\lambda>^a = \kappa^2 a_\lambda = a_\lambda + \rho a_\lambda,$$

so $<\lambda>$ is either a_λ or ρa_λ, depending on how the choices were made between N_k and N'_k. Thus, except for (i) (c), Theorem 8.6 is an immediate consequence of Theorem A8.14.

To make the definition of a_λ into an explicit formula as a linear combination of the c_μ, we need a combinatorial formula for $c_n^\perp(a_\lambda)$. This will be given at the end of Appendix 12.

Notes

There is a more abstract version of the above structure theorem which axiomatizes the idea of a "self dual Hopf algebra over K with canonical basis"; Bean and Hoffman (1988). This follows the ideas of Zelevinsky (1979), who works over $\mathbb{Z}$, with applications to linear representations. The abstract approach has applications to the projective representations of wreath products; Hoffman and Humphreys (1989). In this approach one need not even construct the basic irreducibles. The c_n are inductively characterized by the equation

$$c_1 c_{n-1} = \kappa c_n + u_n$$

where $u_2 = 0$ and u_n is a special irreducible for $n > 2$. Proposition A8.11 then follows, and the arguments to prove A8.14 and A8.15 are still valid in this abstract context. Note that $\oplus T^*\tilde{S}_n$ is not a Hopf algebra over $\mathbb{Z}$. This is one reason to introduce scalars in K. The other reason is the usefulness of the K-valued inner product.

The operators $\mathcal{A}_\ell$ appear in Hoffman (1989).

The definition and properties of $\mathbb{Z}/2$-graded representations were given in Hoffman and Humphreys (1986). See also Sergeev (1984) Proposition 4.

The operation $\boxtimes^-$ and the ring structure on $\oplus T^*\tilde{S}_n$ are defined in Hoffman (1983), Hoffman and Humphreys (1985, 1986). A systematic generalization of $\boxtimes^-$, HOM$^-$, and Theorem A8.6 to the case of several sign homomorphisms, which occur with monomial groups, is given in Hoffman (1990A).

Since the symmetric function algebra $\Delta_{\mathbb{R}}$ is isomorphic to the space $\Theta_{\mathbb{R}}$ of real linear combinations of self-associate characters of $\tilde{S}_n$, Schur undoubtedly realized that $\Theta_{\mathbb{R}}$ has a natural ring structure. The integral version of $\Theta_{\mathbb{R}}$ is $\kappa(\oplus T^\circ\tilde{S}_n)$, which has ring product denoted $\circ$ in Hoffman and Humphreys (1986, p. 1411). Multiplication by κ has kernel equal to $\mathrm{Im}(1-\rho)$. Thus

$$\kappa(\oplus T^\circ\tilde{S}_n) \cong \oplus T^\circ\tilde{S}_n/(1-\rho)\,.$$

The right-hand side has ring structure given by the $\mathbb{Z}/2$-graded tensor product, which is the (0,0)-component of $\boxtimes^-$. In this form, it also appears in Jozefiak (1989). Earlier, El-Sharabasy (1985) had used manipulations in the Clifford algebra (the operator $\tilde{\mathbf{Y}}$ not being at her disposal) to define the same ring product. Independently, Stembridge (1989, p. 104) invented a "reduced Clifford Product" for $\tilde{S}_n$-modules. It is an ℓ-ary operation, for any ℓ, defined using the Clifford algebra. With suitable choices, it coincides with the iteration of $\boxtimes^-$ specialized to objects $\tilde{S}_n$, followed by multiplication by κ^{-t}, where t is the number of self-associate factors in the ℓ-ary product. This is explained further after the table below.

Note that $\oplus T^*\tilde{S}_n$ has simpler structure than $\oplus T^\circ\tilde{S}_n/(1-\rho)$, as well as carrying more information.

The formula for $\nabla^k(v, w)$ in the definition of $\boxtimes^-$ was used in Hoffman and Humphreys (1986, p. 1401), as well as in Jozefiak (1989, p. 240).

The operation $\tilde{\otimes}$ in Chapter 5 of this book is given in terms of $\boxtimes^-$ by the following completion of Table 5.7. Recall that $\tilde{\otimes}$ is only defined on ungraded modules, and is only well-defined up to isomorphism and associates. Below, the $\mathbb{Z}/2$-graded module M' is chosen so that $\kappa M' \cong M$ when M is self-associate, and similarly for N' and N.

M	SA	SA	NSA	NSA
N	SA	NSA	SA	NSA
$M \tilde{\otimes} N$	$\kappa M' \boxtimes^{-} N$	$M' \boxtimes^{-} N$	$M \boxtimes^{-} N'$	$\kappa M \boxtimes^{-} N$

(It is entertaining to reprove the "associativity" of $\tilde{\otimes}$, using this table and the associativity and K-bilinearity of $\boxtimes^{-}$.) Stembridge's ℓ-ary operation coincides, up to isomorphism and associates, with the ℓ-fold iteration of $\tilde{\otimes}$ applied to the case when the groups involved all have the form $\tilde{S}_n$. His formulae are completely different, involving the Clifford algebras explicitly.

9

EXPLICIT Q-FUNCTIONS

More details concerning the Q-functions are given in this chapter. Motivated by Theorem 7.21, we first define a function Q_α for each integer sequence α; but then immediately show that if α is not a strict partition, then Q_α is either zero or is a multiple of Q_λ for a certain λ in $\mathcal{D}$. The Q_α are needed in Chapter 10, but they also play a role in a Laurent identity, (9.5) here. This is a generating "expression" for the Q_α. It could be re-interpreted as a raising operator formula, such as occurs in the theory of Schur functions, but we prefer to use the Laurent formulation. It leads to four more formulae for Q_λ. The first of these, (9.6), is a ratio of series, which is then reduced to a classical expression, (9.10), for Q_λ as the symmetrization of a certain explicit rational function. The third, (9.12), gives Q_λ as a polynomial in the q_i. It leads to Schur's original inductive formulae, (9.14), for Q_λ as the expansion of a certain Pfaffian.

Definition. If $(\alpha_1, \ldots, \alpha_\ell)$ is an integer sequence, define

$$Q_\alpha = \mathcal{B}_{\alpha_1} \circ \ldots \circ \mathcal{B}_{\alpha_\ell}(1),$$

using the operators $\mathcal{B}_n$ defined before 7.21.

The following commutation relations will allow Q_α to be expressed in terms of Q_λ for λ in $\mathcal{D}$.

Theorem 9.1. *For all integers s and t,*

$$\mathcal{B}_s\mathcal{B}_t + \mathcal{B}_t\mathcal{B}_s = (-1)^s 2\delta_{s+t,0}\, Id.$$

It is useful to separate these as follows:

(*i*) $\mathcal{B}_s\mathcal{B}_t = -\mathcal{B}_t\mathcal{B}_s$ *if* $s+t \neq 0$;

(*ii*) $\mathcal{B}_s\mathcal{B}_{-s} + \mathcal{B}_{-s}\mathcal{B}_s = (-1)^s 2Id$;

(*iii*) $\mathcal{B}_s\mathcal{B}_s = 0$ *if* $s \neq 0$;

(*iv*) $\mathcal{B}_0\mathcal{B}_0 = Id$.

Remark. See the discussion after (9.4) for a slicker version of the following proof.

Proof. By the definition of $\mathcal{B}_n$ and (7.19),

$$\mathcal{B}_s\mathcal{B}_t(f) = \sum_{i,j}(-1)^{i+j} q_{s+j}q_j^{\perp}\left(q_{t+i}q_i^{\perp}(f)\right)$$

$$= \sum_{i,j}\sum_{k\leq j}(-1)^{i+j} q_{s+j}q_{j-k}^{\perp}(q_{t+i})q_k^{\perp}q_i^{\perp}(f)\,.$$

Thus

$$(\mathcal{B}_s\mathcal{B}_t + \mathcal{B}_t\mathcal{B}_s)(f)$$

$$= \sum_{i,k}(-1)^i q_i^{\perp}q_k^{\perp}(f)\sum_{j\geq k}(-1)^j\left(q_{s+j}q_{j-k}^{\perp}(q_{t+i}) + q_{t+j}q_{j-k}^{\perp}(q_{s+i})\right)$$

$$= \sum_{i,k}(-1)^i q_i^{\perp}q_k^{\perp}(f)\Bigg((-1)^k(q_{s+k}q_{t+i} + q_{t+k}q_{s+i})$$

$$+ 2\sum_{j>k}(-1)^j(q_{s+j}q_{t+i+k-j} + q_{t+j}q_{s+i+k-j})\Bigg). \qquad (*)$$

The terms in (*) with $i=k$ are the following, letting $M = s+t+2k$:

$$\sum_k(-1)^k q_k^{\perp}q_k^{\perp}(f)\times$$

$$\times\left((-1)^k 2q_{s+k}q_{t+k} + 2\sum_{j>k}(-1)^j(q_{s+j}q_{M-s-j} + q_{M-s-2k+j}q_{s+2k-j})\right).$$

In this expression, use r to replace $s+k$, $s+j$ and $s+2k-j$ respectively in the three products of the form q_aq_b, to obtain, with the help of 7.1,

$$2\sum_k(-1)^k q_k^{\perp}q_k^{\perp}(f)\sum_r(-1)^{r+s}q_rq_{M-r}$$

$$= (-1)^s 2\sum_k(-1)^k q_k^{\perp}q_k^{\perp}(f)\,\delta_{s+t+2k,0}\,. \qquad (**)$$

Writing $N = s+t+i+k$, the remaining terms in (*) are

$$\sum_k \sum_{i \neq k} (-1)^{i+k} q_k^\perp q_i^\perp (f) \left(q_{s+k} q_{t+i} + q_{s+i} q_{t+k} + \sum_{r > s+k} (-1)^{r+s+k} q_r q_{N-r} \right.$$
$$\left. + \sum_{r < s+i} (-1)^{r+s+i} q_r q_{N-r} \right).$$

Using the fact that $q_k^\perp q_i^\perp = q_i^\perp q_k^\perp$, this becomes

$$2 \sum_k \sum_{i>k} (-1)^{i+k} q_k^\perp q_i^\perp (f) \times$$
$$\times \left(q_{s+k} q_{t+i} + \sum_{r>s+k} (-1)^{r+s+k} q_r q_{N-r} + \sum_{r<s+i} (-1)^{r+s+i} q_r q_{N-r} \right.$$
$$\left. + q_{s+i} q_{t+k} + \sum_{r>s+i} (-1)^{r+s+i} q_r q_{N-r} + \sum_{r<s+k} (-1)^{r+s+k} q_r q_{N-r} \right)$$
$$= 2 \sum_k \sum_{i>k} (-1)^{i+k} q_k^\perp q_i^\perp (f) \times$$
$$\times \left(\sum_r (-1)^{r+s+k} q_r q_{N-r} + \sum_r (-1)^{r+s+i} q_r q_{N-r} \right)$$
$$= 2 \sum_k \sum_{i>k} (-1)^{i+k} q_k^\perp q_i^\perp (f)\, \delta_{s+t+k+i,0} \left((-1)^{s+k} + (-1)^{s+i} \right)$$
$$= (-1)^s 2 \sum_k \sum_{i>k} q_k^\perp q_i^\perp (f)\, \delta_{s+t+k+i,0} \left((-1)^i + (-1)^k \right). \qquad (***)$$

Adding (**) and (***), expression (*) for $(\mathcal{B}_s \mathcal{B}_t + \mathcal{B}_t \mathcal{B}_s)(f)$ becomes

$$(-1)^s 2 \left(\sum_k (-1)^k q_k^\perp q_k^\perp (f) \delta_{s+t+2k,0} + \right.$$
$$\left. \sum_k \sum_{i>k} \left[(-1)^i + (-1)^k \right] q_k^\perp q_i^\perp (f) \delta_{s+t+k+i,0} \right)$$
$$= (-1)^s 2 \sum_{i,k} (-1)^k q_k^\perp q_i^\perp (f)\, \delta_{s+t+k+i,0}$$
$$= (-1)^s 2\, \delta_{s+t,0} f,$$

using (7.1). This completes the proof.

Definition. Let α be any finite sequence of integers. If, for some positive integer n, the subsequence of those α_i for which $|\alpha_i| = n$ does *not* have one of the forms

$$n, -n, n, -n, \ldots, n, -n, n$$

or

$$-n, n, -n, \ldots, n, -n, n$$

define $y(\alpha)$ to be zero. Otherwise there is a permutation w which rearranges α into a sequence of the form

$$\lambda, \beta^{(1)}, \ldots, \beta^{(k)}, 0, 0, \ldots, 0\,,$$

where λ is a strict partition, and each $\beta^{(i)}$ has the form $(-n_i, n_i)$ with $n_i > 0$. A description of w is as follows. Move all zeros to the right-hand end, keeping them in order. Move all leftmost $n > 0$ to the left end and rearrange into decreasing order. Rearrange the middle section so that all $(-n, n)$ occur as adjacent pairs, without altering the order of the subsequences $(-n, n, \ldots, -n, n)$. In this case, define

$$\operatorname{str}\alpha := \lambda \quad \text{and} \quad y(\alpha) := \operatorname{sign}(w)\,(-1)^{n_1+\ldots+n_k}\,2^k\,.$$

For given such α, the strict partition λ is unique, the $\beta^{(i)}$ are unique up to re-indexing, and so w is unique up to multiplication by certain even permutations. Thus $y(\alpha)$ and $\operatorname{str}\alpha$ are well defined. Alternatively, this is a consequence of the following theorem.

Theorem 9.2. *For all sequences α of integers,*

$$Q_\alpha = y(\alpha) Q_{\operatorname{str}\alpha}\,.$$

Proof. If $y(\alpha) \neq 0$, factorize the permutation w in the previous definition as a product of transpositions of the form $(i\; i{+}1)$. Now rearrange α into the form given in the definition step-by-step using these transpositions. Theorem 9.1 (i) will apply at each step, yielding

$$Q_\alpha = \operatorname{sign}(w) Q_{\lambda,\beta^{(1)},\ldots,\beta^{(k)},0,0,\ldots,0}\,.$$

Since $\mathcal{B}_0(1) = 1$, the zeroes may be deleted. We may also delete each $\beta^{(i)}$, introducing the factor $(-1)^{n_i}2$, by 9.1 (ii) and since $\mathcal{B}_{-n}(1) = 0$ for $n > 0$. This yields $Q_\alpha = y(\alpha)Q_\lambda$, as required.

If $y(\alpha) = 0$, choose any $n > 0$ for which the condition for $y(\alpha)$ to be non-zero fails. Then, in α, either the rightmost occurrence of $\pm n$ is $-n$, or else, for $m = \pm n$, two copies of m occur such that none of the terms between these occurrences is $-m$. Now use a sequence of adjacent transpositions as in the paragraph above to deduce that $Q_\alpha = \pm Q_\beta$, where β either ends with $-n$ or has adjacent terms m, m. But $\mathcal{B}_{-n}(1) = 0$ and $\mathcal{B}_m\mathcal{B}_m = 0$, so $Q_\beta = 0$, as required.

Examples. (i) For $s>0$, $Q_{-s,s} = (-1)^s 2$ and $Q_{s,-s} = 0$.

(ii) $Q_{2,-3,5,3,1} = -2Q_{5,2,1}$.

(iii) $Q_{4,-2,-3,2,3} = 4q_4$.

(iv) $Q_{2,-3,3,-2} = 0 = Q_{-1,1,1}$.

Proposition 9.3. *For all r and all positive n,*

$$q_n^\perp \mathcal{B}_r = \mathcal{B}_r q_n^\perp + 2\sum_{p=1}^{n} \mathcal{B}_{r-p} q_{n-p}^\perp .$$

Proof. The left side applied to f is, by using (7.19) then (7.20),

$$\sum_i (-1)^i q_n^\perp\left[q_{r+i} q_i^\perp(f)\right]$$

$$= \sum_i \sum_{j=0}^{n} (-1)^i q_{n-j}^\perp(q_{r+i}) q_j^\perp q_i^\perp(f)$$

$$= \sum_i (-1)^i \left[q_{r+i} q_i^\perp q_n^\perp(f) + 2\sum_{j=0}^{n-1} q_{r+i-n+j} q_i^\perp q_j^\perp(f) \right].$$

Letting $p = n-j$, we obtain

$$\sum_i (-1)^i q_{r+i} q_i^\perp q_n^\perp(f) + 2\sum_{p=1}^{n}\sum_i (-1)^i q_{r+i-p} q_i^\perp q_{n-p}^\perp(f)$$

$$= \mathcal{B}_r q_n^\perp(f) + 2\sum_{p=1}^{n} \mathcal{B}_{r-p} q_{n-p}^\perp(f),$$

as required.

The following is immediate by induction on ℓ.

Corollary 9.4. *If* $\alpha \in \mathbb{Z}^\ell$, *then*

$$q_n^\perp(Q_\alpha) = \sum 2^{\eta(\beta)} Q_{\alpha-\beta},$$

summation over all $\beta \in \mathbb{N}^\ell$ *with* $|\beta| = n$, *where* $\eta(\beta)$ *is the number of i for* *which β_i is positive, and $\alpha-\beta$ means termwise subtraction.*

Let $U = \{u_1,\ldots,u_r\}$ be a new, finite, set of variables, disjoint from any other variables in the discussion. The identity in the next theorem takes place in the "ringoid" or "partial ring" of formal "infinite both ways" Laurent expressions

$$R\ll U^{-1}, U\gg \;:=\; \{\sum_{\alpha\in \mathbb{Z}^r} a_\alpha U^\alpha : a_\alpha \in R\},$$

where R is a commutative ring. This abelian group has no ring structure compatible with the two subrings of Laurent series in U and in $\{u^{-1} : u\in U\}$. One must be careful:

(1) to write down only products which are defined;

(2) to specify which embedding is used when a rational function of U is being regarded as such a Laurent expression; and

(3) not to use cancellation recklessly, since $R\ll U^{-1}, U\gg$ is not a "domainoid"!

Definition. Let $F(t)$ denote $1 + 2\sum_{n>0} t^n$, as in Chapter 7. Thus an example of a Laurent expression as discussed above is

$$F(u_i^{-1}u_j) = 1 + 2\sum_{n>0} u_i^{-n}u_j^n,$$

for any distinct indeterminates u_i, u_j in U. This can be written as $(u_i+u_j)/(u_i-u_j)$, since it is the binomial expansion of $(1-u_i^{-1}u_j)^{-1}(1+u_i^{-1}u_j)$. Note however that the other binomial expansion of this rational function (in terms of $u_j^{-1}u_i$) is a *different* Laurent expression, illustrating (2) above.

Our main use of Laurent expressions is in Theorem 9.5 below, and its applications. Such expressions replace the "raising operators" found in some treatments of Schur and Hall–Littlewood functions.

But first let us illustrate the use of these expressions by sketching a sanitized version of the above proof of 9.1, using ideas derived from vertex operator theory and pointed out to us by I.G. Macdonald. Below we define three operators $Q(u)$, $Q^\perp(u)$ and $\mathcal{B}(u)$ whose domains and codomains should be carefully specified for reasons including caution (1) above and the fact that we shall be dealing with composition identities.

Before giving these necessary specifications, here is the core of the proof, demonstrating its formal simplicity. Define

$$Q(u) := \sum_n q_n u^n, \tag{1}$$

regarded both as an "element", and as the operator which is multiplication by that element. Define operators

$$Q^{\perp}(u) := \sum_n q_n^{\perp} u^{-n}, \tag{2}$$

and

$$\mathcal{B}(u) := \sum_n \mathcal{B}_n u^n . \tag{3}$$

Then, directly from the definition of $\mathcal{B}_n$,

$$\mathcal{B}(u) = Q(u) \circ Q^{\perp}(-u) . \tag{4}$$

Equation (7.19) may be re-interpreted as

$$Q^{\perp}(u)(GH) = \left(Q^{\perp}(u)(G)\right)\left(Q^{\perp}(u)(H)\right) . \tag{5}$$

Equation (7.20) is converted to the form

$$Q^{\perp}(v)\left(Q(u)\right) = Q(u)F(v^{-1}u) . \tag{6}$$

Combining (5) and (6) yields

$$Q^{\perp}(v) \circ Q(u) = \left(F(v^{-1}u)Q(u)\right) \circ Q^{\perp}(v) . \tag{7}$$

Letting

$$\phi(u, v) := F(-u^{-1}v) + F(-v^{-1}u) , \tag{8}$$

direct calculations yield

$$\phi(u, v) = 2\sum_n (-u^{-1}v)^n = \phi(-v, -u) , \tag{9}$$

$$\phi(u, v)Q(u)Q(v) = \phi(u, v) , \tag{10}$$

and

$$\phi(u, v) \circ Q^{\perp}(u) \circ Q^{\perp}(v) = \phi(u, v) . \tag{10$^{\perp}$}$$

Using (4) and (7), we find

$$\mathcal{B}(v) \circ \mathcal{B}(u) = Q(v) \circ Q^{\perp}(-v) \circ Q(u) \circ Q^{\perp}(-u)$$

$$= Q(v) \circ \left(F(-v^{-1}u)Q(u)\right) \circ Q^{\perp}(-v) \circ Q^{\perp}(-u)$$

$$= \left(Q(u)Q(v)F(-v^{-1}u)\right) \circ Q^{\perp}(-v) \circ Q^{\perp}(-u) . \tag{11}$$

Now $Q(u)$ commutes with $Q(v)$, and $Q^{\perp}(-u)$ with $Q^{\perp}(-v)$. Thus $\mathcal{B}(u) \circ \mathcal{B}(v)$ is given by the same expression as (11), except for replacing $F(-v^{-1}u)$ by $F(-u^{-1}v)$. Adding and using (9), (10) and $(10)^{\perp}$:

$$\begin{aligned} \mathcal{B}(v) \circ \mathcal{B}(u) + \mathcal{B}(u) \circ \mathcal{B}(v) &= \left(Q(u)Q(v)\phi(u, v)\right) \circ Q^{\perp}(-v) \circ Q^{\perp}(-u) \\ &= \phi(u, v) \circ Q^{\perp}(-v) \circ Q^{\perp}(-u) \\ &= \phi(-v, -u) \circ Q^{\perp}(-v) \circ Q^{\perp}(-u) \\ &= \phi(-v, -u) \\ &= \phi(u, v) . \end{aligned} \tag{12}$$

Now (9.1) is simply the equality of the coefficients of $u^{-s}v^{t}$ in (12). The reader who wishes to fill in the details of this proof will be able to also fill in the foundation by using the following definitions and specifications of domains and codomains. Let

$$\Delta\langle U\rangle = \{\sum f_{\alpha} U^{\alpha} : f_{\alpha} \in \Delta^{(|\alpha|)} \text{ for each } \alpha \in \mathbb{Z}^{U}\} .$$

Let $u \neq v$ be indeterminates, neither being in another set W of indeterminates. Define

$$\Delta\langle u, W\rangle := \Delta\langle\{u\} \cup W\rangle ,$$

and

$$\Delta\langle u^{-}, W\rangle := \{G \in \Delta\langle u, W\rangle : G = \sum_{\substack{n \le 0 \\ \beta \in \mathbb{Z}^{W}}} f_{n,\beta} u^{n} W^{\beta}\} ;$$

that is, u does not "occur" with *positive* powers in G.

Note that for u in U and $\sum g_n u^n$ in $\Delta\langle U\rangle$, the product of $\sum g_n u^n$ with any element of $\Delta\langle U\rangle$ is defined. Furthermore, the element $F(v^{-1}u)$ is in $\Delta\langle u, v^{-}, W\rangle$ and multiplies into any element of $\Delta\langle u, v^{-}, W\rangle$.

Thus, as a multiplication operator, for any subsets S of $\Delta\langle U\rangle$ and T of $\Delta\langle v^{-}, u, W\rangle$, we have

$$Q(u) : S \to \Delta\langle U\rangle$$

or

$$Q(u): T \to \Delta<v^-, u, W> .$$

The operators in (2) and (3) are

$$\begin{array}{rlcl} Q^{\perp}(u): & \Delta<W> & \to & \Delta<u^-, W> \\ & \sum_{\alpha} f_{\alpha} W^{\alpha} & \mapsto & \sum_{n,\alpha} q_n^{\perp}(f_{\alpha}) u^{-n} W^{\alpha} , \end{array}$$

and

$$\begin{array}{rlcl} \mathcal{B}(u): & \Delta<W> & \to & \Delta<u, W> \\ & \sum_{\alpha} f_{\alpha} W^{\alpha} & \mapsto & \sum_{n,\alpha} \mathcal{B}_n(f_{\alpha}) u^{n} W^{\alpha} . \end{array}$$

Equation (4) is commutativity of

$$\begin{array}{ccccc} \Delta<W> & & \xrightarrow{\mathcal{B}(u)} & & \Delta<u, W> \\ & \searrow^{Q^{\perp}(-u)} & & \nearrow_{Q(u)} & \\ & & \Delta<u^-, W> & & \end{array}$$

Equation (5) holds for all those G and H in $\Delta<W>$ for which GH is defined. Equation (7) is commutativity of

$$\begin{array}{ccccc} \Delta<u^-, W> & & \xrightarrow{\quad Q(u) \quad} & & \Delta<u, W> \\ \downarrow{\scriptstyle Q^{\perp}(v)_1} & & & & \downarrow{\scriptstyle Q^{\perp}(v)} \\ \Delta<u^-, v^-, W> & \xrightarrow{Q(u)} & \Delta<u, v^-, W> & \xrightarrow{F(v^{-1}u)} & \Delta<u, v^-, W> \end{array}$$

where $Q^{\perp}(v)_1$ is obtained by restricting $Q^{\perp}(v)$.

Equations (11) and (12), and the latter part of the proof, are obtained from the following *non-commutative* diagram. Of the three commutative squares, the top left is clear, whereas the top right comes from (7), and the bottom left from the symmetry of the diagram about its diagonal (interchanging u and v):

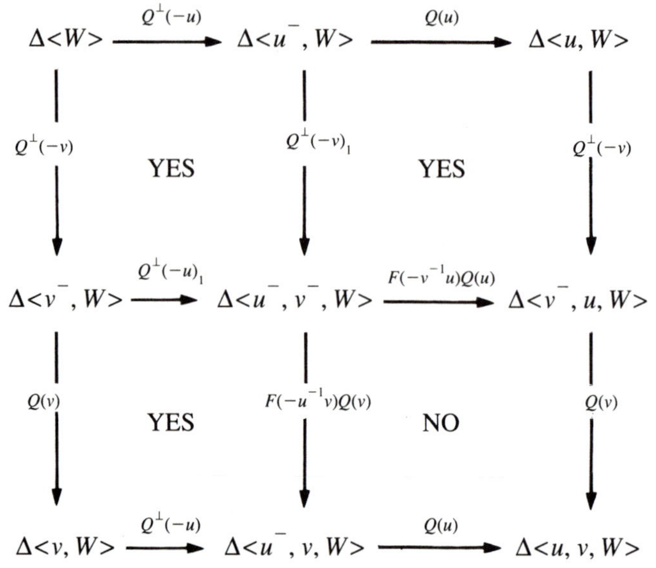

The horizontal composite maps at the top and bottom are two different versions of $\mathcal{B}(u)$. The vertical composites on the left and right are versions of $\mathcal{B}(v)$. Thus, the clean version of 9.1,

$$\mathcal{B}(v) \circ \mathcal{B}(u) + \mathcal{B}(u) \circ \mathcal{B}(v) = \phi(u, v)\,, \tag{12}$$

(obtained by adding the two outside paths in the diagram from its top left to bottom right corners in the case when W is empty), actually involves two different versions of both $\mathcal{B}(u)$ and $\mathcal{B}(v)$.

Theorem 9.5. *The Laurent expression $F(u_i^{-1}u_j)$ has inverse $F(-u_i^{-1}u_j)$, and the following identity is well-defined and valid*:

$$\sum_{\alpha \in \mathbb{Z}^r} Q_\alpha(X)U^\alpha = \prod_{i<j} F(u_i^{-1}u_j)^{-1} \prod_{x,u} F(xu)\,. \tag{9.5}$$

Proof. The inverse of $F(u_i^{-1}u_j)$ is evidently $F(-u_i^{-1}u_j)$, and the product $\prod_{i<j} F(u_i^{-1}u_j)$ is defined. Since $F(xu)$ is a power series, the right-hand side of the identity is well-defined. Taking U to be the set $\{u_1, \ldots, u_r, v_1, \ldots, v_r\}$ and iterating equation (7) above yields

$$Q(u_1)Q^{\perp}(v_1)Q(u_2)Q^{\perp}(v_2)\dots Q^{\perp}(v_r)$$
$$= Q(u_1)Q(u_2)\dots Q(u_r)Q^{\perp}(v_1)\dots Q^{\perp}(v_r)\prod_{i<j} F(v_i^{-1}u_j)\,.$$

Since $Q^{\perp}(v)(1) = 1$, letting $v_i = -u_i$ yields

$$\sum_{\alpha\in\mathbb{Z}^r} Q_\alpha U^\alpha = \mathcal{B}(u_1)\circ\mathcal{B}(u_2)\circ\mathcal{B}(u_r)(1)$$
$$= Q(u_1)\circ\dots\circ Q(u_r)(1)\prod_{i<j} F(-u_i^{-1}u_j)\,.$$

The result now follows, since

$$Q(u)(1) = \prod_x F(xu)\,,$$

which is a restatement of the definition of q_n.

Alternatively, a pedestrian proof proceeds by induction on r, using 9.4 and the definition of q_n. When $r = 1$, the identity is merely the definition of the functions q_n. For the inductive step, write $\alpha = (\alpha_1,\alpha')$ where $\alpha_1 \in \mathbb{Z}$, $\alpha' = (\alpha_2,\alpha_3,\dots)$:

$$\sum_{\alpha\in\mathbb{Z}^r} Q_\alpha(X)U^\alpha$$
$$= \sum_{i\geq 0}\sum_{\alpha\in\mathbb{Z}^r}(-1)^i q_{\alpha_1+i}(X)q_i^{\perp}(Q_{\alpha'})(X)U^\alpha$$
$$= \sum_{\beta\in\mathbb{N}^{r-1}}\sum_{\alpha\in\mathbb{Z}^r}(-1)^{|\beta|}q_{\alpha_1+|\beta|}(X)2^{\eta(\beta)}Q_{\alpha'-\beta}(X)U^\alpha \text{ by (9.4)}$$
$$= \sum_{\beta\in\mathbb{N}^{r-1}}\sum_{\gamma\in\mathbb{Z}^r}(-1)^{|\beta|}q_{\gamma_1}(X)2^{\eta(\beta)}Q_{\gamma'}(X)u_1^{\gamma_1-|\beta|}u_2^{\gamma_2+\beta_1}\dots u_r^{\gamma_r+\beta_{r-1}}$$
$$= \left\{\sum_{\gamma_1\in\mathbb{Z}} q_{\gamma_1}(X)u_1^{\gamma_1}\right\}\left\{\sum_{\mathbb{Z}^{r-1}} Q_{\gamma_2,\dots,\gamma_r}(X)u_2^{\gamma_2}\dots u_r^{\gamma_r}\right\}\times$$
$$\times\left\{\sum_{\mathbb{N}^{r-1}}(-1)^{|\beta|}2^{\eta(\beta)}(u_1^{-1}u_2)^{\beta_1}\dots(u_1^{-1}u_r)^{\beta_{r-1}}\right\}.$$

The last factor is $\prod_{j>1} F(u_1^{-1}u_j)^{-1}$, by multiplying these series together. The middle factor is

$$\prod_{2\le i<j} F(u_i^{-1}u_j)^{-1} \prod_{j>1} \prod_{x\in X} F(xu_j),$$

by the inductive hypothesis. The first factor is $\prod_X F(xu_1)$, by the definition of q_n. This yields the right-hand side of (9.5), as required.

Corollary 9.6. *For any strict* λ *and finite set* $u_1, \ldots, u_r$ *of variables,*

$$Q_\lambda(U) = 2^{\ell(\lambda)} \left(\sum_{\mathrm{str}\,\alpha=\varnothing} y(\alpha)U^\alpha \right)^{-1} \sum_{\mathrm{str}\,\alpha=\lambda} y(\alpha)U^\alpha ,$$

where $\alpha \in \mathbb{Z}^r$ *in both summations (and in those of the proof).*

Proof. Letting X be a sufficiently large set of variables so that the $Q_\lambda(X)$ (for the relevant $|\lambda|$) are linearly independent,

$$\begin{aligned}\sum_{\lambda\in\mathcal{D}} 2^{-\ell(\lambda)} Q_\lambda(U)Q_\lambda(X) &= \prod_{U\times X} F(ux) \quad \text{by (7.16)} \\ &= \prod_{i<j} F(u_i^{-1}u_j) \sum_{\alpha\in\mathbb{Z}^r} Q_\alpha(X)U^\alpha \quad \text{by (9.5)} \\ &= \prod_{i<j} F(u_i^{-1}u_j) \sum_{\substack{\lambda\in\mathcal{D} \\ \mathrm{str}\,\alpha=\lambda}} y(\alpha)U^\alpha Q_\lambda(X) \quad \text{by (9.2)}.\end{aligned}$$

Thus, equating "coefficients" of $Q_\lambda(X)$,

$$Q_\lambda(U) = 2^{\ell(\lambda)} \prod_{i<j} F(u_i^{-1}u_j) \sum_{\mathrm{str}\,\alpha=\lambda} y(\alpha)U^\alpha . \tag{9.7}$$

Since $Q_\varnothing = 1$, for $\lambda = \varnothing$, the empty partition, (9.7) yields

$$\prod_{i<j} F(u_i^{-1}u_j) = \left(\sum_{\mathrm{str}\,\alpha=\varnothing} y(\alpha)(U^\alpha \right)^{-1}, \tag{9.8}$$

completing the proof.

The individual expressions $\sum y(\alpha)U^\alpha$ in 9.6 are seldom polynomials in U, but their quotient in 9.6 is, since $Q_\lambda(U)$ is a polynomial. Next we shall rewrite $\sum y(\alpha)U^\alpha$ so as to make this directly obvious, and at the same time recover a formula for Q_λ which alternatively comes from the theory of Hall–Littlewood polynomials; Macdonald (1979) III.

Let S_r act on $U = \{u_1, ..., u_r\}$ by permuting the variables, so that, when $\ell \leq r$, the Young subgroup $S_1{}^\ell \times S_{r-\ell}$ fixes each of $u_1, ..., u_\ell$.

Lemma 9.9. *Let λ be a strict partition of length $\ell < r$. Then*

$$\sum_{\substack{\alpha\in \mathbb{Z}^r \\ \mathrm{str}\,\alpha=\lambda}} y(\alpha)U^\alpha = \sum_{[w]\in S_r/S_1^\ell\times S_{r-\ell}} \mathrm{sign}\,(w)\, w\left\{u_1^{\lambda_1} \dots u_\ell^{\lambda_\ell} \prod_{\ell<i<j} F(u_i^{-1}u_j)^{-1}\right\}.$$

Proof. The right-hand side is well-defined, since, if $w\in S_1{}^\ell \times S_{r-\ell}$, then w fixes $u_1^{\lambda_1} \dots u_\ell^{\lambda_\ell}$, and maps $\prod_{\ell<i<j}(u_i+u_j)/(u_i-u_j)$ to sign(w) times itself. The cosets $[w]$ in $S_r/S_1{}^\ell \times S_{r-\ell}$ are in $1-1$ correspondence with sequences $I = (i_1, ..., i_\ell)$ of distinct integers between 1 and r, with $[w]$ corresponding to $w(1), ..., w(\ell)$. Thus, for any selection w_I of representatives, the right-hand side is

$$\sum_I \sum_{\substack{\beta\in \mathbb{Z}^{r-\ell} \\ \mathrm{str}\,\beta=\varnothing}} \mathrm{sign}\,(w_I)y(\beta)w_I\left\{u_1^{\lambda_1} \dots u_\ell^{\lambda_\ell} u_{\ell+1}^{\beta_1} \dots u_r^{\beta_{r-\ell}}\right\} \qquad (*)$$

since, by (9.8),

$$\prod_{\ell<i<j} F(u_i^{-1}u_j)^{-1} = \sum_{\substack{\beta\in \mathbb{Z}^{r-\ell} \\ \mathrm{str}\,\beta=\varnothing}} y(\beta)u_{\ell+1}^{\beta_1} \dots u_r^{\beta_{r-\ell}}.$$

On the other hand, by definition, $\mathrm{str}\,\alpha = \lambda$ if and only if α has a subsequence $(\alpha_{i_1}, ..., \alpha_{i_\ell})$ which is a rearrangement of λ such that the complementary subsequence β satisfies $\mathrm{str}(\beta) = \varnothing$. (The subsequence $(\alpha_{i_1},...)$ consists of the leftmost occurrences in α of those $n > 0$ for which n occurs to the left of all occurrences of $-n$). Choose w_I to map $1, ..., \ell$ to $i_1, ..., i_\ell$ respectively, and to be order preserving on $\ell+1, ..., r$. Then

$$y(\alpha) = \mathrm{sign}\,(w_I)y(\beta)$$

and

$$U^{\alpha} = w_I\left\{u_1^{\lambda_1}\ldots u_\ell^{\lambda_\ell}u_{\ell+1}^{\beta_1}\ldots u_r^{\beta_{r-\ell}}\right\}.$$

Thus the left-hand side in the lemma also reduces to (*) for this choice, as required.

Theorem 9.10. *Let λ be a strict partition of length ℓ.*

(i) If $\ell \leq r$, then

$$Q_\lambda(u_1,\ldots,u_r)$$
$$= 2^\ell \sum_{[w]\in S_r/S_1^\ell\times S_{r-\ell}} w\left\{u_1^{\lambda_1}\ldots u_\ell^{\lambda_\ell}\prod_{i=1}^{\ell}\prod_{j=i+1}^{r}(u_i+u_j)/(u_i-u_j)\right\}.$$

(ii) If λ has length greater than r, then $Q_\lambda(u_1, \ldots, u_r) = 0$.

Proof. The second assertion is clear from 9.6, since there are certainly no α in $\mathbb{Z}^r$ with $\operatorname{str}\alpha$ of length greater than r. If $\ell \leq r$, then by 9.7 and 9.9,

$$Q_\lambda(u_1, \ldots, u_r)$$
$$= 2^\ell \prod_{i<j} F(u_i^{-1}u_j) \sum_{\substack{\alpha\in\mathbb{Z}^r\\ \operatorname{str}\alpha=\lambda}} y(\alpha)U^\alpha$$
$$= 2^\ell \prod_{i<j} F(u_i^{-1}u_j) \sum_{[w]\in S_r/S_1^\ell\times S_{r-\ell}} \operatorname{sign}(w)w\left\{u_1^{\lambda_1}\ldots u_\ell^{\lambda_\ell}\prod_{\ell<i<j} F(u_i^{-1}u_j)^{-1}\right\}.$$

But

$$w\left\{\prod_{i<j} F(u_i^{-1}u_j)\right\} = \operatorname{sign}(w)\prod_{i<j} F(u_i^{-1}u_j)$$

for any w, so $\prod_{i<j} F(u_i^{-1}u_j)$ can be absorbed inside the action above, with $\operatorname{sign}(w)$ disappearing, yielding

$$Q_\lambda(u_1, \ldots, u_r) = 2^\ell \sum_{[w]\in S_r/S_1^\ell\times S_{r-\ell}} w\left\{u_1^{\lambda_1}\ldots u_\ell^{\lambda_\ell}\prod_{i=1}^{\ell}\prod_{j>i} F(u_i^{-1}u_j)\right\},$$

as required.

Remarks. (1) The formula in Theorem 9.10 would be a reasonable choice for the official definition of Q_λ. The route from it to the orthogonality property of $\{Q_\lambda\}$, which is central to the material of Chapter 8, is much longer, so we have made a reverse (perhaps perverse) choice for the order of presentation of these functions. The right-hand side within 9.10 (i) is zero when λ is a partition with a repeated part, so 9.10 is in fact valid for all partitions λ.

(2) In general, a definition of the form

$$f(x_1, ..., x_r) = \sum_{[w]\in S_r/G} w\{h(x_1, .., x_r)\},$$

where h is some rational function fixed by all elements of a given subgroup G of S_r, will produce a symmetric rational function f. If h can be written with denominator equal to $\prod_{i<j}(x_i-x_j)$, the Vandermonde determinant, then f is actually a symmetric polynomial, by an elementary argument using the skew symmetry of the Vandermonde and unique factorization in $\mathbb{Z}[x_1, ..., x_r]$.

(3) It can sometimes be more convenient to have an anti-symmetrization or a symmetrization over the whole symmetric group. For example,

$$Q_\lambda(u_1, ..., u_r)$$
$$= 2^{\ell(\lambda)}\sum_{w\in S_r} \operatorname{sign}(w)w\left\{u_1^{\lambda_1}\cdots u_\ell^{\lambda_\ell}\prod_{i=1}^{\ell}\prod_{j=i+1}^{r}(1+u_i^{-1}u_j)\right\}. \tag{9.11}$$

This can be deduced as follows from (9.10), using the expansion of the Vandermonde determinant

$$\prod_{i<j}(u_i-u_j) = \det(u_i^{r-j})_{r\times r}$$
$$= \sum_{w\in S_r} \operatorname{sign}(w)w\left\{u_1^{r-1}u_2^{r-2}\cdots u_{r-1}\right\}.$$

The right-hand side of (9.11) is $2^{\ell(\lambda)}$ multiplied into

$$\sum_{w\in S_r} w\left\{\prod_{i<j}(u_i-u_j)^{-1}u_1^{r-1}u_2^{r-2}\cdots u_{r-1}u_1^{\lambda_1}\cdots u_\ell^{\lambda_\ell}\prod_{j>i\le\ell}(1+u_i^{-1}u_j)\right\}$$

$$= \sum_{w \in S_r} w \left\{ \prod_{\ell<i<j} (u_i - u_j)^{-1} \prod_{j>i\le\ell} (u_i+u_j)/(u_i-u_j) u_1^{\lambda_1} \right.$$

$$\left. \ldots u_\ell^{\lambda_\ell} u_{\ell+1}^{r-\ell-1} u_{\ell+2}^{r-\ell-2} \ldots u_{r-1} \right\}$$

$$= \sum_{[w'] \in S_r/S_1^\ell \times S_{r-\ell}} w' \left\{ u_1^{\lambda_1} \ldots u_\ell^{\lambda_\ell} \prod_{j>i\le\ell} (u_i+u_j)/(u_i-u_j) \times \right.$$

$$\left. \times \sum_{w'' \in S_1^\ell \times S_{r-\ell}} w'' \left(\prod_{\ell<i<j} (u_i-u_j)^{-1} u_{\ell+1}^{r-\ell-1} \ldots u_{r-1} \right) \right\},$$

writing $w = w'w''$, as w' ranges over a fixed set of coset representatives. The inside summation yields 1, by the Vandermonde formula, leaving the expression in (9.10) for Q_λ, as required.

Another corollary to Theorem 9.5 gives Q_λ explicitly as a polynomial in the functions q_n.

Corollary 9.12. *If λ is a strict partition of length ℓ, then*

$$Q_\lambda = \sum_{\operatorname{str}\beta=\varnothing} y(\beta) q_{\lambda-\beta} = \sum_{|\beta|=0} y(\beta) q_{\lambda-\beta},$$

where $\beta \in \mathbb{Z}^\ell$, and $\lambda-\beta$ means termwise subtraction.

Remark. The set $\{\beta \in \mathbb{Z}^\ell : \operatorname{str}\beta = \varnothing$, the empty partition$\}$ is infinite for $\ell>1$, but its intersection with the infinite set $\{\beta : \lambda_i - \beta_i \ge 0$ for all $i\}$ is finite, so that the middle summation is effectively finite. In the summation on the right, the additional terms are all zero, since if $|\beta| = 0$ and $\operatorname{str}\beta$ is undefined, then $y(\beta) = 0$ by definition.

Proof. For any disjoint sets X and $U = \{u_1, \ldots, u_\ell\}$ of variables,

$$\begin{aligned}\sum_{\gamma\in\mathbb{Z}^\ell} Q_\gamma(X)U^\gamma &= \sum_{\mathrm{str}\,\beta=\varnothing} y(\beta)U^\beta \prod_{X\times U} F(xu) && \text{by (9.5) and (9.8)}\\ &= \sum_{\mathrm{str}\,\beta=\varnothing} y(\beta)U^\beta \sum_{\alpha\in\mathbb{Z}^\ell} q_\alpha(X)U^\alpha && \text{by (7.11)}\\ &= \sum_{\gamma\in\mathbb{Z}^\ell}\ \sum_{\mathrm{str}\,\beta=\varnothing} y(\beta)q_{\gamma-\beta}(X)U^\gamma .\end{aligned}$$

Letting $\gamma = \lambda$ and equating coefficients of U^λ yields the result.

To recover Schur's original defining formulae for the Q_λ, we need the following curious identities.

Lemma 9.13. *Let* $\beta\in\mathbb{Z}^\ell$ *with* $|\beta| = 0$. *Then, with* $\hat{\beta}_i$ *meaning "omit* β_i*" and summing over the specified values of* i:

(*i*) *if* $\ell = 2k+1$,

$$y(\beta) = \sum_{\beta_i=0} (-1)^{i+1} y(\beta_1,\ldots,\hat{\beta}_i,\ldots,\beta_{2k+1})\,;$$

(*ii*) *if* $\ell = 2k>0$,

$$y(\beta) = \sum_{\substack{i>1\\ \beta_i=-\beta_1}} (-1)^i y(\beta_1,\beta_i)y(\beta_2,\ldots,\hat{\beta}_i,\ldots,\beta_{2k})\,.$$

Proof. We shall frequently use the fact that, if $|\beta| = 0$ and $y(\beta) \neq 0$, then, for all $n > 0$, $(-n, n, \ldots, -n, n)$ is the form of the subsequence of terms in β equal to $\pm n$.

(i) Let $i_1 > i_2 > \ldots > i_p$ be the list of i for which $\beta_i = 0$. Then, for each s,

$$y(\beta_1,\ldots,\hat{\beta}_{i_s},\ldots,\beta_{2k+1}) = (-1)^{s+i_s} y(\beta)\,,$$

since, in β, we can move β_{i_s} to the right-hand end, changing sign for each $j > i_s$ such that $\beta_j \neq 0$, yielding sign

$$(-1)^{2k+1-i_s+s-1} = (-1)^{s+i_s}\,.$$

Thus the right-hand side in (i) is

$$\sum_{s=1}^{p}(-1)^{s+1}y(\beta)\,.$$

If p is odd, this is $y(\beta)$. If p is even, then $y(\beta) = 0$, since the number, $\ell - p$, of non-zero β_i is odd. In either case, we obtain $y(\beta)$, as required.

(ii) If $\beta_1 > 0$, the identity is trivial, since then $y(\beta) = 0$ and $y(\beta_1, \beta_i) = 0$ for all i such that $\beta_i = -\beta_1$.

If $\beta_1 = 0$, the identity follows by applying part (i) to the sequence $(\beta_2,\ldots, \beta_{2k})$.

If $\beta_1 < 0$, let $\beta_1 = -n$. Suppose $y(\beta) = 0$ because of the terms equal to $\pm m$ in β for some positive $m \neq n$. Then each term $y(\beta_2,\ldots, \hat{\beta}_i,\ldots, \beta_{2k})$ in the identity is zero for the same reason. Suppose instead that $y(\beta) = 0$ only because of the subsequence of terms equal to $\pm n$. If it ends with $-n$ or contains two consecutive copies of $-n$ other than β_1, then each term $y(\beta_2,\ldots, \hat{\beta}_i,\ldots, \beta_{2k})$ in the identity is zero for the same reason. Otherwise, this subsequence has the form

$$(-n, -n), (n, -n, n, -n, \ldots, n, -n), (n, n), (-n, n, \ldots, -n, n)\,.$$

Except for the two values of i corresponding to the adjacent pair (n, n), we have $y(\beta_2,\ldots, \hat{\beta}_i,\ldots, \beta_{2k}) = 0$, and those two values give cancelling terms. Finally, suppose that $y(\beta) \neq 0$. Then for all but the minimum value, i_0, of i, each term $y(\beta_2,\ldots, \hat{\beta}_i,\ldots, \beta_{2k}) = 0$, since it contains two copies of $-n$ "uninterrupted" by any $+n$, or, in one instance, has a $-n$ to the right of all $+n$. The required identity reduces to

$$y(\beta) = (-1)^{i_0}(-1)^{n}2y(\beta_2,\ldots, \hat{\beta}_{i_0},\ldots, \beta_{2k})\,,$$

where

$$\beta = (-n, \beta_2,\ldots, \beta_{i_0-1}, n, \beta_{i_0+1},\ldots, \beta_{2k})\,.$$

This follows immediately from the definition of y, and completes the proof.

Theorem 9.14. *The functions Q_λ satisfy the following two inductive formulae, depending on the parity of the length of λ:*

(i) If $k \geq 0$,

$$Q_{\lambda_1,\ldots,\lambda_{2k+1}} = \sum_{i=1}^{2k+1}(-1)^{i+1}q_{\lambda_i}Q_{\lambda_1,\ldots,\hat{\lambda}_i,\ldots,\lambda_{2k+1}}\,.$$

(ii) If $k > 0$,

$$Q_{\lambda_1,\dots,\lambda_{2k}} = \sum_{i=2}^{2k} (-1)^i Q_{\lambda_1,\lambda_i} Q_{\lambda_2,\dots,\hat{\lambda}_i,\dots,\lambda_{2k}} .$$

Proof. (i) By (9.12) and (9.13) (i),

$$\begin{aligned} Q_{\lambda_1,\dots,\lambda_{2k+1}} &= \sum_{\substack{\beta\in\mathbb{Z}^{2k+1}\\ |\beta|=0}} y(\beta) q_{\lambda-\beta} \\ &= \sum_{\substack{\beta\in\mathbb{Z}^{2k+1}\\ |\beta|=0}} \sum_{\beta_i=0} (-1)^{i+1} y(\beta_1,\dots,\hat{\beta}_i,\dots,\beta_{2k+1}) q_{\lambda-\beta} . \end{aligned}$$

Define

$$B = \{(\beta, i) : \beta\in\mathbb{Z}^{2k+1};\ |\beta|=0;\ 1\le i\le 2k+1;\ \beta_i=0\} ,$$

and

$$C = \{(i,\gamma) : 1\le i\le 2k+1;\ \gamma\in\mathbb{Z}^{2k};\ |\gamma|=0\} ,$$

a Cartesian product. Then the following give mutually inverse bijections between B and C:

$$\begin{aligned} &(\beta, i) \mapsto \big(i, (\beta_1,\dots,\beta_{i-1},\beta_{i+1},\dots,\beta_{2k+1})\big) \\ &(i,\gamma) \mapsto \big((\gamma_1,\dots,\gamma_{i-1}, 0, \gamma_i,\dots,\gamma_{2k}), i\big) . \end{aligned}$$

Using this bijection to change summation variables,

$$\begin{aligned} Q_{\lambda_1,\dots,\lambda_{2k+1}} &= \sum_B (-1)^{i+1} y(\beta_1,\dots,\hat{\beta}_i,\dots,\beta_{2k+1})\, q_{\lambda_i}\, q_{\lambda_1-\beta_1,\dots,\widehat{\lambda_i-\beta_i},\dots} \\ &= \sum_C (-1)^{i+1} q_{\lambda_i}\, y(\gamma)\, q_{(\lambda_1,\dots,\hat{\lambda}_i,\dots)-\gamma} \\ &= \sum_{i=1}^{2k+1} (-1)^{i+1} q_{\lambda_i}\, Q_{\lambda_1,\dots,\hat{\lambda}_i,\dots,\lambda_{2k+1}} \end{aligned}$$

using (9.12) again.

(ii) By (9.12) and (9.13) (ii),

$$\begin{aligned} Q_{\lambda_1,\ldots,\lambda_{2k}} &= \sum_{\substack{\beta\in\mathbb{Z}^{2k}\\ |\beta|=0}} y(\beta) q_{\lambda-\beta} \\ &= \sum_{\substack{\beta\in\mathbb{Z}^{2k}\\ |\beta|=0}} \sum_{\substack{1<i\leq 2k\\ \beta_i+\beta_1=0}} (-1)^i\, y(\beta_1,\beta_i)\, y(\beta_2,\ldots,\hat{\beta}_i,\ldots,\beta_{2k})\, q_{\lambda-\beta}\,. \end{aligned}$$

Define

$$\begin{aligned} D &= \{(\beta,i) : \beta\in\mathbb{Z}^{2k};\ |\beta|=0;\ 1<i\leq 2k;\ \beta_i=-\beta_1\} \\ E &= \{(i,b,\gamma) : 1<i\leq 2k:\ b\in\mathbb{Z};\ \gamma\in\mathbb{Z}^{2k-2};\ |\gamma|=0\}\,. \end{aligned}$$

Then

$$\begin{aligned} (\beta,i) &\mapsto \big(i,\ \beta_i,\ (\beta_2,\ldots,\hat{\beta}_i,\ldots,\beta_{2k})\big) \\ (i,b,\gamma) &\mapsto \big((-b,\gamma_1,\ldots,\gamma_{i-2},b,\gamma_{i-1},\ldots,\gamma_{2k-2}),\ i\big) \end{aligned}$$

yield mutually inverse bijections between D and E. Thus

$$\begin{aligned} Q_{\lambda_1,\ldots,\lambda_{2k}} &= \sum_{D} (-1)^i\, y(\beta_1,\beta_i)\, y(\beta_2,\ldots,\hat{\beta}_i,\ldots,\beta_{2k})\, q_{\lambda-\beta} \\ &= \sum_{E} (-1)^i\, y(-b,b)\, y(\gamma)\, q_{\lambda_1+b,\,\lambda_i-b}\, q_{(\lambda_2,\ldots,\hat{\lambda}_i,\ldots)-\gamma} \\ &= \sum_{i=2}^{2k} (-1)^i \left\{\sum_{b\in\mathbb{Z}} y(-b,b)\, q_{\lambda_1+b,\lambda_i-b}\right\} \left\{\sum_{\substack{\gamma\in\mathbb{Z}^{2k-2}\\ |\gamma|=0}} y(\gamma)\, q_{(\lambda_2,\ldots,\hat{\lambda}_i,\ldots)-\gamma}\right\} \\ &= \sum_{i=2}^{2k} (-1)^i\, Q_{\lambda_1,\lambda_i}\, Q_{\lambda_2,\ldots,\hat{\lambda}_i,\ldots,\lambda_{2k}}\,, \end{aligned}$$

again using (9.12), as required.

The formulae in (9.14) are expansions of certain Pfaffians. More precisely,

$$Q_{\lambda_1,\dots,\lambda_{2k}} = Pf(Q_{\lambda_i,\lambda_j})_{2k\times 2k} \tag{9.15}$$

and

$$Q_{\lambda_1,\dots,\lambda_{2k+1}} = Pf\begin{pmatrix} & & q_{\lambda_1} \\ & & \cdot \\ & Q_{\lambda_i,\lambda_j} & \cdot \\ & & \cdot \\ & & q_{\lambda_{2k+1}} \\ -q_{\lambda_1} \cdots & & 0 \end{pmatrix}_{(2k+2)\times(2k+2)} . \tag{9.15$'$}$$

For expansions of Pfaffians, see DeConcini and Procesi (1976). Together with $Q_{(a)} = q_a$ and

$$Q_{(a,b)} = q_a q_b + 2\sum_{n>0}(-1)^n q_{a+n} q_{b-n} ,$$

these formulae were Schur's inductive definition of the Q_λ; Schur (1911). This last formula is the case $\ell=2$ of (9.12).

Here is a direct deduction of (9.15) and (9.15)′ from (9.5). One needs to know that the function Pf, on anti-symmetric matrices of even degree m, has the form

$$Pf(A) = \sum_I n_I A_{i_1,i_2} A_{i_3,i_4} \dots A_{i_{m-1},i_m}$$

for certain n_I (actually ± 1), with summation over certain rearrangements $I = (i_1, \dots, i_m)$ of $(1, \dots, m)$; and also that

$$Pf\left(\frac{u_i - u_j}{u_i + u_j}\right) = \prod_{i<j} \frac{u_i - u_j}{u_i + u_j} . \tag{*}$$

See Jozefiak (1989), p. 231, for the latter. By (9.5)

$$\sum_{\alpha \in \mathbb{Z}^m} Q_\alpha U^\alpha$$

$$= \prod_{i<j} F(-u_i^{-1} u_j)\, Q(u_1) \dots Q(u_m)$$

$$= Pf[F(-u_i^{-1}u_j)]\, Q(u_1) \dots Q(u_m), \text{ by } (*),$$

$$= \sum_I n_I F(-u_{i_1}^{-1} u_{i_2})\, Q(u_{i_1})\, Q(u_{i_2}) \dots F(-u_{i_{m-1}}^{-1} u_{i_m})\, Q(u_{i_{m-1}})\, Q(u_{i_m})$$

$$= \sum_I n_I \left(\sum_{\alpha_{i_1}, \alpha_{i_2}} Q_{\alpha_{i_1}, \alpha_{i_2}} u_{i_1}^{\alpha_{i_1}} u_{i_2}^{\alpha_{i_2}} \right) \left(\sum_{\alpha_{i_3}, \alpha_{i_4}} \dots \right) \dots, \text{ by (9.5) again,}$$

$$= \sum_{\alpha, I} n_I Q_{\alpha_{i_1}, \alpha_{i_2}} \dots Q_{\alpha_{i_{m-1}}, \alpha_{i_m}} u_1^{\alpha_1} \dots u_m^{\alpha_m}$$

$$= \sum_{\alpha} Pf(Q_{\alpha_i, \alpha_j}) U^{\alpha} .$$

Thus

$$Q_\alpha = Pf(Q_{\alpha_i, \alpha_j})$$

as required.

Notes

Schur considered Q_α for α with positive parts, not necessarily in decreasing order. When α has negative parts, the functions were introduced in Morris (1962). Rules for reducing Q_α to Q_λ were also given there. Note that rule 4 (p. 63) must be restricted, since it would imply, for example, that $Q_{1,-1,1}$ is zero. Also rules 1 and 2 need qualification once parts equal to zero are allowed. See Humphreys (1986) and Morris and Olsson (1988) for more details. The treatment using $\mathcal{B}_n$ and $y(\alpha)$ is in Hoffman (1990), as are 9.5, 9.6 and 9.12. Formula (9.5), when rewritten as a raising operator formula, becomes the specialization at $t = -1$ of III (2.5) in Macdonald (1979). The lack of ring structure on the group of Laurent expressions was pointed out to us by John Lawrence. The "vertex operator" $\mathcal{B}(t)$ and the associated proofs of 9.1 and 9.5 were shown to us by I.G. Macdonald. Formula (9.10) is the specialization at $t = -1$ of one of the two equivalent defining formulae for Hall–Littlewood polynomials; Macdonald (1979) III (2.1), (2.2). The other becomes 0/0 at $t = -1$. See also Appendix 9. The Pfaffians (9.15), (9.15)$'$ are from Schur (1911); our first treatment is from Hoffman (1990). The deduction from 9.5 was shown to us by I.G. Macdonald. Other proofs of the equivalence of Schur's definition to the specialization of the Hall–Littlewood functions are given in Humphreys (1983), and sketched in the appendix to Pragacz (1987). The proof of this equivalence apparently does not occur earlier in the literature, though it was undoubtedly known to Schur.

Appendix 9

Hall–Littlewood and Schur functions

The functions referred to in the title are treated thoroughly in Macdonald (1979). We shall give a brief description of their relation to the Q-functions. The generalities in Remark (2) after Theorem 9.10 referred to rational functions with integer coefficients. They are clearly valid for functions with coefficients in any commutative ring R with 1. Below we shall take R to be the polynomial ring $\mathbb{Z}[t]$.

Let n be a positive integer and let $\lambda = (\lambda_1, \lambda_2, \ldots, \lambda_n)$ be a weakly decreasing sequence of non-negative integers (i.e. a partition followed possibly by some zeroes). Recall that

$$G_\lambda = \{\beta \in S_n : \lambda_{\beta(i)} = \lambda_i \text{ for all } i\}.$$

Define an element of $\mathbb{Z}[t]\,(x_1, .., x_n)$ as follows:

$$R_\lambda(x_1, \ldots, x_n; t) = x_1^{\lambda_1} \ldots x_n^{\lambda_n} \prod_{\{(i,j):\lambda_i > \lambda_j\}} \frac{x_i - tx_j}{x_i - x_j}.$$

It is not difficult to see that $\beta R_\lambda = R_\lambda$ for all β in G_λ. Then, by the remark above, the following defines a symmetric rational function with coefficients in $\mathbb{Z}[t]$:

$$P_\lambda(x_1, \ldots, x_n; t) = \sum_{[w] \in S_n/G_\lambda} w\{R_\lambda(x_1, \ldots, x_n; t)\}.$$

It is clear that R_λ can be written as the quotient of a polynomial divided by the Vandermonde determinant $V_n(x_1, \ldots, x_n)$. Thus P_λ is actually a polynomial in $(x_1, \ldots, x_n)$. Since R_λ is homogeneous of degree $\lambda_1 + \ldots + \lambda_n$, so is P_λ. It is not difficult to see that

$$P_{\lambda_1, \ldots, \lambda_n, 0}(x_1, \ldots, x_n, 0; t) = P_{\lambda_1, \ldots, \lambda_n}(x_1, \ldots, x_n; t).$$

This yields a symmetric function $P_\lambda(X;t)$ in $\Lambda \otimes \mathbb{Z}[t]$, where X may be thought of as any finite or countable set of variables. The subscript λ is now simply a partition (we may omit the zeroes, if any); and $P_{\lambda_1, \ldots, \lambda_\ell}(x_1, \ldots, x_k; t)$ makes sense even when $k < \ell$. The function $P_\lambda(X; t)$ is called a *Hall–Littlewood symmetric function.*

Directly from its definition,

$$P_\lambda(x_1, ..., x_n; 1) = m_\lambda(x_1, ..., x_n).$$

When $\lambda = (1^r) = (1, 1, ..., 1)$, we find

$$P_{1^r}(x_1, ..., x_r; t) = R_{1^r}(x_1, ..., x_r; t) = x_1 x_2 \ldots x_r.$$

Thus for any number of variables, $P_{1^r}(X; t)$ is $e_r(X)$, the rth elementary symmetric function, and so is independent of t.

Define $P_\lambda(x_1, ..., x_n)$ to be $2^{-\ell(\lambda)}Q_\lambda(x_1, ..., x_n)$, where $\ell(\lambda)$ is the number of *non-zero* terms in the sequence λ. It has integer coefficients, by Theorem 9.10. We shall not use this function, nor does Macdonald, but it does occur in the literature, so the following comments may be useful to the reader.

The relationship between the Hall–Littlewood functions and the Q-functions may be stated loosely as: "Take $t = -1$". One must exercise some caution, however. For example,

$$P_{1^r}(x_1, ..., x_n) \neq P_{1^r}(x_1, ..., x_n; -1)$$

for $1 < r \leq n$, since the left side is zero and the right side is $e_r(x_1, ..., x_n)$. On the other hand, if λ is a strict partition, then $P_\lambda(X) = P_\lambda(X; -1)$. (Substitute $t = -1$ in the definition and compare 9.10.) Furthermore, the function $Q_\lambda(X;t)$, defined in Macdonald (1979) III (2.11) as $b_\lambda(t)P_\lambda(X; t)$ for a certain function $b_\lambda(t)$, does specialize to $Q_\lambda(X)$ when $t = -1$ for *any* partition λ. This is simply because $b_\lambda(-1)$ is zero if λ is not strict (and is the correct power of 2 when λ is strict). Note that $\{P_\lambda : \lambda \in \mathcal{D}\}$ is the basis for $\Delta_{\mathbb{R}}$ which is dual to the basis $\{Q_\lambda : \lambda \in \mathcal{D}\}$ with respect to the inner product of (7.7).

The $Q_\lambda(X; t)$ can be defined for any sequence λ of integers; Macdonald (1979) (p. 109, ex. 2). The operators $\mathcal{B}_n$ of this book also generalize almost verbatim to the Hall–Littlewood functions; Hoffman (1990). See also Jing (1989B).

Recall the inner product on the vector space $\Delta_{\mathbb{R}}$ with basis $\{Q_\lambda\}$, which makes that basis orthogonal. The corresponding theory for the Hall–Littlewood functions is treated in Macdonald (1979) III 4. The fundamental identity of Schur's theory, which we give in Corollary 7.16, has an analogue for Hall–Littlewood functions; see Macdonald (1979) (4.1) (4.2) (4.4). It is possible to deduce Corollary 7.16 from this analogue by "taking $t = -1$". Note, however, that "taking $t = -1$" does not produce an isometry with respect to the above inner products, since it is not injective.

The question of counting chains of subgroups of a finite abelian p-group has an explicit solution in terms of products of the functions $Q_\lambda(X; t)$ and the substitution $t = p^{-1}$; see Macdonald (1979) (II, III). There is also a close relationship between the Hall–Littlewood functions and the representation theory of the finite general linear groups; Macdonald (1979) (IV).

To finish, we describe the Schur functions and their connection with the Hall–Littlewood functions. For any sequence $\alpha = (\alpha_1, ..., \alpha_n)$ of non-negative integers, let

$$g_\alpha(x_1, ..., x_n) = \sum_{\sigma \in S_n} \operatorname{sgn} w\, w(x_1^{\alpha_1} \ldots x_n^{\alpha_n}).$$

For example, let $\delta = (n-1, n-2, ..., 2, 1, 0)$. Then g_δ is the Vandermonde determinant. For a weakly decreasing sequence $\lambda = (\lambda_1, ..., \lambda_n)$ of non-negative integers, set

$$s_\lambda(x_1, ..., x_n) = g_\delta(x_1, ..., x_n)^{-1} g_{\lambda+\delta}(x_1, ..., x_n).$$

The arguments after Theorem 9.10 show that this is a symmetric polynomial, which can be used to define a symmetric function s_λ, called *the Schur function*, where now λ is simply a partition.

In Macdonald (1979)(I 7) there is a succinct treatment relating the s_λ to the linear representations of the symmetric groups. In Chapter 8 we gave the analogous relation of the Q_λ to their projective representations.

The Schur and Hall–Littlewood functions are related as follows:

$$s_\lambda(X) = P_\lambda(X; 0) = Q_\lambda(X; 0).$$

The second equality comes from the fact that $b_\lambda(0) = 1$ for all λ. Working directly from our definition of $P_\lambda(X; t)$, the first equality reduces to showing that

$$g_{\lambda+\delta} = \sum_{[w] \in S_n/G_\lambda} \operatorname{sgn}(w)\, w \left\{ \prod_{i=1}^{n} x_i^{\lambda_i + |\{j : \lambda_j > \lambda_i\}|} \prod_{\substack{i<j \\ \text{and } \lambda_i < \lambda_j}} (x_i - x_j) \right\},$$

which is not obvious, but is straightforward.

A generalization of this yields another formula for the Hall–Littlewood functions, Macdonald (1979) III (2.1), (where $\lambda_1 \geq \lambda_2 \geq ... \geq \lambda_n \geq 0$): letting

$$m(i, \lambda) = |\{j : \lambda_j = i\}|,$$

$$w_m(t) = (1-t)^{-m} \prod_{i=1}^{m} (1-t^i)$$

and

$$v_\lambda(t) = \prod_i w_{m(i,\lambda)}(t),$$

we have

$$P_\lambda(x_1, \ldots, x_n; t) = v_\lambda(t)^{-1} \sum_{w \in S_n} w \left\{ x_1^{\lambda_1} \ldots x_n^{\lambda_n} \prod_{\{(i,j): i<j\}} \frac{x_i - tx_j}{x_i - x_j} \right\}.$$

When t is replaced by -1, this yields $0^{-1}0$. However, (9.11) gives an antisymmetrization formula for $Q_\lambda(x_1, \ldots, x_n)$ over all of S_n, which, if desired, may be made into a symmetrization by introducing a Vandermonde factor.

10

REDUCTION, BRANCHING AND DEGREE FORMULAE

The first objective of this chapter is an inductive method of calculation for all those values of the irreducible negative characters of $\tilde{S}_n$ which are not determined explicitly by Theorem 8.7. This result is analogous to the Murnaghan–Nakayama formula for linear characters of S_n, James (1978, p. 79), which reduces the calculation of the values of such a character to the calculation of some character values in S_k for $k<n$. Given partitions λ in $\mathcal{D}(n)$ and π in $\mathcal{P}^0(n)$, and a part r of π, the formula given in Theorem 10.1 enables one to express $<\lambda>(\pi)$ in terms of character values on $\tilde{S}_{n-r}$. By Theorem 8.7, if g has cycle type π not in $\mathcal{P}^0(n)$, then $<\lambda>(g)$ is zero except (possibly) when $\pi = \lambda$. Since the latter case is given by Theorem 8.7, the complete table of character values can be determined in this way.

Using this reduction formula, we then derive a "branching rule", the formula for restriction of representations from $\tilde{S}_n$ to $\tilde{S}_{n-1}$.

The relationship between combinatorics and linear representations of the symmetric group has proved to be of mutual benefit to both. Much recent progress on the projective representation theory of the symmetric group depends on a combinatorial framework which allows one to develop a similar relationship. The main idea is that of a shifted diagram, which we introduce next. The reduction formula and branching rule are then interpreted combinatorially using diagrams. Finally, the degrees of the irreducible projective representations of S_n are calculated and related to the idea of a standard shifted tableau.

Let r be an odd positive integer, and let $\lambda \in \mathcal{D}(n)$ have length ℓ. Below we define:

(i) a subset, $I_+ \cup I_0 \cup I_- = I(\lambda,r)$, of integers between 1 and ℓ; and

(ii) for each $i \in I(\lambda, r)$, a strict partition $\lambda(i,r)$ in $\mathcal{D}(n-r)$. (Despite the notation, $\lambda(i,r)$ is a function of λ, as well as of (i,r).) Let

$$I_+ = \{i : \lambda_j < \lambda_i - r < \lambda_{j+1} \text{ for some } j \leq \ell, \text{ taking } \lambda_{\ell+1} = 0\} .$$

If $i \in I_+$, then $\lambda_i > r$, and we define $\lambda(i,r)$ to be the partition obtained from λ by removing λ_i and inserting $\lambda_i - r$ between λ_j and λ_{j+1}. Let

$$I_0 = \{i : \lambda_i = r\} ,$$

which is empty or a singleton. For $i \in I_0$, remove λ_i from λ to obtain $\lambda(i,r)$. Let

$$I_- = \{i : r - \lambda_i = \lambda_j \text{ for some } j \text{ with } i < j \leq \ell\} .$$

If $i \in I_-$, then $\lambda_i < r$, and $\lambda(i,r)$ is formed by removing both λ_i and λ_j from λ.

The motivation for the strict partitions $\lambda(i,r)$ is that, for each i in $I(\lambda, r)$,

$$\mathrm{str}(\lambda_1,\ldots,\lambda_i - r,\ldots,\lambda_\ell) = \lambda(i,r) .$$

for str as defined before Theorem 9.2. Thus

$$Q_{\lambda_1,\ldots,\lambda_i - r,\ldots,\lambda_\ell} = y(\lambda_1,\ldots,\lambda_i - r,\ldots,\lambda_\ell)\, Q_{\lambda(i,r)} .$$

The coefficient comes immediately from the definition of y before Theorem 9.2:

$$y(\lambda_1,\ldots,\lambda_i - r,\ldots,\lambda_\ell) = \begin{cases} (-1)^{j-i} & \text{if } i \in I_+; \\ (-1)^{\ell-i} & \text{if } i \in I_0; \\ (-1)^{j-i+\lambda_i} 2 & \text{if } i \in I_-. \end{cases}$$

Recall that, if $\pi \in \mathcal{P}^0(n)$, then $<\lambda>(\pi)$ means $<\lambda>(g)$, where $g \in \tilde{S}_n$ is determined up to conjugacy by projecting to cycle type π and satisfying $\chi_n(g) > 0$.

Theorem 10.1. *Let $\pi \in \mathcal{P}^0(n)$ be a partition containing r at least once. Define $\pi' \in \mathcal{P}^0(n-r)$ by removing a copy of r from π. Then*

$$<\lambda>(\pi) = \sum_{i \in I(\lambda, r)} n_i <\lambda(i, r)>(\pi') ,$$

where

$$n_i = \begin{cases} (-1)^{j-i} 2^{1-\varepsilon(\lambda)} & \text{if } i \in I_+; \\ (-1)^{\ell-i} & \text{if } i \in I_0; \\ (-1)^{j-i+\lambda_i} 2^{1-\varepsilon(\lambda)} & \text{if } i \in I_-. \end{cases}$$

(The integer j is that occurring in the definitions of $I_\pm$, and $\varepsilon(\lambda)$ is the parity of λ; i.e. 0 or 1.)

Proof. The partitions λ and π below are always in $\mathcal{D}(n)$ and $\mathcal{P}^0(n)$ respectively. By Theorem 8.6,

$$Q_\lambda = d(\lambda)^{-1}\mathrm{och}\begin{cases} <\lambda> & \text{if } \varepsilon(\lambda)=0; \\ (<\lambda> + <\lambda>^a) & \text{if } \varepsilon(\lambda)=1; \end{cases}$$

where $d(\lambda) = 2^{[\varepsilon(\lambda)-\ell(\lambda)]/2}$. If λ is odd, since $<\lambda>^a(\pi) = <\lambda>(\pi)$, we have

$$(<\lambda> + <\lambda>^a)(\pi) = 2^{\varepsilon(\lambda)} <\lambda>(\pi)\,.$$

Thus, using the definition of och,

$$\begin{aligned} Q_\lambda &= d(\lambda)^{-1} \sum_{\pi\in\mathcal{P}^0(n)} 2^{\ell(\pi)/2} 2^{\varepsilon(\lambda)} <\lambda>(\pi)\, p_\pi/z_\pi \\ &= \sum_{\pi\in\mathcal{P}^0(n)} 2^{c(\pi,\lambda)} <\lambda>(\pi)\, p_\pi/z_\pi\,, \end{aligned} \tag{1}$$

where

$$c(\pi,\lambda) = [\ell(\lambda)+\ell(\pi)+\varepsilon(\lambda)]/2\,.$$

Letting $f_\pi(r)$ be the frequency of r in π, and regarding $p_1, p_2, \ldots$ as variables,

$$\begin{aligned} \partial Q_\lambda/\partial p_r &= \sum_{\pi\in\mathcal{P}^0(n)} 2^{c(\pi,\lambda)} <\lambda>(\pi) f_\pi(r)\, p_\pi/(p_r z_\pi) \\ &= \sum_{\beta\in\mathcal{P}^0(n-r)} 2^{c(\beta\cup r,\lambda)} <\lambda>(\beta\cup r)\, p_\beta/(r z_\beta)\,. \end{aligned} \tag{2}$$

In this formula, define the partition $\beta \cup r$ by inserting an extra part equal to r into the partition β. We have used the fact that

$$z_{\beta\cup r} = [f_\beta(r)+1] r z_\beta\,. \tag{3}$$

But $\partial Q_\lambda/\partial p_r$ can be calculated in another way. Recall formulae (7.8) and (9.5):

$$\prod_{x\in X}(1+xu)/(1-xu) = \sum_{\beta\in\mathcal{P}^0} 2^{\ell(\beta)} z_\beta^{-1} p_\beta(X)\, u^{|\beta|}\,;$$

$$\sum_{\alpha\in\mathbb{Z}^\ell} Q_\alpha(X) U^\alpha = \prod_{i<j} F(u_i^{-1}u_j)^{-1} \prod_{x,u}(1+xu)/(1-xu)\,.$$

Differentiating the latter after substituting the former,

$$\sum_{\alpha}(\partial Q_\alpha/\partial p_r)U^\alpha = \prod_{i<j} F(u_i^{-1}u_j)^{-1}\frac{\partial}{\partial p_r}\left[\prod_{a=1}^{\ell}\sum_{\beta\in\mathcal{P}^0} 2^{\ell(\beta)}z_\beta^{-1}p_\beta u_a^{|\beta|}\right]$$

$$= \prod_{i<j} F(u_i^{-1}u_j)^{-1} \times$$

$$\times \sum_{b=1}^{\ell}\left[\left\{\sum_{\gamma\in\mathcal{P}^0} 2^{\ell(\gamma)}f_\gamma(r)u_b^{|\gamma|}p_\gamma/(z_\gamma p_r)\right\}\left\{\prod_{a\neq b}\sum_{\beta\in\mathcal{P}^0} 2^{\ell(\beta)}z_\beta^{-1}p_\beta u_a^{|\beta|}\right\}\right]$$

$$= \sum_{b=1}^{\ell}\left[\prod_{i<j} F(u_i^{-1}u_j)^{-1}\left\{\prod_{a=1}^{\ell}\sum_{\beta\in\mathcal{P}^0} 2^{\ell(\beta)}z_\beta^{-1}p_\beta u_a^{|\beta|}\right\}2r^{-1}u_b^r\right] \text{ using (3)}$$

$$= \sum_{b=1}^{\ell} 2r^{-1}u_b^r \sum_{\beta} Q_\beta U^\beta \text{, again using (9.5) with (7.8) substituted,}$$

$$= \sum_{\alpha} 2r^{-1}\sum_{b=1}^{\ell} Q_{\alpha_1,\dots,\alpha_b-r,\dots,\alpha_\ell} U^\alpha .$$

Thus, for all odd r and all α in $\mathbb{Z}^\ell$,

$$\partial Q_\alpha/\partial p_r = 2r^{-1}\sum_{b=1}^{\ell} Q_{\alpha_1,\,\dots,\,\alpha_b-r,\,\dots,\,\alpha_\ell} . \tag{4}$$

Now let $\alpha = \lambda$, a strict partition of weight n and length ℓ. By Theorem 9.2,

$$Q_{\lambda_1,\dots,\lambda_b-r,\dots,\lambda_\ell} = \begin{cases} y(\lambda_1, \dots, \lambda_b-r, \dots, \lambda_\ell)\, Q_{\lambda(b,r)} & \text{if } b\in I(\lambda, r); \\ 0 & \text{otherwise.}\end{cases}$$

Substituting this into (4) and using (1),

$$\partial Q_\lambda/\partial p_r = 2r^{-1}\sum_{i\in I(\lambda,r)} y(\lambda_1, \dots, \lambda_i-r, \dots, \lambda_\ell)\, Q_{\lambda(i,r)}$$

$$= 2r^{-1}\sum_{\beta\in\mathcal{P}^0(n-r)}\ \sum_{i\in I(\lambda,r)} y(\lambda_1, \dots, \lambda_i-r, \dots, \lambda_\ell)2^{c[\beta,\lambda(i,r)]} \times$$

$$\times <\lambda(i,r)>(\beta)\, p_\beta/z_\beta .$$

Comparing this to (2), the algebraic independence of $\{p_i\}$ allows one to "equate coefficients", yielding

$$<\lambda>(\pi) = \sum_{i \in I(\lambda,r)} m_i <\lambda(i,r)>(\pi') ,$$

where

$$m_i = 2^{1+c[\pi',\lambda(i,r)]-c(\pi,\lambda)} y(\lambda_1,...,\lambda_i - r,...,\lambda_\ell) .$$

An easy exercise with the calculation of y given before the theorem shows that m_i agrees with n_i given in the theorem. This is done by separating into cases according to which one of I_+, I_0 or I_- contains i, and noting that in the cases of $I_\pm$ we have that $\varepsilon[\lambda(i,r)] \neq \varepsilon(\lambda)$, so that

$$\varepsilon[\lambda(i,r)] - \varepsilon(\lambda) = 1 - 2\varepsilon(\lambda) .$$

This completes the proof.

Examples. (i) When $\lambda = (4, 2, 1)$ and $r = 3$, both I_+ and I_0 are empty, whereas $I_- = \{2\}$ and $\lambda(2, 3) = (4)$. Thus

$$<4, 2, 1>(3, 3, 1) = (-1)^3 2<4>(3, 1) .$$

Taking $r = 3$ again yields $<4>(3, 1) = <1>(1) = 1$. Thus $<4, 2, 1>(3, 3, 1) = -2$.

(ii) Now let $\lambda = (5, 2)$ and let $\pi = (3, 3, 1)$. If $r = 1$ is used, Theorem 4.1 yields

$$<5, 2>(3, 3, 1) = <5, 1>(3, 3) + <4, 2>(3, 3) .$$

Taking $r = 3$ for both applications,

$$<5, 1>(3, 3) = 2<2, 1>(3) ;$$

$$<4, 2>(3, 3) = -2<2, 1>(3) .$$

Thus $<5, 2>(3, 3, 1) = 0$. But if we had taken $r = 3$ at the beginning, this would have followed immediately, since $I\big((5, 2), 3\big)$ is empty.

Definitions. Let $\lambda \in \mathcal{D}(n)$. Define $M(\lambda)$ to be the subset of $\mathcal{D}(n-1)$ consisting of those strict partitions which can be obtained by subtracting 1 from one of the parts of λ. If $\lambda = (\lambda_1, ..., \lambda_{\ell-1}, 1)$, let $\lambda^- = (\lambda_1, ..., \lambda_{\ell-1})$, a member of $M(\lambda)$. Define $N(\lambda)$ to be the subset of $\mathcal{D}(n+1)$ consisting of those strict partitions which can be obtained by adding 1 to one of the parts of λ. If $\lambda = (\lambda_1,...,\lambda_\ell)$ with $\lambda_\ell > 1$, let $\lambda^+ = (\lambda_1, ..., \lambda_\ell, 1)$, a non-member of $N(\lambda)$. Let

$$<\lambda>^* = \begin{cases} <\lambda> & \text{if } \lambda \text{ is even;} \\ <\lambda> + <\lambda>^a & \text{if } \lambda \text{ is odd.} \end{cases}$$

With this notation, the next theorem states the "branching rules", which give both the decomposition of the restriction to $\tilde{S}_{n-1}$ of an irreducible $<\lambda>$ of $\tilde{S}_n$, and the decomposition of the character obtained by inducing $<\lambda>$ to $\tilde{S}_{n+1}$. Recall that, although $<\lambda>$ is fixed once χ_λ has been specified, the $\tilde{\otimes}$ operation of Chapter 5 only specifies χ_λ up to replacement by its associate, when $\ell > 1$.

Theorem 10.2. *For a strict partition λ of length ℓ, assume that if $\lambda_\ell = 1$, the choices of χ_{λ^-} and χ_λ, when λ is odd, are correlated so that*

$$\chi_\lambda(g, 1) = \chi_{\lambda^-}(g)\,,$$

where g projects to cycle type λ^- (similarly for λ and λ^+, where $\lambda_\ell > 1$). Then

(*i*) $<\lambda>\downarrow \tilde{S}_{n-1} = \delta_{1,\lambda_\ell}<\lambda^- > + \sum_{\lambda^- \neq \mu \in M(\lambda)} <\mu>^*$, *and*

(*ii*) $<\lambda>\uparrow \tilde{S}_{n+1} = (1-\delta_{1,\lambda_\ell})<\lambda^+ > + \sum_{\mu \in N(\lambda)} <\mu>^*$.

(If the other correlations were used, $<\lambda^- >$ and $<\lambda^+ >$ should be replaced by their associates.)

Proof. Let $g \in \tilde{S}_{n-1}$ project to cycle type β in $\mathcal{P}(n-1)$. We prove (i) by evaluating both sides at g. Let (i)(g) denote this required identity. Since the inclusion $\tilde{S}_{n-1} \to \tilde{S}_n$ is $g \mapsto (g, 1)$, where $(g, 1) \in \tilde{S}_{n-1} \,\tilde{\curlyvee}\, \tilde{S}_1 \subset \tilde{S}_n$ has cycle type $\beta \cup 1$, it is clear that

$$(<\lambda>\downarrow \tilde{S}_{n-1})(g) = <\lambda>(g, 1)\,.$$

Suppose first that β is not in $\mathcal{P}^0(n-1)$, so that β has an even part. If $\beta \cup 1$ is different from λ, the left-hand side of (i)(g) is zero by Theorem 8.7. Also $<\mu>(g)$ is zero unless $\mu = \beta$ is odd, but in that case $<\mu>^*(g) = 0$. Thus (i)(g) holds when $\beta \cup 1 \neq \lambda$. If $\beta \cup 1 = \lambda$, then (i)(g) reduces to $0 = 0$ when λ is even, and to $<\lambda>(g, 1) = <\lambda^- >(g)$ when λ is odd. The latter holds by Theorem 8.7 (i) and the correlation of χ_{λ^-} to χ_λ.

We may therefore suppose that β is in $\mathcal{P}^0(n-1)$. It suffices to take g with $\chi_\beta(g) > 0$. Then (i)(g) becomes the case $r = 1$, $\pi = \beta \cup 1$ of Theorem 10.1, as we now explain. In that theorem, when $r = 1$ it is clear that I_- is empty and that

$$I_0 = \begin{cases} \text{empty set} & \text{if } \lambda_\ell > 1; \\ \{\ell\} & \text{if } \lambda_\ell = 1. \end{cases}$$

In the latter case, $n_\ell = 1$ and $\lambda(\ell, 1) = \lambda^-$, yielding the first term $<\lambda^->(\beta)$ on the right-hand side of (i)(g). The remaining terms arise from the partitions μ in

$$M(\lambda)\backslash\{\lambda^-\} = \{\lambda(i, 1): i \in I_+\}.$$

For each $i \in I_+$, the corresponding j is equal to i, giving sign $+1$. Finally, letting $\mu = \lambda(i, 1)$,

$$2^{1-\varepsilon(\lambda)} <\lambda(i, 1)>(\beta) = 2^{\varepsilon(\mu)} <\mu>(\beta) = <\mu>^*(\beta),$$

so the remaining terms are as stated.

The formula for the induced character in (ii) can now be obtained using Frobenius reciprocity ((C6) of Chapter 4). To do this, note that the irreducibles which occur as constituents of the induced character $<\lambda>\uparrow\tilde{S}_{n+1}$, are precisely those $<\nu>$ such that $<\lambda>$ occurs as a constituent of $<\nu>\downarrow\tilde{S}_n$. These are the irreducibles corresponding to those strict partitions ν which are either λ^+ or are in $N(\lambda)$. The odd ones occur with their associates except for $<\lambda^+>$.

Example. In Chapter 6, we determined the table of irreducible negative characters of $\tilde{S}_5$. Inducing these to $\tilde{S}_6$ gives the following array. (It should be noted that although the value of <4, 1> and its associate on the class (4, 1) is non-zero, the value of the induced character is zero on (4, 1, 1) since this partition has a repeated part.) Here $\uparrow$ means $\uparrow \tilde{S}_6$.

	1	(3)	(5)	(3,3)
$<5>\uparrow$	24	6	1	0
$<4,1>\uparrow$	36	0	-1	0
$<4,1>^a\uparrow$	36	0	-1	0
$<3,2>\uparrow$	24	-3	1	$\sqrt{3}$
$<3,2>^a\uparrow$	24	-3	1	$-\sqrt{3}$

We now determine the table of irreducible negative characters of $\tilde{S}_6$. The value of the basic character <6> is given by Theorem 6.8. It is not self-associate. Its non-zero values are as shown:

	1	(3)	(5)	(3,3)	(6)
$<6>$	4	2	1	1	$\sqrt{-3}$

By Theorem 10.2, the induced characters have the following decompositions:

$$\begin{aligned}<5>\uparrow \tilde{S}_6 &= <5,1>^* + <6>^*;\\ <4,1>\uparrow \tilde{S}_6 &= <5,1>^* + <4,2>^*;\\ <3,2>\uparrow \tilde{S}_6 &= <4,2>^* + <3,2,1> .\end{aligned}$$

It is therefore an easy task to obtain the table as shown.

	1	(3)	(5)	(3,3)	(6)	(3,2,1)
<6>	4	2	1	1	$\sqrt{-3}$	0
$<6>^a$	4	2	1	1	$-\sqrt{-3}$	0
<5,1>	16	2	−1	−2	0	0
<4,2>	20	−2	0	2	0	0
<3,2,1>	4	−1	1	−2	0	$\sqrt{3}$
$<3,2,1>^a$	4	−1	1	−2	0	$-\sqrt{3}$

We have illustrated the branching rule by calculating a complete table of negative characters. However, one of the main uses of Theorem 10.1 is to calculate single character values without knowing complete character tables.

It is convenient at this point to introduce a number of combinatorial objects.

Definition. Let $\lambda = (\lambda_1, \lambda_2, ..., \lambda_\ell)$ be in $\mathcal{D}(n)$ so that $\lambda_1 > \lambda_2 > \ldots > \lambda_\ell > 0$. The *shifted diagram*, $S(\lambda)$, of shape λ, is obtained by forming ℓ rows of nodes, with λ_i nodes in the ith row such that, for all $i > 1$, the first node in row i is placed underneath the second node in row $(i-1)$.

Example. Consider the partition $\lambda = (7, 5, 3, 2, 1)$ of 18. Its shifted diagram is as shown.

$$\begin{array}{ccccccc} \times & \times & \times & \times & \times & \times & \times \\ & \times & \times & \times & \times & \times & \\ & & \times & \times & \times & & \\ & & & \times & \times & & \\ & & & & \times & & \end{array}$$

Definition. The (unshifted) *Young diagram* corresponding to a partition λ is a *left justified* diagram with λ_i nodes in the ith row (so that the nodes at the left-hand ends of the rows are all in the same column). Such diagrams will be familiar to readers who have studied linear representations of the symmetric groups.

Definition. Another combinatorial structure associated with a strict partition λ is the *shift-symmetric diagram.* This is a diagram $Y(\lambda)$ with $2n$ nodes produced from $S(\lambda)$ by a adding a new (zeroth) column consisting of λ_1 nodes to the left of $S(\lambda)$, and then, for $i \geq 2$, inserting λ_i nodes under the bottom node in the $(i-1)$st column of $S(\lambda)$. We shall distinguish the shifted diagram $S(\lambda)$ within $Y(\lambda)$, using crosses, $\times$, to mark the nodes in $S(\lambda)$, and circles, $\circ$, to denote the remaining nodes.

Example. The shift symmetric diagram associated with the partition $(7, 5, 3, 2, 1)$ is

$$\begin{array}{cccccccc}
\circ & \times & \times & \times & \times & \times & \times & \times \\
\circ & \circ & \times & \times & \times & \times & \times & \\
\circ & \circ & \circ & \times & \times & \times & & \\
\circ & \circ & \circ & \circ & \times & \times & & \\
\circ & \circ & \circ & \circ & \circ & \times & & \\
\circ & \circ & & & & & & \\
\circ & & & & & & &
\end{array}$$

Remark. The shift-symmetric diagram $Y(\lambda)$ is the Young diagram of a partition $\tilde{\lambda}$ of $2n$. The easiest way to describe $\tilde{\lambda}$ is to use Frobenius' notation for partitions. In this notation, a partition μ of k is described by two sequences of non-negative integers

$$\{\alpha_1,\ldots,\alpha_\ell \mid \beta_1,\ldots,\beta_\ell\}\ ,$$

where the Young diagram of μ has α_i nodes in the ith row after position i, and has β_i nodes in the ith column below position i. Then $\tilde{\lambda}$ is the partition $\{\lambda_1,\ldots,\lambda_\ell \mid \lambda_1-1,\ldots,\lambda_\ell-1\}$. It should be noted that although the Young diagram associated with $\tilde{\lambda}$ has the same number of nodes in each row as $Y(\lambda)$, the numbering of columns is different. The leftmost column of $Y(\lambda)$ was defined to be the zeroth column.

Definitions. In a (non-shifted) Young diagram, the *hook* at a given node has, by definition, an *arm*, consisting of all the nodes to its right, and a *leg*, consisting of all the nodes below the given node. The *hook length* of a hook is the total number of nodes in the hook. It is the sum of the *arm length* plus the *leg length* plus 1 (to count the node itself). The hook length and leg length of a node in a shifted diagram $S(\lambda)$ are both defined with reference to the hook at the corresponding node in the associated shift-symmetric diagram $Y(\lambda)$.

Example. For the example above, the hook lengths are as recorded at the nodes of $S(\lambda)$:

$$12 \ 10 \ 9 \ 8 \ 7 \ 3 \ 1$$
$$8 \ 7 \ 6 \ 5 \ 1$$
$$5 \ 4 \ 3$$
$$3 \ 2$$
$$1$$

The leg lengths are as follows:

$$5 \ 4 \ 4 \ 4 \ 4 \ 1 \ 0$$
$$3 \ 3 \ 3 \ 3 \ 0$$
$$2 \ 2 \ 2$$
$$1 \ 1$$
$$0$$

Definition. Let λ be a strict partition of length ℓ, and let $S(\lambda)$ be the corresponding shifted diagram. The nodes in $S(\lambda)$ are partitioned into three types as follows. The nodes in the ℓth column of $S(\lambda)$ are the *type 2* nodes. The nodes of *type 1* are those in all the columns to the right of the type 1 nodes, and the remaining nodes are of *type 3*.

Example. To illustrate this definition, we consider the partition $(8, 6, 4, 3, 2)$.

$$
\begin{array}{ccccccccc}
\times & \times & \times & \times & | & \times & | & \times & \times & \times \\
 & \times & \times & \times & | & \times & | & \times & \times \\
 & & \times & \times & | & \times & | & \times \\
 & & & \times & | & \times & | & \times \\
 & & \text{type 3} & & | & \times & | & \times & \text{type 1} \\
 & & & & & \text{type 2} & & &
\end{array}
$$

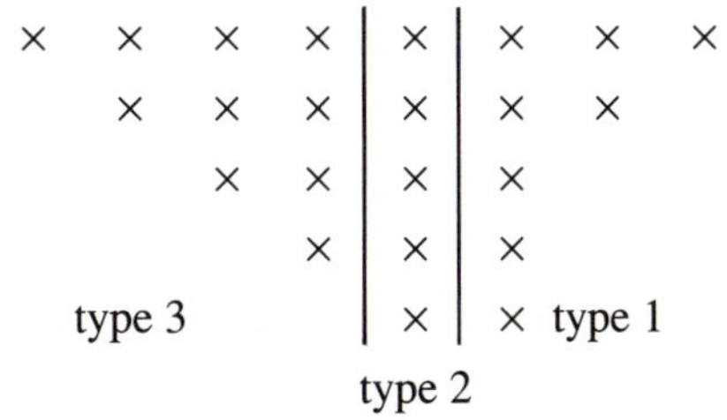

Its shifted diagram has been divided into regions to show the location of the three types of node. Note that a node in $S(\lambda)$ has circle nodes in $Y(\lambda)$ directly below it if and only if it is a node of type 3. Consider nodes in the ith row of $S(\lambda)$. At a node of type 1, the cardinality h of the hook in $S(\lambda)$ is smaller than λ_i. Furthermore, $h = \lambda_i$ for nodes of type 2, and $h > \lambda_i$ for nodes of type 3.

Let us relate these notions to the reduction formula in Theorem 10.1. Given λ in $\mathcal{D}(n)$ and odd $r > 0$, recall the set $I(\lambda, r)$. Think of its members as row numbers in $S(\lambda)$ and $Y(\lambda)$.

Proposition 10.3. *For each* $i \in I(\lambda, r)$ *there is a unique node* $\times_i$ *in the i-th row of* $S(\lambda)$ *for which the hook at* $\times_i$ *has hook length r. This node is of type 1, 2 or 3 according to whether i is in* I_+, I_0 *or* I_- *respectively. The sign of the corresponding term in the reduction formula of Theorem 10.1 is* $(-1)^s$, *where s is the leg length (in* $Y(\lambda)$*) of the hook at* $\times_i$.

Proof. Uniqueness is clear since hook lengths strictly increase as we move from right to left along a given row.

Suppose $i \in I_+$ with $\lambda_j > \lambda_i - r > \lambda_{j+1}$. Let $\times_i$ be the $(r+i-j)$th node from the right in the ith row of $S(\lambda)$. Then $\times_i$ is directly above the rightmost node in the jth row of $S(\lambda)$, so it is of type 1 with leg length $j-i$ and hook length $(r+i-j)+(j-i) = r$. Since $(-1)^{j-i}$ is the sign in Theorem 10.1, this case is proved.

If $i \in I_0$, let $\times_i$ be the $(r+i-\ell)$th node from the right in the ith row, which happens to be the $(\ell-i+1)$st node from the left, since $\lambda_i = r$. This node lies directly above the leftmost (diagonal) node in the bottom row of $S(\lambda)$, so it is a type 2 node with leg length $\ell-i$, as required, and hook length $(r+i-\ell)+(\ell-i) = r$, as required.

Finally, if $i \in I_-$, let $\times_i$ be the $(\lambda_i+i-j+1)$st node from the right (which is also the $(j-i)$th node from the left) in the ith row of $S(\lambda)$, where $\lambda_j+\lambda_i = r$. It is directly above the leftmost (diagonal) node of the $(j-1)$st row of $S(\lambda)$, and so is a type 3 node with an additional λ_j circle nodes directly below it in $Y(\lambda)$. It therefore has hook length

$$(\lambda_i+i-j+1)+(j-i-1)+\lambda_j = \lambda_i+\lambda_j = r$$

as required. Its leg length is

$$(j-i-1)+\lambda_j = j-i-1+r-\lambda_i \equiv j-i+\lambda_i \pmod 2,$$

as required, since r is odd. This completes the proof.

There is another type of subset of $S(\lambda)$ which will occur in Appendix 10. With notation as above, for each $i \in I(\lambda, r)$, the associated *r-bar* is given as follows. If i is in I_+ or I_0, the r-bar consists of the rightmost "r" nodes in the ith row of $S(\lambda)$; i.e. when $i \in I_0$, the entire ith row. If i is in I_-, the r-bar consists of all the nodes in both the ith and jth rows, a total of "r" nodes. The relationship of this to our work so far is that the shifted diagram of $\text{str}(\lambda_1, ..., \lambda_i - r, ..., \lambda_\ell)$ (the strict partition $\lambda(i,r)$ such that $Q_{\lambda_1, ..., \lambda_i - r, ..., \lambda_\ell} = \pm Q_{\lambda(i,r)}$) is given by removing the associated r-bar from $S(\lambda)$, and shuffling the resulting rows so as to form a shifted diagram.

Example. Let $\lambda = (7, 5, 3, 2, 1)$ and let $r = 3$. Then $I(\lambda, r) = \{1, 3, 4\}$, with each of I_+, I_0 and I_- being a singleton. The vertices $\times_i$ are labelled in $S(\lambda)$ below, and each is within its corresponding bar as noted. To avoid cluttering, we have left the reader to draw the corresponding hooks in $Y(\lambda)$, which have leg lengths 1, 2 and 1. This example gives, for each $\beta \in \mathcal{P}^0(15)$, the reduction

$$\langle 75321\rangle(\beta \cup 3) = -\langle 54321\rangle(\beta) + \langle 7521\rangle(\beta) - \langle 753\rangle(\beta)$$

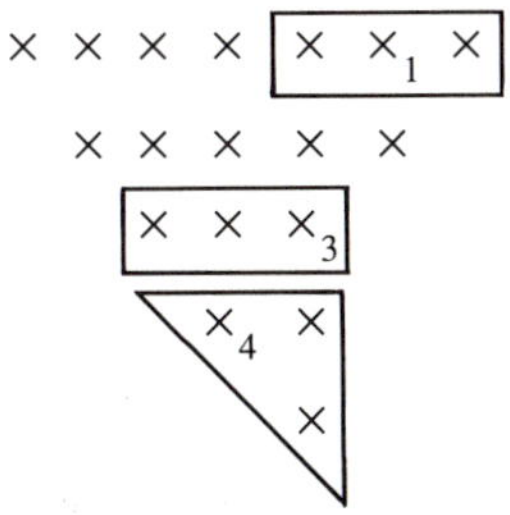

The combinatorial interpretation of Theorem 10.2 is rather obvious: the diagrams of strict partitions μ such that $\langle\mu\rangle$ occurs in $\langle\lambda\rangle\downarrow \tilde{S}_{n-1}$ are obtained by the various ways of removing a single node from $S(\lambda)$ so as to leave a shifted diagram. As for $\langle\lambda\rangle\uparrow\tilde{S}_{n+1}$, one works out all the ways in which a single node can be added to $S(\lambda)$ to produce a shifted diagram.

Definition. Let λ be a strict partition. A *standard shifted tableau* T of shape λ is obtained by replacing each of the nodes in $S(\lambda)$ by one of the integers $1, 2, \ldots, |\lambda|$,

$$\begin{array}{cccc} T[1,1] & T[1,2] & T[1,3] & \cdots \\ & T[2,2] & T[2,3] & \cdots \\ & & T[3,3] & \cdots \\ & & & \cdot \\ & & & \quad\cdot \\ & & & \qquad\cdot \end{array}$$

such that the entries of T increase down the columns and from left to right along the rows. Thus both $T[i+1, j]$ and $T[i, j+1]$ are larger than $T[i, j]$.

The following is an example of a standard shifted tableau of shape $(7, 5, 3, 2, 1)$:

1	2	4	6	8	10	18
	3	5	7	9	17	
		11	12	13		
			14	15		
				16		

Given a strict partition λ, let g_λ denote the number of standard shifted tableaux of shape λ.

Proposition 10.4. *Let* $\lambda = (\lambda_1, \lambda_2, \ldots, \lambda_\ell)$ *be a strict partition of n. Then*

$$g_\lambda = \frac{n!}{\lambda!} \prod_{1 \le i < j \le \ell} \frac{\lambda_i - \lambda_j}{\lambda_i + \lambda_j},$$

where $\lambda!$ *denotes the product* $\lambda_1!\lambda_2! \ldots \lambda_\ell!$.

Proof. First notice that in a standard shifted tableau with n nodes, the integer n must occur at the end of a row, since we demand increase along rows. It must also occur at the bottom of a column. Recall that $M(\lambda)$ is the set of partitions in $\mathcal{D}(n-1)$ obtained by subtracting 1 from one of the parts of λ. It is clear from this discussion that

$$g_\lambda = \sum_{\mu \in M(\lambda)} g_\mu . \tag{10.5}$$

Given this recursive formula for g_λ, it is natural to establish the required expression for g_λ by induction on n. The statement clearly holds when $n = 1$.

For each integer i, let $\lambda^{(i)}$ be the partition of $n-1$ obtained by subtracting 1 from the ith part of λ. Thus $\lambda^{(i)}! = \lambda!/\lambda_i$. The set $M(\lambda)$ consists of those $\lambda^{(i)}$ which are strict. If $\ell(\lambda^{(i)}) = \ell$, it follows by induction that, for those i such that $\lambda^{(i)}$ is strict,

$$g_{\lambda^{(i)}} = \frac{(n-1)!}{\lambda!} \lambda_i \prod_{1 \le k < j \le \ell} \frac{\lambda_k^{(i)} - \lambda_j^{(i)}}{\lambda_k^{(i)} + \lambda_j^{(i)}} \tag{*}$$

where $\lambda_j^{(i)} = \lambda_j - \delta_{ij}$. Furthermore, there is at most one case when the length of $\lambda^{(i)}$ is not ℓ. This arises if $\lambda_\ell = 1$, so that the length of $\lambda^{(\ell)}$ is $\ell-1$. In this case, setting $\lambda_\ell^{(\ell)} = 0$, we again see by induction that (*) holds. Thus, for those i such that $\lambda^{(i)}$ is strict, letting h_λ be the right-hand side of the formula in the proposition,

$$g_{\lambda^{(i)}}/h_\lambda$$

$$= \frac{\lambda_i}{n} \prod_{i<j\le\ell} \frac{(\lambda_i-\lambda_j-1)(\lambda_i+\lambda_j)}{(\lambda_i+\lambda_j-1)(\lambda_i-\lambda_j)} \prod_{1\le k<i} \frac{(\lambda_k-\lambda_i+1)(\lambda_k+\lambda_i)}{(\lambda_k+\lambda_i-1)(\lambda_k-\lambda_i)}$$

$$= \frac{\lambda_i}{n} \prod_{\substack{1\le j\le\ell \\ i\ne j}} \frac{(\lambda_i-\lambda_j-1)(\lambda_i+\lambda_j)}{(\lambda_i+\lambda_j-1)(\lambda_i-\lambda_j)} .$$

When $\lambda^{(i)}$ is not strict, the last expression is zero. Thus, if $g_{\lambda^{(i)}}$ is defined to be zero for such i, the inductive formula for g_λ is equivalent to

$$g_\lambda = \sum_{i=1}^{\ell} g_{\lambda^{(i)}} .$$

Therefore, to prove, as required, that $g_\lambda = h_\lambda$, it remains to prove that

$$n = \sum_{i=1}^{\ell} \lambda_i = \sum_{i=1}^{\ell} \left(\lambda_i \prod_{\substack{1\le j\le\ell \\ i\ne j}} \frac{(\lambda_i-\lambda_j-1)(\lambda_i+\lambda_j)}{(\lambda_i+\lambda_j-1)(\lambda_i-\lambda_j)} \right). \tag{1}$$

In order to do this, consider the decomposition into partial fractions of the rational function

$$f(x) = (2x-1) \prod_{i=1}^{\ell} \frac{(x-\lambda_i-1)(x+\lambda_i)}{(x+\lambda_i-1)(x-\lambda_i)} . \tag{2}$$

It can be easily checked that the numbers $A_1, A_2, \ldots, A_\ell$ and $B_1, B_2, \ldots, B_\ell$ such that

$$f(x) = (2x-1) + \sum_{i=1}^{\ell} \frac{A_i}{(x+\lambda_i-1)} + \sum_{i=1}^{\ell} \frac{B_i}{(x-\lambda_i)} \tag{3}$$

are given by

$$A_i = B_i = -2\lambda_i \prod_{\substack{j=1 \\ i\ne j}}^{\ell} \frac{(\lambda_i-\lambda_j-1)(\lambda_i+\lambda_j)}{(\lambda_i+\lambda_j-1)(\lambda_i-\lambda_j)} . \tag{4}$$

Now expand expressions (2) and (3) for $f(x)$ in powers of x^{-1} using the binomial theorem. In

$$\prod_{i=1}^{\ell}\{1-(\lambda_i+1)x^{-1}\}\{1+\lambda_i x^{-1}\}\{1+(\lambda_i-1)x^{-1}\}^{-1}\{1-\lambda_i x^{-1}\}^{-1},$$

the coefficient of x^{-1} is zero, and that of x^{-2} is $-2(\lambda_1+\lambda_2+\ldots\lambda_\ell)$. Thus the coefficient of x^{-1} in (2) is $-4(\lambda_1+\lambda_2+\ldots\lambda_\ell)$. On the other hand, the coefficient of x^{-1} in expression (3) is

$$A_1+A_2+\ldots+A_\ell+B_1+B_2+\ldots B_\ell.$$

Equating coefficients of x^{-1} in expressions (2) and (3) gives equation (1), as required.

Proposition 10.6. *For any partition* λ *in* $\mathcal{D}(n)$,

$$g_\lambda = \frac{n!}{h(\lambda)},$$

where $h(\lambda)$ *denotes the product of the hook lengths in* $Y(\lambda)$ *at all nodes in the shifted diagram* $S(\lambda)$.

Proof. By Proposition 10.4, what we must prove is that, with $\ell = \ell(\lambda)$,

$$h(\lambda) = \prod_{1\le i\le \ell}\left((\lambda_i)!\prod_{i<j\le \ell}\frac{(\lambda_i+\lambda_j)}{(\lambda_i-\lambda_j)}\right).$$

More precisely, we shall show that the product of the hook lengths for nodes in the ith row is the ith factor above.

The hook lengths of the type 3 nodes in row i are $\lambda_i+\lambda_{i+1}, \lambda_i+\lambda_{i+2}, \ldots, \lambda_i+\lambda_\ell$. The hook length of the type 2 node is λ_i. It remains to show that the hook lengths of the type 1 nodes are all the numbers $1, 2, \ldots, \lambda_i-1$ except the $\lambda_i-\lambda_t$ for $i+1\le t\le \ell$. In the diagram below are pictured, for given s, the longest and shortest hooks, at type 1 nodes on row i, which end on row s. The lengths of these two hooks are $\lambda_i-\lambda_{s+1}-1$ and $\lambda_i-\lambda_s+1$, as one sees by inspecting the diagram. This includes the cases $s = i$ and $s = \ell$ (where $\lambda_{\ell+1}$ is taken as zero), in which the diagram collapses somewhat.

This completes the proof.

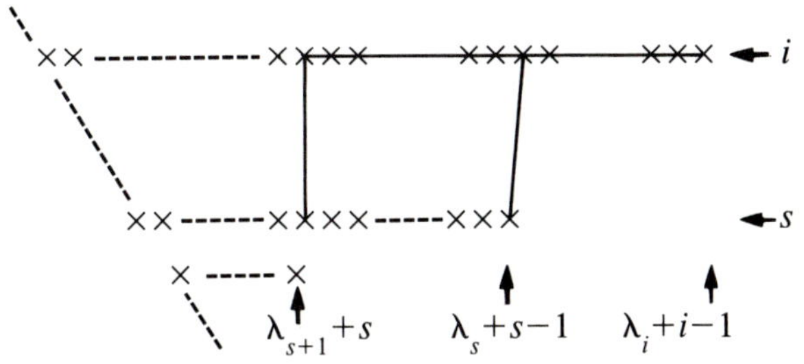

Remark. The proof has the following consequence, which will be used in Appendix 10. For each node in $S(\lambda)$, let r be the length of the hook in $Y(\lambda)$ at the given node. (Earlier, we began with a given odd r; here r may be even as well.) There is thus an r-bar associated with the given node. By the above proof, removal of this bar will always result in a set of rows of distinct lengths. These may therefore be shuffled to obtain the shifted diagram of a strict partition in $\mathcal{D}(n-r)$.

Theorem 10.7. *Let* $\lambda \in \mathcal{D}(n)$ *have length* ℓ. *Then the degree of* $<\lambda>$ *is* $2^{[n-\ell-\varepsilon(\lambda)]/2} g_\lambda$, *where* g_λ *is the number of standard shifted tableaux of shape* λ.

Proof. The proof is by induction on n. Let $\delta = \delta_{1,\lambda_\ell}$. By Theorem 10.2,

$$<\lambda>(1) = \delta <\lambda^- >(1) + \sum_{\lambda^- \neq \mu \in M(\lambda)} <\mu>^*(1) .$$

Furthermore,

$$<\mu>^*(1) = 2^{\varepsilon(\mu)} <\mu>(1) .$$

Now by induction, if μ is in $M(\lambda)$ and $\mu \neq \lambda^-$, then

$$<\mu>(1) = 2^{[n-1-\ell-\varepsilon(\mu)]/2} g_\mu .$$

Furthermore,

$$<\lambda^- >(1) = 2^{[n-1-(\ell-1)-\varepsilon(\lambda^-)]/2} g_{\lambda^-} .$$

Since $\varepsilon(\mu) \neq \varepsilon(\lambda)$ for μ as above,

$$[n-\ell-\varepsilon(\lambda)]/2 = \varepsilon(\mu) + [n-1-\ell-\varepsilon(\mu)]/2 .$$

Thus

$$<\lambda>(1) = \delta<\lambda^- >(1) + \sum_{\lambda^- \neq \mu \in M(\lambda)} 2^{\varepsilon(\mu)} <\mu>(1)$$

$$= \delta 2^{[n-\ell-\varepsilon(\lambda)]/2} g_{\lambda^-} + \sum_{\lambda^- \neq \mu \in M(\lambda)} 2^{\varepsilon(\mu)} 2^{[n-1-\ell-\varepsilon(\mu)]/2} g_\mu$$

$$= 2^{[n-\ell-\varepsilon(\lambda)]/2} \sum_{\mu \in M(\lambda)} g_\mu$$

$$= 2^{[n-\ell-\varepsilon(\lambda)]/2} g_\lambda$$

by (10.5), as required.

Corollary 10.8. *With summation over all* λ *in* $\mathcal{D}(n)$,

$$\sum 2^{n-\ell(\lambda)} g_\lambda^2 = n!$$

Proof. By Proposition 4.1, the sum of the squares of the degrees of the negative irreducible representations of $\tilde{S}_n$ is $n!2/2 = n!$. Since these representations are the $<\lambda>$, together with the $<\lambda>^a$ for those λ with $\varepsilon(\lambda) = 1$, that sum of squares is

$$\sum_\lambda 2^{\varepsilon(\lambda)} \{2^{[n-\ell(\lambda)-\varepsilon(\lambda)]/2} g_\lambda\}^2 = \sum_\lambda 2^{n-\ell(\lambda)} g_\lambda^2,$$

as required.

A combinatorial proof of this corollary will be given at the end of Chapter 13.

Notes

Theorem 10.1, as well as Proposition 10.3, are due to Morris (1962, 1962A, 1965). Our proof of 10.1 uses the Laurent expression in place of raising operators, and $[y(\alpha), \mathrm{str}\,\alpha]$ in place of "rules", but is essentially the same. Theorem 10.2 is due to Dehuai and Wybourne (1986). See also Salam and Wybourne (1989). Shifted diagrams appeared in Thrall (1952). See also Morris and Yaseen (1986) and Appendix 10 overleaf. The proofs of 10.4 and 10.5 follow the outline given in Macdonald (1979, pp. 135-6). The formula for $c_1^{\perp}$ at the end of Appendix 12 implies immediately the appropriate analogues of 10.2 and 10.7 for the algebra $\oplus T^*\tilde{S}_n$ of Appendix 8. For other proofs of 10.1, see Appendix 12 of this book, Hoffman (1988) and Stembridge (1988). The degree formula (10.7), hook formula (10.6) and sum of squares corollary (10.8) are analogous to corresponding formulae for linear representations of S_n, in which Young diagrams, rather than shifted diagrams, are used. Unshifted diagrams were used in classical algebra and combinatorics long before Young, but we have continued the custom of naming them after him. The idea of using shifted and shift-symmetric diagrams in this subject originated with I.G. Macdonald.

Appendix 10

Modular projective representations of S_n

This appendix is a survey of results in the theory of projective representations of S_n over algebraically closed fields of non-zero characteristic. It is not intended to give an expository introduction to the subject. It is assumed that the reader of this appendix will be familiar with the problems and techniques of modular representation theory. (The last chapter of Isaacs (1976) gives an introduction to the subject.) The first objective is to explain how to assign the complex irreducible negative representations of $\tilde{S}_n$ to p-blocks. This can be done by reference to the strict partition associated with the representation. There is then a brief discussion of decomposition numbers and related problems.

The first idea to discuss is that of the p-bar core of a shifted diagram. As we have seen in Chapter 10, there is a bar corresponding to each node of the shifted diagram $S(\lambda)$ associated with a strict partition λ of n. If the bar at a node has length p, removal of this bar (as discussed after the proof of Proposition 10.6) will result in a diagram which is the shifted diagam corresponding to a strict partition of $n-p$. Continuing to remove p-bars, we obtain a shifted diagram of a partition μ of $n-kp$, for some k, with the property that it is not possible to remove a p-bar from $S(\mu)$. It can be shown that μ, the p-bar core of λ, is independent of the order in which the p-bars are removed.

In the case of linear representations of S_n, the irreducible representations are in bijective correspondence with the partitions of λ. The combinatorial object corresponding to a bar at a node of the shifted diagram is a hook at a node of a Young diagram. A hook is the set of nodes weakly to the right of, and strictly below the given node. In (1940) Nakayama conjectured that two irreducible linear representations of S_n corresponding to partitions λ and μ are in the same p-block if and only if λ and μ have the same p-core. The corresponding result for projective representations (or negative representations of $\tilde{S}_n$) was conjectured by Morris (1965). In fact the following result holds.

Theorem A10.1. *Let p be an odd prime. Let P and Q be irreducible negative representations of $\tilde{S}_n$ corresponding to strict partitions λ and μ respectively. If P and Q are in the same p-block then λ and μ have the same p-bar core. Conversely, if λ and μ have the same p-bar core, then either P and Q are in the* *same p-block or P and Q are associate representations corresponding to a partition λ ($=\mu$) which has no p-bars.* (*In the latter case, λ is its own p-bar core and P, Q are definitely not in the same block.*)

The first proof of this result was given by Humphreys (1986), based on the proof of the Nakayama conjecture for linear representations given by Meier and Tappe (1976). Subsequent proofs have been given by Cabanes (1988) and Olsson (1989), this latter being the closest to the original proof of the Nakayama conjecture given in Brauer and Robinson (1947). More recently, Olsson has established the following facts for a block B of negative representations of $\tilde{S}_n$ with defect group D.

(1) Let $k(B)$ denote the number of irreducible characters in B. Then $k(B) < |D|$.

(2) Let $k_0(B)$ denote the number of irreducible characters in B of height zero. Then $k_0(B) < |[D, D]|$.

(3) The integers $k(B)$ and $k_0(B)$ are equal if and only if D is abelian.

When $p = 2$, the p-bar core does not determine the p-blocks. However, the following conjecture has been made by Knörr and Olsson in Olsson (1987).

Conjecture. *Let λ be a strict partition. Then the irreducible negative representation corresponding to λ is in that 2-block of S_n which contains the irreducible representation indexed by the partition*

$$\lceil\lambda_1 + 1/2\rceil\lceil\lambda_1/2\rceil\,\lceil\lambda_2 + 1/2\rceil\,\lceil\lambda_2/2\rceil\ldots\lceil\lambda_\ell + 1/2\rceil\,\lceil\lambda_\ell/2\rceil,$$

where $\lceil\ \rceil$ denotes the integer part function.

An alternative way to visualize the process of forming the p-bar core, at least when p is odd, is to use a p-abacus as follows. The abacus has p wires or runners numbered 0 to $p-1$ from left to right. The rows of the abacus are numbered 0, 1, 2, ... A strict partition λ of length ℓ is represented by placing ℓ beads on the abacus. A bead is placed in the jth row of the ith runner if and only if $pj+i$ is one of the parts of λ. For $1\leq i\leq (p-1)/2$, the runners i and $p-i$ are *conjugate*.

The removal of a p-bar from λ is registered on the abacus as follows:

Type 1: slide a bead one position up on the same runner.

Type 2: remove a bead on the zeroth row of the zeroth runner.

Type 3: remove beads on conjugate runners of the zeroth row.

Using this abacus, it can be seen that the p-bar core is attained by the following process. First move all the beads as far up the runners as possible and then perform all possible moves of type 2 and 3. Repeat this procedure until no further moves are possible.

Example. Let $\lambda = (17, 13, 12, 9, 6, 4, 2)$ and $p = 5$. The configuration of the 5-abacus corresponding to λ is (with bead positions underlined):

$$\begin{array}{ccccc} 0 & 1 & \underline{2} & 3 & \underline{4} \\ 5 & \underline{6} & 7 & 8 & \underline{9} \\ 10 & 11 & \underline{12} & \underline{13} & 14 \\ 15 & 16 & \underline{17} & 18 & 19 \end{array}$$

Thus the 5-core of λ is $(7, 4, 2)$.

The *p-bar quotient* of λ is a sequence of $(p+1)/2$ partitions defined as follows. The partition $\lambda^{(0)}$ shows the location of the beads on the zeroth runner of the abacus. Thus if the beads on runner 0 are in rows $\alpha_1, \alpha_2, \ldots, \alpha_r$, where $\alpha_1 > \alpha_2 > \ldots > \alpha_r$, then

$$\lambda^{(0)} = (\alpha_1, \alpha_2, \ldots, \alpha_r)\,.$$

To determine the other $(p-1)/2$ partitions in the p-quotient of λ, suppose that $1 \leq i \leq (p-1)/2$, and that the locations of the beads on runners i and $p-i$ are $\alpha_1, \alpha_2, \ldots, \alpha_s$ and $\beta_1, \beta_2, \ldots, \beta_r$. Assume without loss of generality that $u = s-r$ is positive. The partition $\lambda^{(i)}$ is then written as the concatenation of two partitions, the first being

$$(\alpha_1 - u+1, \alpha_2 - u+2, \ldots, \alpha_u)$$

and the second, written in Frobenius notation, being

$$(\alpha_{u+1}, \ldots, \alpha_s; \beta_1, \ldots, \beta_r)\,.$$

Example. When $\lambda = (13, 11, 10, 9, 7, 6, 4, 2)$ and $p = 3$, the 3-quotient consists of two partitions. These are

$$\begin{aligned} \lambda^{(0)} &= (3\ 2);\ \text{and} \\ \lambda^{(1)} &= (3, 3; (2\ 1; 3\ 0)) = (3^4\ 1^2)\,. \end{aligned}$$

The following result is proved in Olsson (1987).

Theorem A10.2. *A strict partition determines and is uniquely determined by its p-bar core and p-bar quotient.*

Morris and Olsson (1988) have used these ideas to define a p-bar quotient character associated with $\langle\lambda\rangle$. The determination of the values of this character is a step in their proof of Theorem A10.1.

Michler and Olsson (1990) have established the Alperin–McKay conjecture for $\tilde{S}_n$ by proving the following.

Theorem A10.3. *For each odd prime p, there is a* $1-1$ *correspondence between the irreducible characters of height zero in a p-block B of* $\tilde{S}_n$ *(or of* $\tilde{A}_n$*) with defect group D and the irreducible characters of height zero of the Brauer correspondent b of B in the normalizer of D.*

The conjecture on modular representations of finite groups which has received most attention in recent years is Alperin's weight conjecture. Given a prime p, a p-subgroup R of a finite group of G is a radical subgroup if R is the largest normal p-subgroup of $N_G(R)$. An ordinary irreducible character ζ of $N_G(R)$ is a weight character if ζ has R in its kernel and is of zero defect as a character of $N_G(R)/R$. The G-conjugacy class (R, ζ) is a weight of G. Alperin's conjecture (which also has a blockwise version) is that the number of weights of G is equal to the number of p-modular irreducible representations of G. This has recently been established for the covering groups of the symmetric and alternating groups by Michler and Olsson (1990A).

A useful application of negative representations of $\tilde{S}_n$ arises in the study of 2-modular linear representations. Since $\tilde{S}_n$ has a normal subgroup of order 2, each negative representation of $\tilde{S}_n$ may be regarded as a 2-modular representation of S_n. (The Brauer character obtained by restricting the character of a negative representation of $\tilde{S}_n$ to 2-regular classes is a Brauer character of S_n.) Thus the irreducible negative representations provide an interesting family of 2-modular representations, but with some unexpected results. Although the basic irreducible representation $<n>$ behaves as the "identity" projective representation in many respects, $<n>$ is not in the principal 2-block of S_n when n is of the form $4k+3$. (The Knörr–Olsson conjecture concerning the 2-blocks has been verified for the representation $<n>$.)

Benson (1988) has a general result on the composition factors of the reduction of $<\lambda>$ modulo 2. This involves the idea of the double, dbl(λ), of a partition λ. This is obtained from λ by replacing each odd part $\lambda_i = 2k+1$ by two parts, k and $k+1$ and replacing each even part $\lambda_i = 2\ell$ by the parts $\ell-1$ and $\ell+1$. There is an irreducible 2-modular character D^λ corresponding to each strict partition λ. Given two partitions λ, μ of n, we say that λ is less than μ in the strict dominance partial order on the set of partitions of n if $\lambda \neq \mu$ and for all i,

$$\sum_{j=1}^{i} \lambda_j \leq \sum_{j=1}^{i} \mu_j .$$

We can now state Benson's result.

Theorem A10.4. *Let* λ *be a strict partition with t odd parts. The reduction of* $<\lambda>$ *modulo 2 contains the irreducible* $D^{\mathrm{dbl}(\lambda)}$ *with multiplicity* $2^{\lceil t/2 \rceil}$*, together with possibly some other composition factors, each of which is of the form* D^{μ}*, where* μ *is a partition with* μ *less than* dbl(λ) *in the strict dominance order.*

When p is odd very little is known about the decomposition numbers. Wales (1977) discusses the modular decomposition of the projective representations $<n>$ and $<n-1, 1>$. Some calcuations of the decomposition numbers for small values of n have been considered by Morris and Yaseen (1988). However, when $p = 3$, there are problems which seem to be unresolvable by algebraic methods even when $n = 9$. At present, the only solution of these ambiguties uses Richard Parker's MEATAXE computer package.

11

CONSTRUCTION OF THE IRREDUCIBLE NEGATIVE REPRESENTATIONS

In this chapter, we give a construction due to Nazarov (1989) of the irreducible negative representations of $\tilde{S}_n$. They are obtained from those of the group $\hat{S}_n$ (as defined in Remark (3) after Theorem 2.11) by multiplying the matrix for each generator s_k by i to give the matrix for t_k. We shall follow Nazarov closely, giving the matrices for $\hat{S}_n$. Replacing the generators s_k by $r_k = z^k s_k$, the group $\hat{S}_n$ has a presentation with generators $z, r_1, \ldots, r_{n-1}$, and relations: z is central of order 2, and

$$\begin{aligned} r_j^2 &= 1 && \text{for } 1 \le j \le n-1; \\ (r_j r_{j+1})^3 &= z && \text{for } 1 \le j \le n-2; \\ r_j r_k &= z r_k r_j && \text{for } |j-k| > 1 \text{ and } 1 \le j,k \le n-1. \end{aligned}$$

We shall define matrices corresponding to these generators, and prove that the resulting representations are a complete set of irreducibles, and are mutually inequivalent. The verification that the above relations are satisfied by the matrices will be referred to the literature.

Recall that a partition λ in $\mathcal{P}(n)$ of length ℓ is said to be even or odd according as $n-\ell$ is even or odd. Let $\varepsilon(\lambda)$ be 1 if λ is odd, and be 0 if λ is even. The following apply in this chapter only.

Definitions. Let $m(\lambda) = (n-\ell-\varepsilon(\lambda))/2$. Let $\mathcal{S}(\lambda)$ denote the set of all standard shifted tableaux of shape λ, as defined before Proposition 10.4. For any $T \in \mathcal{S}(\lambda)$, and any k with $1 \le k \le n$, let $g(T, k)$ denote the number of integers in the set $\{1, ..., k+1\}$ which do not occur on the main diagonal of T. Also, if k is the entry in position $[i, j]$, define $h(T, k)$ to be $j-i$. Thus $h(T, k)$ determines which north-west to south-east diagonal contains the entry k.

Proposition 11.1. *Let k and k' be integers with $1 \leq k \leq k' \leq n$.*
(*a*) *Then $g(T, k') \geq g(T, k)$.*
(*b*) *If $k' > k$, then $g(T, k') > g(T, k)$ unless $h(T, k'+1) = 0$ with $k' < n$.*

Proof. Part (a) is trivial. If $g(T, k') = g(T, k)$, then all x with $k+2 \leq x \leq k'+1$ are on the main diagonal of T. In particular, $k' < n$ and $h(T, k'+1) = 0$, as required for (b).

Remark. In (b) we must also have $k = k'-1$ if $g(T, k') = g(T, k)$, since no two consecutive integers can appear on any diagonal of T.

Given a strict partition λ of n, we shall construct the irreducible representation(s) of $\tilde{S}_n$ with character(s) $<\lambda>$ (and $<\lambda>^a$). Let δ denote either of the integers 1 or -1. Two representations $R(\lambda, \delta)$ of $\hat{S}_n$ corresponding to λ will be defined on the space $V_\lambda = U_\lambda \otimes W_\lambda$, where W_λ is the tensor power of $m(\lambda)$ copies of the complex vector space $\mathbb{C}^2$, and U_λ is the complex vector space with basis $\{u_T : T \in \mathcal{S}(\lambda)\}$. These representations will be determined by explicitly writing down their values on r_k ($k=1, 2, ..., n-1$). It will be shown later that $R(\lambda, 1)$ and $R(\lambda, -1)$ are equivalent if λ is even and are inequivalent associate representations if λ is odd.

From here till the examples below, fix a pair (λ, δ) as above. Consider the following 2×2 matrices:

$$E = \begin{pmatrix} 1 & 0 \\ 0 & 1 \end{pmatrix}, \quad A = \begin{pmatrix} 0 & -i \\ i & 0 \end{pmatrix}, \quad B = \begin{pmatrix} 1 & 0 \\ 0 & -1 \end{pmatrix}, \quad C = \begin{pmatrix} 0 & 1 \\ 1 & 0 \end{pmatrix}.$$

Define the matrices $M_1, ..., M_{2m(\lambda)+1}$, regarded as operators on W_λ, as follows:

$$M_1 = \delta B^{\otimes m(\lambda)},$$

and, for $1 \leq k \leq m(\lambda)$,

$$M_{2k} = \delta E^{\otimes(k-1)} \otimes C \otimes B^{\otimes[m(\lambda)-k]},$$

$$M_{2k+1} = \delta E^{\otimes(k-1)} \otimes A \otimes B^{\otimes[m(\lambda)-k]}.$$

It may be easily checked that $M_j^2 = I$ and $M_j M_k = -M_k M_j$ if $j \neq k$ (compare Proposition 6.1). Note that the M_i are not the same as the matrices given the same names in Theorem 2.8 and Chapter 6.

Definition. For any distinct non-negative integers p, q, let

$$\phi(p,q) = \left(2q(q+1)\right)^{1/2}\left((p-q)(p+q+1)\right)^{-1},$$

and

$$\rho(p,q) = \left\{\left(1-(p-q)^{-2}\right)\left(1-(p+q+1)^{-2}\right)/2\right\}^{1/2}.$$

Some values of ϕ and ρ for small p and q are shown in the following table:

(p,q)	(1,0)	(2,0)	(3,0)	(2,1)	(3,2)
$\phi(p,q)$	0	0	0	1/2	$1/\sqrt{3}$
$\phi(q,p)$	-1	$-1/\sqrt{3}$	$-1/\sqrt{6}$	$-\sqrt{3}/2$	$-\sqrt{(2/3)}$
$\rho(p,q)$	0	$1/\sqrt{3}$	$\sqrt{(5/12)}$	0	0

Next, given T in $\mathcal{S}(\lambda)$, and k an integer with $1 \leq k \leq n-1$, we define the vector space $U_{T,k}$. Let s_kT be the tableau obtained by interchanging k and $k+1$ in T. If s_kT is also a standard tableau, then $U_{T,k}$ is defined to be the $\mathbb{C}$-vector space with basis $\{u_T, u_{s_kT}\}$. If s_kT is not standard, then $U_{T,k}$ is defined to be the space spanned by u_T. Thus, for $1 \leq k \leq n-1$, we have direct sum decompositions

$$U_\lambda = \bigoplus_{T\in\mathcal{S}(\lambda)} U_{T,k}.$$

Put $p = h(T, k)$ and $q = h(T, k+1)$, which are unequal by the remark after 11.1. Define matrices $A_{T,k}$, $B_{T,k}$ and $C_{T,k}$, to be regarded as operators on $U_{T,k}$ using its given basis, as follows.

If $U_{T,k}$ is two-dimensional, let

$$A_{T,k} = \begin{pmatrix} \phi(p,q) & \rho(p,q) \\ \rho(p,q) & \phi(q,p) \end{pmatrix}, \quad B_{T,k} = \begin{pmatrix} -\phi(q,p) & \rho(p,q) \\ \rho(p,q) & -\phi(p,q) \end{pmatrix},$$

$$C_{T,k} = \begin{pmatrix} \phi(p,q)-\phi(q,p) & \rho(p,q)\sqrt{2} \\ \rho(p,q)\sqrt{2} & \phi(q,p)-\phi(p,q) \end{pmatrix}.$$

If $U_{T,k}$ is one-dimensional, instead of the given matrices, take their (1, 1) entries:

$$A_{T,k} = \phi(p,q),\ B_{T,k} = -\phi(q,p),\ C_{T,k} = \phi(p,q)-\phi(q,p).$$

Definition. Define the action of the generator r_k of $\hat{S}_n$ on the subspace $U_{T,k}\otimes W_\lambda$ to be given by the matrix:

$$\begin{cases} A_{T,k} \otimes M_{g(T,k)} + B_{T,k} \otimes M_{g(T,k)-1}, & \text{if } h(T, k)h(T, k+1) \neq 0; \\ C_{T,k} \otimes M_{g(T,k)}, & \text{otherwise.} \end{cases}$$

Taking the direct sum over $T \in \mathcal{S}(\lambda)$, this produces the matrix $R(\lambda, \delta)(r_k)$ as an operator on V_λ.

Examples. (1) Take $\lambda = (n)$. There is only one standard tableau of shape λ, namely

$$T = 1\,2 \ldots n\,.$$

It follows that each $U_{T,k}$ is one-dimensional, $h(T, k) = k-1$ and $g(T, k) = k$. Hence for the two representations corresponding to (n), we obtain the following matrices:

(a) If $k = 1$, then $p = 0$, $q = 1$ and $\phi(p, q) = -1$, so

$$R[(n), \delta](r_1) = -M_1\,.$$

(b) If $k > 1$, then $q = p+1 = k$, hence

$$R[(n), \delta](r_k) = -[(k+1)/2k]^{1/2} M_k - [(k-1)/2k)]^{1/2} M_{k-1}\,.$$

This is evidently closely related to the basic spin representation (of $\tilde{S}_n$ rather than $\hat{S}_n$) discussed in Chapters 2 and 6.

(2) Consider the partition $\lambda = (4, 2)$ of 6. Here there are five standard tableaux:

$$T_1 = \begin{matrix} 1\,2\,3\,4 \\ 5\,6 \end{matrix}, \quad T_2 = \begin{matrix} 1\,2\,3\,5 \\ 4\,6 \end{matrix}, \quad T_3 = \begin{matrix} 1\,2\,3\,6 \\ 4\,5 \end{matrix},$$

$$T_4 = \begin{matrix} 1\,2\,4\,5 \\ 3\,6 \end{matrix}, \quad T_5 = \begin{matrix} 1\,2\,4\,6 \\ 3\,5 \end{matrix}\,.$$

The partition λ is even and $m(\lambda) = 2$. Thus $M_1 = \delta B^{\otimes 2}$, $M_2 = \delta C \otimes B$, $M_3 = \delta A \otimes B$, $M_4 = \delta E \otimes C$ and $M_5 = \delta E \otimes A$.

When $k = 1$, each $U_{T,1}$ is one-dimensional, $p = h(T, 1) = 0$ and $g(T, 1) = 1 = h(T, 2) = q$. Thus r_1 is represented as five 4×4 blocks each equal to $-M_1$.

When $k = 2$, each $U_{T,2}$ is again one-dimensional, but $p = h(T, 2) = 1$. For $i = 1, 2$ and 3,

$$g(T_i, 2) = 2 \text{ and } q = h(T_i, 3) = 2\,.$$

However,

$$g(T_4, 2) = g(T_5, 2) = 1\,,$$

and

$$h(T_4, 3) = h(T_5, 3) = 0\,.$$

Thus $R[(4, 2), \delta]$ represents r_2 as a sum of five 4×4 blocks (but using a different ordering of the basis than for r_1). Three of the blocks are $-(\sqrt{3}/2)M_2 - (1/2)M_1$, and the other two are $-M_1$.

As an example where not all U are one-dimensional, take $k = 4$. Then $U_{T_1,4}$ is two-dimensional, $p = 3$ and $q = 0$, giving rise to an 8×8 block $X \otimes M_3$, where

$$X = \begin{pmatrix} 1/\sqrt{6} & \sqrt{(5/6)} \\ \sqrt{(5/6)} & -1/\sqrt{6} \end{pmatrix}.$$

The remaining blocks of this matrix, as well as the matrices representing r_3 and r_5, are left as an exercise.

Theorem 11.2. *The operators defined above determine a negative representation of* $\hat{S}_n$ *on the space* V_λ.

The proof proceeds by showing that the matrices we have defined do indeed satisfy the relations for the group $\hat{S}_n$, as given in the first paragraph of this chapter. The details of this, although straightforward, involve manipulations with 6×6 matrices with entries of the form $\phi(p, q)$ and $\rho(p, q)$. We refer the reader to Nazarov (1989) for details of the checks required.

Note that $R(\lambda, 1)$ and $R(\lambda, -1)$ are associates, since their values on r_k are negatives of each other.

The basic objective is to show that these representations are irreducible. In order to do that, we first consider their restrictions to $\hat{S}_{n-1}$.

Again fix (λ, δ) till Proposition 11.5. Let $M(\lambda, \delta)$ be the set of all pairs (ω, δ'), where ω is a strict partition of $n-1$ obtained by decreasing a part of λ by 1 (so that $\omega \in M(\lambda)$, as defined in Chapter 10), and δ' takes the value(s) $(\pm 1)^{[\varepsilon(\delta)-1]\varepsilon(\omega)}\delta$. For each pair (ω, δ') in $M(\lambda, \delta)$, we shall define an inclusion of vector spaces

$$\iota_{\omega,\delta'} : V_\omega \to V_\lambda\,.$$

To do this, first denote the standard basis of $\mathbb{C}^2$ by $\{w_1, w_{-1}\}$. Then the space $U_\omega \otimes W_\omega$ has basis

$$\left\{ u_S \otimes w_e : S \in \mathcal{S}(\omega),\, e \in \{1, -1\}^{m(\omega)} \right\},$$

where

$$w_e = w_{e_1} \otimes w_{e_2} \otimes \ldots w_{e_{m(\omega)}} .$$

Suppose that ω is obtained from λ by decreasing the part $q+1$ to q. Let S be a tableau in $\mathcal{S}(\omega)$. Let T be the tableau in $\mathcal{S}(\lambda)$ obtained by adding to S an entry equal to n at the end of the row of length q. Now define $\iota_{\omega,\delta'}$ as follows:

If $\varepsilon(\lambda) = 0$ and $\varepsilon(\omega) = 1$, then $m(\omega) = m(\lambda) - 1$, and define

$$\iota_{\omega,\delta'}(u_S \otimes w_e) = u_T \otimes w_e \otimes w_{\delta\delta'} .$$

If $\varepsilon(\lambda) = 1$ or $\varepsilon(\omega) = 0$, then $m(\omega) = m(\lambda)$, and define

$$\iota_{\omega,\delta'}(u_S \otimes w_e) = u_T \otimes w_e .$$

Denote by U_λ^q the subspace of U_λ with basis

$$\{u_T : T \in \mathcal{S}(\lambda) \text{ and } h(T, n) = q\} .$$

It follows that

$$\iota_{\omega,\delta'}(V_\omega) = U_\lambda^q \otimes \begin{cases} (\mathbb{C}^2)^{\otimes[m(\lambda)-1]} \otimes \mathbb{C} w_{\delta\delta'}, & \text{if } m(\omega) = m(\lambda) - 1, \\ W_\lambda & \text{if } m(\omega) = m(\lambda). \end{cases}$$

By the definition of $M(\lambda, \delta)$, we obtain a decomposition

$$V_\lambda = \bigoplus_{(\omega,\, \delta') \in M(\lambda,\delta)} \iota_{\omega,\delta'}(V_\omega) .$$

Theorem 11.3: *(a) For $1 \leq k \leq n-2$, the operator $R(\lambda, \delta)(r_k)$ preserves each subspace $\iota_{\omega,\delta'}(V_\omega)$, and acts on it as $R(\omega, \delta')(r_k)$. Thus the embedding $\iota_{\omega,\delta'}$ is an $\hat{S}_{n-1}$-map between the $\hat{S}_{n-1}$-modules given by $R(\omega, \delta')$ and $R(\lambda, \delta)\downarrow \hat{S}_{n-1}$.*

(b) Let M be any non-empty, proper subset of $M(\lambda, \delta)$. Then the subspace

$$\bigoplus_{(\omega,\delta') \in M} \iota_{\omega,\delta'}(V_\omega)$$

is not invariant under the operator $R(\lambda, \delta)(r_{n-1})$.

Proof. (a) Fix a pair (ω, δ') in $M(\lambda, \delta)$ and a tableau S in $\mathcal{S}(\omega)$. We shall show that $R(\lambda, \delta)(r_k)$ preserves the subspace $\iota_{\omega,\delta'}(U_{S,k} \otimes W_\omega)$, and acts on it as $R(\omega, \delta')(r_k)$ acts on $U_{S,k} \otimes W_\omega$.

Let T be the tableau in $\mathcal{S}(\lambda)$ obtained from S by adding n to a row of length q. Then, since $1 \le k \le n-2$, we have $h(T, k) = h(S, k)$, $h(T, k+1) = h(S, k+1)$, and $g(T, k) = g(S, k)$. Considering the possibilities that 1 is a part of λ or not, and also the cases when λ is even or odd, it can be seen that, in three of the four cases, $m(\lambda) = m(\omega)$. The only other possibility, namely that $m(\omega) = m(\lambda)-1$, occurs when λ is even and 1 is not a part of λ (so that ω is odd).

Suppose that $m(\omega) = m(\lambda)$. Then $[\varepsilon(\lambda)-1]\varepsilon(\omega)$ is zero, and so $\delta' = \delta$ and

$$\iota_{\omega,\delta'}(U_{S,k} \otimes W_{\omega}) = U_{T,k} \otimes W_{\lambda} .$$

Now (a) follows from the definitions of $R(\lambda, \delta)(r_k)$ and $R(\omega, \delta')(r_k)$.

Alternatively, if $m(\omega) = m(\lambda)-1$, then $\delta' = \pm\delta$ and

$$\iota_{\omega,\delta'}(U_{S,k} \otimes W_{\omega}) = U_{T,k} \otimes (\mathbb{C}^2)^{\otimes[m(\lambda)-1]} \otimes \mathbb{C} w_{\delta\delta'} .$$

The fact that $[\varepsilon(\lambda)-1]\varepsilon(\omega)$ is non-zero implies both that $\varepsilon(\lambda) = 0$ and that $h(T, n) \neq 0$. Now apply Proposition 11.1 (b), with $k' = n-1$, to obtain that $g(T, k) < g(T, n-1)$. Also, since λ is even,

$$g(T, n-1) = n-\ell = 2m(\lambda) .$$

Therefore the operators $M_{g(T,k)}$ and $M_{g(T,k)-1}$ preserve the subspace $(\mathbb{C}^2)^{\otimes[m(\lambda)-1]} \otimes \mathbb{C} w_{\delta\delta'}$ of W_{λ}. Now (a) follows from the definition of $R(\lambda, \delta)(r_k)$ and $R(\omega, \delta')(r_k)$.

(b) Since M is non-empty and is a proper subset of $M(\lambda, \delta)$, there exist (ω_1, δ'_1) in $M(\lambda, \delta) \setminus M$, and (ω, δ') in M, where $\delta'_1 = (\pm 1)^{[\varepsilon(\delta)-1]\varepsilon(\omega_1)}\delta$. We shall construct an element v of $\iota_{\omega,\delta'}(V_{\omega})$ such that the component v_1 of $R(\lambda, \delta)(r_{n-1})v$ in the direct summand $\iota_{\omega_1,\delta'_1}(W_{\omega_1})$ is non-zero.

Put $w = w_1^{\otimes[m(\lambda)-1]}$. Suppose that the strict partitions ω and ω_1 are obtained from λ by decreasing the parts $q+1$ and $p+1$, respectively, by 1.

Suppose firstly that $\omega = \omega_1$ (so that $p = q$) and hence $\delta' = -\delta'_1$. Then $\varepsilon(\lambda)-1$ and $\varepsilon(\omega)$ are both non-zero, so $m(\omega) = m(\lambda)-1$ and $p = q \neq 0$. Let $u_T \in U_{\lambda}^{q}$ be fixed. Put $v = u_T \otimes w \otimes w_{\delta\delta'}$. Then

$$v_1 = \delta\phi\big(h(T, n-1), q\big) u_T \otimes w \otimes w_{\delta\delta'} \neq 0 .$$

Now suppose that $\omega \neq \omega_1$. Then $|p-q| > 1$ and $\rho(p, q) \neq 0$. There is a tableau T in $\mathcal{S}(\lambda)$ such that $h(T, n-1) = p$ and $h(T, n) = q$. Put $T' = s_{n-1}T$, so that $T' \in \mathcal{S}(\lambda)$. Each of $m(\omega)$ and $m(\omega_1)$ may be equal to $m(\lambda)$ or to $m(\lambda)-1$, giving rise to the following four possibilities.

(i) $m(\omega) = m(\omega_1) = m(\lambda)$: then $\varepsilon(\lambda) = 1$. Put $v = u_T \otimes w \otimes w_1$. Then

$$v_1 = \delta\rho(p, q)u_{T'}\otimes w\otimes w_{-1}z,$$

where $z = i\sqrt{2}$ if $pq = 0$, and $z = 1+i$ otherwise.

(ii) $m(\omega) = m(\lambda)-1$, $m(\omega_1) = m(\lambda)$: then $\varepsilon(\lambda) = 0$ and $p = 0$, but q is non-zero. Put $v = u_T\otimes w\otimes w_{\delta\delta'}$. Then

$$v_1 = \delta\rho(p, q)\sqrt{2}\, u_{T'}\otimes w\otimes w_{-\delta\delta'}\,.$$

(iii) $m(\omega) = m(\lambda)$, $m(\omega_1) = m(\lambda)-1$: then $\varepsilon(\lambda) = 0$, p is non-zero and $q = 0$. Put $v = u_T\otimes w\otimes w_{-\delta\delta'_1}$. Then

$$v_1 = \delta(\lambda)\rho(p, q)\sqrt{2}\, u_{T'}\otimes w\otimes w_{\delta\delta'_1}\,.$$

(iv) $m(\omega) = m(\omega_1) = m(\lambda)-1$: then $\varepsilon(\lambda) = 0$ and pq is non-zero. Now put $v = u_T\otimes w\otimes w_{\delta\delta'}$. Then

$$v_1 = \rho(p, q)u_{T'}\otimes w_1^{\otimes[m(\lambda)-2]}\otimes x$$

where $x = i\delta' w_{-1}\otimes w_{\delta\delta'}$ if $\delta' = \delta_1'$, and $x = \delta w_1\otimes w_{\delta\delta'_1}$ otherwise.

Since $\rho(p, q) \neq 0$, in each case $v_1 \neq 0$.

This completes the proof.

For 11.4 and 11.5 below, we assume that our fixed partition λ is odd. The objective is to evaluate the character of the representation $R(\lambda, \delta)$ on a certain conjugacy class. Let $K(\lambda)$ be the set of integers k with $1 \leq k \leq n$ and such that

$$k \in \{\lambda_1, \lambda_1+\lambda_2, \ldots, \lambda_1+\ldots+\lambda_\ell\}.$$

For any tableau T in $\mathcal{S}(\lambda)$, let $F(T)$ be the set of all functions $f: K(\lambda)\to\mathbb{N}$ such that

$$f(k) = \begin{cases} g(T, k) \text{ or } g(T, k)-1 & \text{if } h(T, k)h(T, k+1) \neq 0; \\ g(T, k) & \text{otherwise.} \end{cases}$$

It follows from Proposition 11.1 (b) that each function in $F(T)$ is non-decreasing. The set $K(\lambda)$ has cardinality $n-\ell = 2m(\lambda)+1$, and we define f_0 to be the unique non-decreasing function on $K(\lambda)$ taking each of the values $1, 2, \ldots, 2m(\lambda)+1$. Let T_0 be the element of $\mathcal{S}(\lambda)$ obtained by entering the integers $1, \ldots, n$ in order, along rows.

Lemma 11.4. *With the above notation, f_0 is in $F(T)$ if and only if $T = T_0$.*

Proof. If $T = T_0$ then $f_0(k) = g(T_0, k)$ for all k in $K(\lambda)$, and so f_0 is in $F(T_0)$. To prove the converse, suppose that T is not equal to T_0. Let $k+1$ be the smallest integer having different positions in the tableaux T and T_0. If $k+1$ is in the $[i, j]$ position of T_0, then the entry $T[i, j]$ is greater than $k+1$, and so $k+1$ is not in the ith row of T. Thus $k+1$ is in the $[i+1, i+1]$ position of T and so $g(T, k) = g(T_0, k)-1$. In particular, $k+1$ cannot be on the diagonal of T_0 and so k is in $K(\lambda)$. Also $h(T, k+1) = 0$, so for any f in $F(T)$, we have that

$$f(k) = g(T, k) = g(T_0, k)-1 = f_0(k)-1 \neq f_0(k).$$

Thus $f \neq f_0$, completing the proof.

Below, products of group elements and matrices which are indexed by integers k will be taken with k increasing from left to right. Now let r_λ be the element $\prod_{k \in K(\lambda)} r_k$ of $\hat{S}_n$. We next calculate the character of the representation $R(\lambda, \delta)$ at r_λ.

Proposition 11.5. *If* λ *is odd, then*

$$\mathrm{tr}[R(\lambda, \delta)(r_\lambda)] = -\delta i^{m(\lambda)}(\lambda_1\lambda_2 \ldots \lambda_\ell/2)^{1/2}.$$

Proof. Consider the decomposition of the representation space

$$V_\lambda = \bigoplus_{T \in \mathcal{S}(\lambda)} \mathbb{C}u_T \otimes W_\lambda.$$

Denote by $D(T)$, and $D(T, k)$, the diagonal blocks of the matrices $R(\lambda, \delta)(r_\lambda)$, and $R(\lambda, \delta)(r_k)$, respectively, corresponding to the direct summand in the above decomposition indexed by T. Let $p = h(T, k)$, $q = h(T, k+1)$ and $g = g(T, k)$. Then

$$D(T, k) = \begin{cases} \phi(p, q)M_g - \phi(q, p)M_{g-1} & \text{if } pq \neq 0; \\ [\phi(p, q) - \phi(q, p)]M_g & \text{if } pq = 0. \end{cases}$$

Since the product $\prod r_k$, as k ranges over any non-empty subset of $K(\lambda)$, is never 1, we have that

$$D(T) = \prod_{k \in K(\lambda)} D(T, k) = \sum_{f \in F(T)} a_f \prod M_{f(k)},$$

where $a_f \in \mathbb{R}$. The trace of the product $\prod M_{f(k)}$ is zero unless $f = f_0$. It now follows from Lemma 11.4 that $\mathrm{tr}[D(T)]$ is zero unless $T = T_0$. Finally, it may be checked that

$$\mathrm{tr}[D(T_0)] = \left(\prod_{i=1}^{\ell} \prod_{j=1}^{\lambda_i+i-2} \phi(j-i, j-i+1) \right) \mathrm{tr}(M_1 M_2 \ldots M_{2m(\lambda)+1})$$

$$= -\delta i^{m(\lambda)} (\lambda_1 \lambda_2 \ldots \lambda_\ell / 2)^{1/2},$$

as required. (For these traces, compare Proposition 6.1, but recall that the M_k there are different.)

Corollary 11.6. *If λ is odd, then the representations $R(\lambda, 1)$ and $R(\lambda, -1)$ are inequivalent.*

Proposition 11.7. *If λ is even, then $R(\lambda, 1)$ and $R(\lambda, -1)$ are equivalent.*

Proof. We shall define an invertible operator H_λ, on the space $U_\lambda \otimes W_\lambda$, which is a map of $\hat{S}_n$-modules, where the action is $R(\lambda,1)$ on the domain, and $R(\lambda,-1)$ on the codomain. To do this, put $H_\lambda = I \otimes E^{\otimes[m(\lambda)-1]} \otimes A$; that is, H_λ is the matrix $\delta I \otimes M_{2m(\lambda)+1}$. Now apply Proposition 11.1(a) with $k' = n-1$ to obtain, for $1 \leq k \leq n-1$,

$$g(T, k) \leq g(T, n-1) = n - \ell = 2m(\lambda).$$

Thus

$$H_\lambda R(\lambda, 1)(r_k) = -R(\lambda, 1)(r_k) H_\lambda = R(\lambda, -1)(r_k) H_\lambda,$$

as required, the first equality arising because the M_j anti-commute with each other.

Before proving the irreducibility of $R(\lambda, \delta)$, there is one more preliminary result.

Lemma 11.8. *Let λ and $\bar{\lambda}$ be strict partitions with $M(\lambda) = M(\bar{\lambda})$. Then $\lambda = \bar{\lambda}$ unless $M(\lambda)$ is the singleton $\{(2)\}$.*

Proof. We consider three possibilities.

(a) $M(\lambda)$ consists of a single partition ρ of length 1. In this case, since $\lambda = (2)$ is excluded, λ must also be of length 1, and so is uniquely determined by $M(\lambda)$.

(b) $M(\lambda)$ consists of a single partition ρ of length greater than 1. In this case, the parts of λ must be a sequence of successive integers, and ρ is obtained by decreasing the smallest part by 1. Thus λ is uniquely determined by ρ.

(c) Finally, we may suppose that $M(\lambda)$ contains two distinct partitions σ and ρ, where $\ell(\sigma) \leq \ell(\rho)$. If $\ell(\sigma) = \ell(\rho)$, then $\ell(\lambda) = \ell(\rho)$ and $\lambda_i = \max\{\sigma_i, \rho_i\}$. If $\ell(\sigma) = \ell(\rho) - 1$, then $\lambda = (\sigma_1, \ldots, \sigma_{\ell(\sigma)}, 1)$, completing the proof.

We can now prove the main result of this chapter.

Theorem 11.9. *The representations $R(\lambda, \delta)$, where λ is a strict partition of n, and $\delta = (\pm 1)^{\varepsilon(\lambda)}$, are irreducible and pairwise inequivalent (and so they are all the irreducible negative representations of $\hat{S}_n$.)*

Proof. We shall prove by induction on n that, if $R(\lambda, \delta)$ and $R(\bar{\lambda}, \bar{\delta})$ are equivalent, then $(\lambda, \delta) = (\bar{\lambda}, \bar{\delta})$, and simultaneously that $R(\lambda, \delta)$ is irreducible.

If $n = 1$, then $[\lambda, \delta] = [(1), 1] = [\bar{\lambda}, \bar{\delta}]$. The representation $R(\lambda, \delta)$ is one-dimensional and therefore irreducible.

For the inductive step, consider the decomposition of V_λ into $\hat{S}_{n-1}$-invariant subspaces given by Theorem 11.3 (a). These subspaces are irreducible $\hat{S}_{n-1}$-modules by the inductive hypothesis, using Proposition 11.7 in the case of pairs $(\omega, -1) \in M(\lambda, \delta)$ with ω even. These subspaces are also pairwise inequivalent by the inductive hypothesis, since we cannot have both $(\omega, 1)$ and $(\omega, -1)$ in $M(\lambda, \delta)$ if ω is even. It is immediate that any $\hat{S}_n$-invariant subspace of V_λ must be a direct sum of a subcollection of these subspaces. Theorem 11.3 (b) now implies that V_λ is $\hat{S}_n$-irreducible; i.e. $R(\lambda, \delta)$ is irreducible.

To complete the induction, we show that $(\lambda, \delta) = (\bar{\lambda}, \bar{\delta})$. First note that the equivalence of $R(\lambda, \delta)$ and $R(\bar{\lambda}, \bar{\delta})$ over $\hat{S}_n$ implies (by restriction) the equivalence over $\hat{S}_{n-1}$ of the following direct sums of irreducibles:

$$\bigoplus_{M(\lambda, \delta)} R(\omega, \delta') \cong \bigoplus_{M(\bar{\lambda}, \bar{\delta})} R(\bar{\omega}, \bar{\delta}') .$$

Now let $\omega \in M(\lambda)$. Then $(\omega, \delta') \in M(\lambda, \delta)$ for some choice of δ'. Thus $R(\omega, \delta')$ is equivalent to $R(\bar{\omega}, \bar{\delta}')$ for some $(\bar{\omega}, \bar{\delta}') \in M(\bar{\lambda}, \bar{\delta})$. Using Proposition 11.7 when either of these pairs has the form (even partition, -1), it follows from the inductive hypothesis that $\omega = \bar{\omega}$. In particular, $\omega \in M(\bar{\lambda})$. Thus $M(\lambda) \subset M(\bar{\lambda})$. By symmetry, $M(\lambda) = M(\bar{\lambda})$. By Lemma 11.8, either $\lambda = \bar{\lambda}$ or $M(\lambda) = \{(2)\}$. In the latter case, λ and $\bar{\lambda}$ are either (3) or (2 1). But $R[(3), \delta]$ and $R[(2\ 1), \bar{\delta}]$ have degrees 2 and 1, respectively, so, once again, $\lambda = \bar{\lambda}$. Since $R(\lambda, \delta)$ and $R(\lambda, \bar{\delta})$ are equivalent, if λ is odd Corollary 11.6 gives that $\delta = \bar{\delta}$, whereas if λ is even, then $\delta = 1 = \bar{\delta}$ by hypothesis.

This completes the proof.

Our indexings of representations and of characters agree with each other; that is, the character of the $\tilde{S}_n$-representation arising from the $\hat{S}_n$-representation

$R(\lambda, \delta)$ is either $<\lambda>$ or $<\lambda>^a$. This follows by induction on n, comparing the branching rule, Theorem 10.2 (i), to the decomposition of the restriction of V_λ given in Theorem 11.3 (a).

Notes

The results and techniques of this chapter are all due to Nazarov (1989). Other than the last paragraph, they do not depend on Chapter 10, except for motivation. Independent proofs of the branching rule and degree formula of Chapter 10 therefore follow from Nazarov's construction, modulo an independent check that indexings by λ agree.

12

COMBINATORIAL AND SKEW Q-FUNCTIONS

The first objective of this chapter is to find combinatorial descriptions of two sets of numbers: those which occur (i) as inner products $<q_\nu, Q_\lambda>$, and (ii) as coefficients in the expression of the product $q_\nu Q_\mu$ in terms of the basis $\{Q_\lambda\}$. The former is essentially a special case of the latter, and leads to an enumerative description of the coefficients in the polynomial $Q_\lambda(X)$; in particular, these coefficients are non-negative integers. It also leads to a strengthening of the unitriangularity phenomena, in both the definition of Q_λ, and in Theorem 8.6 relating the irreducibles $<\lambda>$ to the induced representations ξ_λ. Schur's fundamental orthogonality identity then yields an important enumerative result connecting the marked shifted tableaux occurring in the above formulae with integer matrices which have "marked" entries.

Next we define the skew Q-functions and find the combinatorial description of their coefficients. The final part of the chapter gives some formulae for the skew Q-functions, including a Laurent generating expression, an explicit formula as polynomials in the q_i, and a Pfaffian expansion.

The first result is a rather technical commutation relation for operators on Δ.

Proposition 12.1. *For $r > s > 0$ and $n > 0$, and with summation over positive j less than the smaller of n and $r-s$,*

$$q_n^\perp \mathcal{B}_r \mathcal{B}_s$$
$$= \mathcal{B}_r q_n^\perp \mathcal{B}_s + 2\sum \mathcal{B}_{r-j} q_{n-j}^\perp \mathcal{B}_s + \mathcal{B}_s q_{n+s-r}^\perp \mathcal{B}_s + 2\mathcal{B}_{r-n}\mathcal{B}_s \delta_{\lceil n/(r-s)\rceil,0},$$

where $\lceil\ \rceil$ denotes the "rounding down" function. Note that the third term is zero for $n \leq r-s$, and the fourth term is zero for $n \geq r-s$.

Proof. Compose the equation of Proposition 9.3 on the right with $\mathcal{B}_s$, to obtain

$$q_n^\perp \mathcal{B}_r \mathcal{B}_s = \mathcal{B}_r q_n^\perp \mathcal{B}_s + 2\sum_{p=1}^{n} \mathcal{B}_{r-p} q_{n-p}^\perp \mathcal{B}_s .$$

The result follows directly when $n \le r-s$. When $n > r-s$, we need to show that

$$\mathcal{B}_s q^\perp_{n+s-r} \mathcal{B}_s = 2 \sum_{i=r-s}^{n} \mathcal{B}_{r-i} q^\perp_{n-i} \mathcal{B}_s .$$

To do this, let $k=r-s$, and consider their difference:

$$2\sum_{i=k}^{n} \mathcal{B}_{s+k-i} q^\perp_{n-i} \mathcal{B}_s - \mathcal{B}_s q^\perp_{n-k} \mathcal{B}_s$$

$$= \mathcal{B}_s q^\perp_{n-k} \mathcal{B}_s + 2\mathcal{B}_{s+k-n} \mathcal{B}_s + 2 \sum_{i=k+1}^{n-1} \mathcal{B}_{s+k-i} q^\perp_{n-i} \mathcal{B}_s$$

$$= \mathcal{B}_s (\mathcal{B}_s q^\perp_{n-k} + 2\sum_{p=1}^{n-k} \mathcal{B}_{s-p} q^\perp_{n-k-p}) + 2\mathcal{B}_{s+k-n} \mathcal{B}_s$$

$$+ 2 \sum_{i=k+1}^{n-1} \mathcal{B}_{s+k-i} (\mathcal{B}_s q^\perp_{n-i} + 2\sum_{p=1}^{n-i} \mathcal{B}_{s-p} q^\perp_{n-i-p}) .$$

Since $\mathcal{B}_s \mathcal{B}_s$ is zero by Theorem 9.1, the first of these five terms is zero. In the remaining four terms the coefficient of $q^\perp_{n-u}$ is

$$2\mathcal{B}_s \mathcal{B}_{s+k-u} + 2\mathcal{B}_{s+k-u} \mathcal{B}_s + 4 \sum_{i=k+1}^{u-1} \mathcal{B}_{s+k-i} \mathcal{B}_{s+i-u} .$$

This is zero by Theorem 9.1, completing the proof.

Some notation is needed for the next result.

Definition. Let λ be a strict partition of length ℓ, and let n be a positive integer. For convenience, set $\lambda_{\ell+1} = 0$. Define $\mathcal{S}(\lambda, n)$ to be the set of all sequences α of weight n in $\mathbb{Z}^\ell$ such that

$$\lambda_1 \ge \lambda_1 - \alpha_1 \ge \lambda_2 \ge \lambda_2 - \alpha_2 \ge \ldots \ge \lambda_\ell \ge \lambda_\ell - \alpha_\ell \ge 0 ,$$

and such that the sequence $\lambda - \alpha$, whose ith term is $\lambda_i - \alpha_i$, is a strict partition (or else a strict partition followed by a zero, if $\alpha_\ell = \lambda_\ell$).

Example. Let λ be the partition (r, s), and let n be a positive integer. Then

$$\mathcal{S}(\lambda, n)$$
$$= \{(a, b) : 0 \le a \le r-s;\ 0 \le b \le s;\ a+b=n;\ \text{if } b=0, \text{ then } a<r-s\} .$$

Definition. Given a sequence α in $\mathcal{S}(\lambda, n)$ define $\mathrm{fr}(\lambda, \alpha)$ to be the number of indices j such that both $\alpha_j > 0$ and, if $j < \ell$, then $\alpha_j < \lambda_j - \lambda_{j+1}$; (that is, such that $\lambda_j > \lambda_j - \alpha_j > \lambda_{j+1}$, ignoring the second inequality when $j = \ell$).

In the above example, $\mathrm{fr}(\lambda, \alpha)$ is always 1 or 2. The cases where it is 1 are those when $\alpha = (a, b)$ with $a = 0, r - s$ or n.

Theorem 12.2. *With the above notation,*

$$q_n^\perp(Q_\lambda) = \sum_{\alpha \in \mathcal{S}(\lambda, n)} 2^{\mathrm{fr}(\lambda, \alpha)} Q_{\lambda \ \alpha} \ .$$

Proof. The proof is by induction on the length of λ. When λ has a single part, say $\lambda = (r)$, by (7.20)

$$q_n^\perp(Q_\lambda) = q_n^\perp(q_r) = 2q_{r-n} \ .$$

However, $\mathcal{S}(\lambda, n)$ consists of the single sequence (n) and, since $\ell = 1$ and n is positive, $\mathrm{fr}(\lambda, \alpha) = 1$, as required.

For the inductive step, let $\ell(\lambda) > 1$, $\lambda_1 = r$, and $\lambda_2 = s$. Let μ be the partition $(\lambda_2, \lambda_3, ..., \lambda_\ell)$, so that $\lambda = (r, \mu)$. Then $\mathcal{S}(\lambda, n)$ is the union of two or three of the following sets (depending on the size of n as compared to $r - s$):

(i) $\{(0, \beta) : \beta \text{ is in } \mathcal{S}(\mu, n)\}$;

(ii) $\{(a, \beta) : \beta \in \mathcal{S}(\mu, n-a) \text{ and } 0 < a < \min(n, r-s)\}$;

(iii) $\{(r-s, \beta) : \beta \in \mathcal{S}(\mu, n+s-r) \text{ and } \beta_1 \neq 0\}$, when $n > r-s$; and

(iv) $\{(n, \beta) : \beta = (0, ..., 0) \in \mathbb{Z}^{\ell-1}\}$, when $n < r-s$.

Clearly, $\mathrm{fr}(\lambda, \alpha) = \mathrm{fr}(\mu, \beta)$ for α in (i) or (iii), whereas $\mathrm{fr}(\lambda, \alpha) = 1 + \mathrm{fr}(\mu, \beta)$ for α in (ii) or (iv). Now let ν be the partition obtained by deleting the parts r and s from λ, so that $(r, s, \nu) = (r, \mu) = \lambda$. Then, using Proposition 12.1 (so $0 < j < \min(n, r-s)$ in the summation):

$$q_n^\perp(Q_\lambda) = q_n^\perp\left(\mathcal{B}_\lambda(1)\right)$$

$$= q_n^\perp \mathcal{B}_r \mathcal{B}_s \mathcal{B}_\nu(1)$$

$$= \mathcal{B}_r q_n^\perp \mathcal{B}_\mu(1) + 2 \sum \mathcal{B}_{r-j} q_{n-j}^\perp \mathcal{B}_\mu(1) + \mathcal{B}_s q_{n+s-r}^\perp \mathcal{B}_\mu(1)$$

$$+ 2\mathcal{B}_{r-n} \mathcal{B}_\mu(1) \delta_{\lceil n/(r-s)\rceil, 0} \ .$$

The result now follows by induction, since the four terms in the above sum correspond in the same order to the four subsets, (i) to (iv), whose union is $\mathcal{S}(\lambda, n)$.

Theorem 12.2 is probably best expressed combinatorially. A subset Z of the shifted diagram $S(\lambda)$ will be called a *zig-zag strip* when it intersects each north-west to south-east diagonal in at most one node. We shall say that a node in Z *can be marked* when no node in Z is immediately to its left; the node *must be marked* when there is a node in Z immediately below it; and the node is *free* when it is among those nodes which can be marked but not among those which must be marked.

Examples. When $\lambda = (7, 5, 3, 2, 1)$, the following is a zig-zag strip in $S(\lambda)$. The free node is denoted by f, and the node which must be marked by m:

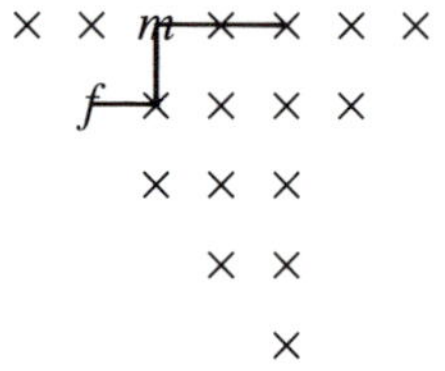

For the second example, the only two nodes in the zig-zag strip are denoted by f, since both are free.

$$\begin{array}{ccccccc} \times & \times & \times & \times & \times & \times & \times \\ & \times & \times & f & \times & \times & \\ & & f & \times & \times & & \\ & & & \times & \times & & \\ & & & & \times & & \end{array}$$

Fix $n > 0$ and consider the set $\mathscr{Z}(\lambda, n)$ of those zig-zag strips Z containing n nodes in $S(\lambda)$ such that $S(\lambda)\backslash Z$ is the shifted diagram of another strict partition, denoted $\mu(Z)$. Any Z in $\mathscr{Z}(\lambda, n)$ corresponds to a sequence α in $\mathscr{S}(\lambda, n)$, where α_i is the number of nodes in Z which are in the ith row. (They must be the right-most α_i nodes in that row of $S(\lambda)$.) Conversely, given α, one recovers Z as a union over i of the rightmost α_i nodes in the ith row. Since $\operatorname{fr}(\lambda, Z)$, the number of free nodes in Z, is equal to $\operatorname{fr}(\lambda, \alpha)$, we may rewrite Theorem 12.2 as

$$q_n^{\perp}(Q_\lambda) = \sum_Z 2^{\operatorname{fr}(\lambda, Z)} Q_{\mu(Z)},$$

where the summation is over all Z in $\mathscr{Z}(\lambda, n)$. Note that any zig-zag strip Z, such that $S(\lambda)\backslash Z$ is the shifted diagram of a partition, is a subset of the "south-east" boundary (or *rim*) of $S(\lambda)$. Conversely, all subsets of the rim of $S(\lambda)$ are necessarily zig-zag strips.

A *marked zig-zag strip* Z' will be a zig-zag strip together with marks on all the "must-be-marked" nodes and on none, some or all of the free nodes. The formula is then

$$q_n^{\perp}(Q_\lambda) = \sum_{Z'} Q_{\mu(Z')}, \tag{12.3}$$

where the summation is over all those marked zig-zag strips Z' whose underlying zig-zag strips are in $\mathcal{Z}(\lambda, n)$.

Now we can iterate. For any sequence $\nu \in \mathbb{N}^k$,

$$q_\nu^{\perp}(Q_\lambda) = q_{\nu_1}^{\perp} q_{\nu_2}^{\perp} \cdots q_{\nu_k}^{\perp}(Q_\lambda).$$

This will be a sum over certain sequences $Z'_k, \ldots, Z'_2, Z'_1$ of marked zig-zag strips of sizes $\nu_k, \ldots, \nu_1$. To distinguish nodes in different strips, we can think of inserting the integer p at all nodes in Z'_p, and marking the *integer* (as p') for those nodes which are marked. This motivates the following.

Definition. Let λ and μ be strict partitions, and let $\nu \in \mathbb{N}^k$. A *marked shifted tableau* of shape λ/μ and content ν is a sequence $Z'_k, \ldots, Z'_1$ of pairwise disjoint marked zig-zag strips of sizes $\nu_k, \ldots, \nu_1$, respectively, in the shifted Young diagram $S(\lambda)$, such that for each i, the array $S(\lambda) \setminus (Z'_i \cup \ldots \cup Z'_k)$ is the shifted Young diagram of some strict partition which, when $i = 1$, is the partition μ. Equivalently, a marked shifted tableau of shape λ/μ and content ν is the result of using ν_1 copies of "1", ν_2 copies of "2", . . ., ν_k copies of "k", to replace nodes of $S(\lambda)$, some of the integers being marked, such that:

(1) the remaining nodes form $S(\mu)$;

(2) with respect to the ordering

$$1' < 1 < 2' < 2 < 3' < \ldots,$$

both rows and columns are weakly increasing; and

(3) each (unmarked) p occurs at most once in each column, and each (marked) p' occurs at most once in each row.

Example. Let $\lambda = (7, 5, 3, 2, 1)$ and $\mu = (5, 1)$. An example of a marked shifted tableau of shape λ/μ is

$$\begin{array}{ccccccccc}
\times & \times & \times & \times & \times & 3' & 4 & & \\
 & \times & 1' & 1 & 2 & 3 & & & \\
 & & 1 & 3 & 3 & & & & \\
 & & & 4 & 5' & & & & \\
 & & & & 5' & & & &
\end{array}$$

The content of this tableau is the sequence (3, 1, 4, 2, 2).

By iteration from 12.3, we immediately obtain

$$q_\nu^\perp(Q_\lambda) = \sum Q_{\mu(T)}, \qquad (*)$$

where the summation is over all marked shifted tableaux T in $S(\lambda)$ of content $\nu = (\nu_1, ..., \nu_k)$, where the shape is $\lambda/\mu(T)$. Let $MST(\lambda/\mu; \nu)$ denote the set of all marked shifted tableaux of shape λ/μ and content ν. In particular, $MST(\lambda/\mu; \nu)$ is empty unless $S(\mu) \subset S(\lambda)$ and $|\mu| + |\nu| = |\lambda|$.

Since $<q_\nu Q_\mu, Q_\lambda> = <Q_\mu, q_\nu^\perp(Q_\lambda)>$ and $<Q_\mu, Q_\mu> = 2^{\ell(\mu)}$, equation (*) is equivalent to either equation in the following theorem:

Theorem 12.4. *With summations over* $\mathcal{D}$,

$$q_\nu^\perp(Q_\lambda) = \sum_\mu |MST(\lambda/\mu; \nu)| Q_\mu$$

and

$$q_\nu Q_\mu = \sum_\lambda 2^{\ell(\mu)-\ell(\lambda)} |MST(\lambda/\mu; \nu)| Q_\lambda \, .$$

The latter equation is an analogue of the Pieri formula (also known as Young's rule) for Schur functions.

Corollary 12.5. *The cardinality* $|MST(\lambda/\mu; \nu)|$ *does not change when the terms in the sequence* ν *are permuted.*

When μ is empty, one refers to the "non-skew" case and denotes the set as $MST(\lambda; \nu)$. It can only be non-empty if $|\nu| = |\lambda|$, in which case (*) yields

$$<q_\nu, Q_\lambda> = q_\nu^\perp(Q_\lambda) = |MST(\lambda; \nu)| \, . \qquad (12.6)$$

For example, when $\nu = (1, 1, ..., 1)$ with $|\nu| = |\lambda|$, the set $MST(\lambda; 1^{|\lambda|})$ consists of the standard shifted tableaux occurring in Chapter 10, modified by marking some of the entries. Since no integer appears more than once, any subset of entries can be marked. Thus

$$|MST(\lambda; 1^{|\lambda|})| = 2^{|\lambda|} g_\lambda \, .$$

The subset of those tableaux with unmarked diagonal entries therefore has cardinality $2^{|\lambda|-\ell(\lambda)} g_\lambda$, the number which appeared in Corollary 10.8, and which will appear in the above form at the end of Chapter 13.

In view of Theorem 7.7 (i), formula (12.6) yields

Corollary 12.7. *For* λ *in* $\mathcal{D}(n)$ *and each set* $X = \{x_1,...,x_k\}$ *of variables,*

$$Q_\lambda(X) = \sum_{\alpha \in \mathbb{N}^k} |MST(\lambda;\, \alpha)| X^\alpha .$$

Equivalently,

$$Q_\lambda - \sum_{\nu \in \mathcal{P}} |MST(\lambda;\, \nu)| m_\nu .$$

This combinatorial formula for the coefficients of Q_λ is sometimes used to define the Q_λ. Their symmetry, Corollary 12.5, must then be proved directly. None of the earlier formulae revealed in any obvious way the fact that the coefficients of Q_λ are non-negative.

The second version in 12.7 gives the matrix expressing $\{Q_\lambda : \lambda \in \mathcal{D}\}$ in terms of $\{m_\nu : \nu \in \mathcal{P}\}$. It is not square, but does have a certain triangularity property given in the next result.

Definition. If λ and ν are any partitions of n, define "λ precedes ν" to mean that, for all i,

$$\lambda_1 + ... + \lambda_i \geq \nu_1 + ... + \nu_i$$

(where $\lambda_i := 0$ if $i > \ell(\lambda)$ and similarly for ν). The relation "precedes" is a partial order with which the reverse lexicographic (total) order is clearly compatible; i.e. if λ precedes ν, then $\lambda \leq \nu$. The relation "precedes" is called the natural order in Macdonald (1979).

Proposition 12.8. *If* λ *does not precede* ν, *then* $MST(\lambda;\, \nu)$ *is empty* (*where* λ *is strict and* ν *is any partition*).

Proof. For a contradiction, assume that λ does not precede ν and $T \in MST(\lambda;\, \nu)$. Choose some k such that

$$\lambda_1 + . . . + \lambda_k < \nu_1 + . . . + \nu_k .$$

Then the entries equal to $1', 1, 2', 2, ..., k', k$ in T will not fit onto the first k rows of T. Choose any entry m or m' with $m \leq k$ on the jth row of T for some $j > k$. By the definition of marked shifted tableaux, the entry immediately north-west of this is p or p' where $p < m$. Continuing up this diagonal in a northwesterly direction, we obtain a contradiction, since that north-west portion of the diagonal has j entries, and $j > m$, and since entries strictly decrease (ignoring markings).

Now for any f in Δ^n,

$$f = \sum_{\lambda \in \mathcal{D}(n)} 2^{-\ell(\lambda)} <f, Q_\lambda> Q_\lambda \, ,$$

since they have the same inner product with Q_μ for all μ. This holds in particular when $f = q_\nu$, so that (12.6) gives the transition matrix from the basis $\{Q_\lambda : \lambda \in \mathcal{D}(n)\}$ to the basis $\{q_\nu : \nu \in \mathcal{D}(n)\}$. Combining (12.6) and 12.8 yields

Corollary 12.9. *If $\mathcal{D}(n)$ is ordered by any total order which is compatible with the partial order "precedes", then the transition matrix between $\{Q_\lambda\}$ and $\{q_\nu\}$ is unitriangular.*

Since the map och in Chapter 8 essentially matches up the Q_λ with the irreducibles $<\lambda>$, and the q_λ with the induced representations ξ_λ, we obtain a strengthening of Theorem 8.6 (i) (b).

Corollary 12.10. *If the irreducible $<\mu>$ or $<\mu>^a$ occurs in the decomposition of ξ_λ, then μ precedes λ.*

The basic identity in Corollary 7.16,

$$\sum_\lambda 2^{-\ell(\lambda)} Q_\lambda(X) Q_\lambda(Y) = \prod_{\substack{x \in X \\ y \in Y}} (1+xy)/(1-xy) \, ,$$

now leads to a fundamental enumerative result for shifted tableaux.

Definition. Let $MST^u(\lambda/\mu; \alpha)$ denote the set of those tableaux in $MST(\lambda/\mu; \alpha)$ whose main diagonal entries are all unmarked. There are exactly $\ell(\lambda) - \ell(\mu)$ entries on the main diagonal in a marked shifted tableaux of shape λ/μ. Since these are free,

$$|MST^u(\lambda/\mu; \alpha)| = 2^{\ell(\mu)-\ell(\lambda)} |MST(\lambda/\mu; \alpha)| \, .$$

Definition. For finite sequences ρ, γ from $\mathbb{N}$, let $MM(\rho, \gamma)$ denote the set of "marked" matrices with entries in $\mathbb{N}$, whose row sums are ρ, whose column sums are γ, and any of whose non-zero entries may be marked. Thus each ordinary matrix gives rise to 2^k marked matrices, where k is its number of non-zero entries.

Theorem 12.11. *For any* ρ *and* γ, *there is a bijection from* $MM(\rho, \gamma)$ *to the union, over all strict partitions* λ, *of the sets*

$$MST(\lambda; \rho) \times MST^{u}(\lambda; \gamma)\,.$$

Equivalently,

$$|MM(\rho; \gamma)| = \sum_{\lambda} |MST(\lambda; \rho)|\ |MST^{u}(\lambda; \gamma)|\,.$$

Proof. We have

$$\sum_{\lambda} 2^{-\ell(\lambda)} Q_{\lambda}(x_1, x_2, \ldots) Q_{\lambda}(y_1, y_2, \ldots) = \prod_{i,j} (1+x_i y_j)/(1-x_i y_j)\,.$$

Using Corollary 12.7, the left-hand side is

$$\sum_{\lambda} \sum_{\rho} \sum_{\gamma} |MST(\lambda; \rho)|\ |MST^{u}(\lambda; \gamma)|\ x_1^{\rho_1} \cdots y_1^{\gamma_1} \cdots\,.$$

The right-hand side is

$$\begin{aligned} &\prod_{i,j} (1 + x_i y_j)(1 + x_i y_j + x_i^2 y_j^2 + \ldots) \\ &= \prod_{i,j} (1 + 2x_i y_j + 2x_i^2 y_j^2 + \ldots) \\ &= \sum \prod_{i,j} 2^{\eta(a_{ij})} (x_i y_j)^{a_{ij}}\,, \end{aligned}$$

where the summation is over arrays (a_{ij}) of non-negative integers, almost all of which are zero; and where $\eta(a) = 0$ if $a = 0$, and $\eta(a) = 1$ if $a > 0$. For any $\mathbb{N}$-matrix A, let $\eta(A)$ be the number of non-zero entries in A, let $\rho_i(A)$ be the ith row sum of A, and let $\gamma_j(A)$ be the jth column sum of A. Then the above expression becomes

$$\begin{aligned} &\sum_{A} 2^{\eta(A)} x_1^{\rho_1(A)} x_2^{\rho_2(A)} \cdots y_1^{\gamma_1(A)} y_2^{\gamma_2(A)} \cdots \\ &= \sum_{\rho,\gamma} |MM(\rho, \gamma)|\ x_1^{\rho_1} x_2^{\rho_2} \cdots y_1^{\gamma_1} y_2^{\gamma_2} \cdots\,, \end{aligned}$$

giving the required result by equating coefficients.

Remark. In the more combinatorial approach to the subject where Corollary 12.7 is taken as the definition of the Q-functions, a construction of algorithms is given to provide an explicit bijection for Theorem 12.11. Since this is of considerable interest, we give some details concerning it in Chapter 13.

Definition. Given strict partitions λ and μ, define the skew Q-function $Q_{\lambda/\mu}$ as follows:

$$Q_{\lambda/\mu} = 2^{-\ell(\mu)} Q_{\mu}^{\perp}(Q_{\lambda}) .$$

Note that $Q_{\lambda/\varnothing} = Q_{\lambda}$ where $\varnothing$ is the empty partition, so the "non-skew" case is included in this definition. It is clear that the skew Q-functions are in $\Delta_{\mathbb{Q}}$, and, in particular, are symmetric. By the following theorem, the coefficients of $Q_{\lambda/\mu}$ lie in $2^{\ell(\lambda)-\ell(\mu)}\mathbb{N}$. Later we shall see that $Q_{\lambda/\mu}$ is in Δ; see Corollary 14.5 (iii).

Theorem 12.12. *For all integer sequences ν and strict partitions λ and μ,*

$$\langle Q_{\lambda/\mu}, q_{\nu}\rangle = |MST(\lambda/\mu; \nu)| .$$

Thus, for a set $X = \{x_1, ..., x_k\}$ of variables,

$$Q_{\lambda/\mu}(X) = \sum_{\nu \in \mathbb{N}^k} |MST(\lambda/\mu; \nu)|\, X^{\nu} .$$

In particular, $Q_{\lambda/\mu} \neq 0$ if and only if the shifted diagrams satisfy $S(\mu) \subset S(\lambda)$.

Proof. Using Theorem 12.4,

$$\begin{aligned} 2^{\ell(\mu)}\langle Q_{\lambda/\mu}, q_{\nu}\rangle &= \langle Q_{\mu}^{\perp}(Q_{\lambda}), q_{\nu}\rangle \\ &= \langle Q_{\lambda}, Q_{\mu}q_{\nu}\rangle \\ &= \langle q_{\nu}^{\perp}(Q_{\lambda}), Q_{\mu}\rangle \\ &= \sum_{\omega \in \mathcal{D}} |MST(\lambda/\omega; \nu)|\, \langle Q_{\omega}, Q_{\mu}\rangle \\ &= 2^{\ell(\mu)}|MST(\lambda/\mu; \nu)| . \end{aligned}$$

The second statement (a generalization of Corollary 12.7) now follows directly from Theorem 7.7 (i).

Remarks. Theorem 12.12 and the symmetry of $Q_{\lambda/\mu}$ provide another proof of Corollary 12.5. If T is marked shifted tableau of content γ, write X^T for X^{γ}. Then the equation in Theorem 12.12 becomes

$$Q_{\lambda/\mu}(X) = \sum_T X^T,$$

where T ranges over all marked shifted tableaux of shape λ/μ.

The remainder of this chapter consists of a number of formulae involving skew Q-functions, including a Laurent generating expression, a polynomial in the q_i, and a Pfaffian expansion. Since these will not be used later in the book, we shall leave some details to the reader.

Firstly, by induction on i using (7.1) and (7.19), for all f and h in Δ,

$$[q_i^{\perp}(f)]^{\perp}(h) = \sum_s (-1)^{s+i} q_{i-s} f^{\perp}(q_s h). \tag{12.13}$$

Now using the definition of $\mathcal{B}_n$ and substituting (12.13),

$$[\mathcal{B}_n(f)]^{\perp}(h) = \sum_{s,t} (-1)^s q_t f^{\perp}[q_s q_{n+s+t}^{\perp}(h)]. \tag{12.14}$$

Next, for all β in $\mathbb{Z}^k$,

$$Q_\beta^{\perp}(q_m f) = \sum_{\omega \in \mathbb{N}^k} 2^{\eta(\omega)} q_{m-|\omega|} Q_{\beta-\omega}^{\perp}(f), \tag{12.15}$$

recalling that $\eta(\omega)$ is the number of i with $\omega_i > 0$. This is proved by induction on k, the initial step, $k = 1$, being immediate from (7.19) and (7.20). (In fact, we could start with $k = 0$, where (12.15) holds with the usual conventions.) Letting $\beta' = (\beta_2, \ldots, \beta_k)$, the inductive step proceeds as follows:

$$Q_\beta^{\perp}(q_m f) = \left[\mathcal{B}_{\beta_1}(Q_{\beta'})\right]^{\perp}(q_m f)$$

$$= \sum_{s,t} (-1)^s q_t Q_{\beta'}^{\perp}\left[q_s q_{\beta_1+s+t}^{\perp}(q_m f)\right] \quad \text{by (12.14)}$$

$$= \sum_{s,t} (-1)^s q_t Q_{\beta'}^{\perp}\left[\sum_{\omega_1 \in \mathbb{N}} q_s 2^{\eta(\omega_1)} q_{m-\omega_1} q_{\beta_1-\omega_1+s+t}^{\perp}(f)\right]$$

by (7.19), (7.20)

$$= \sum_{s,t} (-1)^s q_t \sum_{\omega_1 \in \mathbb{N}} \sum_{\omega' \in \mathbb{N}^{k-1}} 2^{\eta(\omega_1)+\eta(\omega')} q_{m-\omega_1-|\omega'|} \times$$

$$\times Q_{\beta'-\omega'}^{\perp}\left[q_s q_{\beta_1-\omega_1+s+t}^{\perp}(f)\right]$$

by the inductive hypothesis ,

$$= \sum_{\omega} 2^{\eta(\omega)} q_{m-|\omega|} \left[\mathcal{B}_{\beta_1-\omega_1}(Q_{\beta'-\omega'}) \right]^{\perp}(f) \quad \text{by (12.14),}$$

completing the induction,.

Now $Q_\beta^\perp(Q_\alpha)$, and in particular $Q_{\lambda/\mu}$, will be expressed as a linear combination of Q_ψ, but over *sequences* ψ (not just over strict partitions, which is a rather difficult question treated in Chapter 14). Let $\alpha \in \mathbb{Z}^\ell$ and $\beta \in \mathbb{Z}^k$. In the summation below, M ranges over all $k\times\ell$ matrices with entries from $\mathbb{N}$ which add up to $|\beta|$. Recall that the sequence of row sums is denoted $\rho(M) \in \mathbb{Z}^k$, the column sums $\gamma(M) \in \mathbb{Z}^\ell$, and $\eta(M)$ is the number of non-zero entries. Then, with the function y as in 9.2,

$$Q_\beta^\perp(Q_\alpha) = \sum_M 2^{\eta(M)} y(\beta - \rho(M)) Q_{\alpha-\gamma(M)} \,. \tag{12.16}$$

The proof is by induction on ℓ, with the initial step, $\ell = 1$, being given by (12.15) with $f = 1$ and $\alpha = (m)$. (In fact, (12.16) holds for $\ell = 0$ with the usual conventions, so the induction could start there.) For the inductive step,

$$\begin{aligned}
Q_\beta^\perp(Q_\alpha) &= Q_\beta^\perp\left[\mathcal{B}_{\alpha_1}(Q_{\alpha'})\right] \\
&= \sum_i (-1)^i Q_\beta^\perp \left[q_{\alpha_1+i} q_i^\perp (Q_{\alpha'}) \right] \\
&= \sum_i \sum_{\omega\in\mathbb{N}^k} (-1)^i 2^{\eta(\omega)} q_{\alpha_1+i-|\omega|} Q_{\beta-\omega}^\perp Q_i^\perp (Q_{\alpha'}) \quad \text{by (12.15)} \\
&= \sum_{\omega\in\mathbb{N}^k} 2^{\eta(\omega)} \mathcal{B}_{\alpha_1-|\omega|} \left[Q_{\beta-\omega}^\perp (Q_{\alpha'}) \right] \\
&= \sum_{\substack{(\omega,M')\in\mathbb{N}^k\times\mathbb{N}^{k\times(\ell-1)} \\ |M'|=|\beta-\omega|}} 2^{\eta(\omega)+\eta(M')} \mathcal{B}_{\alpha_1-|\omega|} \left(y[\beta-\omega-\rho(M')] Q_{\alpha'-\gamma(M')} \right) \\
&= \sum_{\substack{M\in\mathbb{N}^{k\times\ell} \\ |M|=|\beta|}} 2^{\eta(M)} y[\beta-\rho(M)] Q_{\alpha-\gamma(M)} \,,
\end{aligned}$$

as required, taking

$$M = \begin{pmatrix} \omega_1 & \\ \cdot & \\ \cdot & M' \\ \cdot & \\ \omega_k & \end{pmatrix}.$$

A Laurent generating expression for the $Q^{\perp}_{\beta}(Q_{\alpha})$ (in particular, for the $Q_{\lambda/\mu}$) is as follows:

$$\sum_{(\alpha,\beta)\in \mathbb{Z}^{\ell}\times\mathbb{Z}^{k}} \left[Q^{\perp}_{\beta}(Q_{\alpha})\right](X)U^{\alpha}V^{\beta}$$

$$= \prod_{i<j\leq\ell} F(u_i^{-1}u_j)^{-1} \prod_{i<j\leq k} F(v_i^{-1}v_j)^{-1} \prod_{U\times V} F(uv) \prod_{U\times X} F(ux), \qquad (12.17)$$

where U and V are sets of variables with cardinalities ℓ and k, disjoint from each other and from X. This can, and should not, be reinterpreted as a raising operator formula. To prove it, the left side is

$$\sum_{\substack{\alpha,\beta,M\\ |M|=|\beta|}} 2^{\eta(M)} Q_{\beta-\rho(M)} Q_{\alpha-\gamma(M)}(X)U^{\alpha}V^{\beta} \quad \text{by (12.16)}$$

$$= \sum_{\substack{M,\theta,\psi\\ |\psi|=0}} 2^{\eta(M)} Q_{\psi} Q_{\theta}(X) U^{\theta+\rho(M)} V^{\psi+\gamma(M)}$$

$$= \left\{\sum_{M} 2^{\eta(M)} U^{\rho(M)} V^{\gamma(M)}\right\} \left\{\sum_{|\psi|=0} Q_{\psi} V^{\psi}\right\} \left\{\sum_{\theta} Q_{\theta}(X) U^{\theta}\right\}.$$

The right-hand factor is $\prod_{i<j} F(u_i^{-1}u_j)^{-1} \prod_{U\times X} F(ux)$, by (9.5). By setting all x in X equal to zero in (9.5), the middle factor is $\prod_{i<j} F(v_i^{-1}v_j)^{-1}$. The left-hand factor is $\prod_{U\times V} F(uv)$, as explained in the proof of (12.11).

This Laurent identity (12.17) becomes (9.5) by taking $k=0$ or by inspecting the coefficient of $V^{0,\ldots,0}$.

Now the first three factors on the right-hand side of (12.17) can be re-expressed:

$$\prod_{i<j} F(u_i^{-1}u_j)^{-1} \prod_{i<j} F(v_i^{-1}v_j)^{-1} \prod_{U\times V} F(uv)$$

$$= \sum_{\lambda\in\mathcal{D}} 2^{-\ell(\lambda)} Q_\lambda(U) \prod_{i<j} F(u_i^{-1}u_j)^{-1} Q_\lambda(V) \prod_{i<j} F(v_i^{-1}v_j)^{-1} \quad \text{by (7.16)}$$

$$= \sum_{\lambda\in\mathcal{D}} 2^{-\ell(\lambda)} \left\{ \sum_{\substack{\gamma\in\mathbb{Z}^\ell \\ \operatorname{str}\gamma=\lambda}} 2^{\ell(\lambda)} y(\gamma) U^\gamma \right\} \left\{ \sum_{\substack{\beta\in\mathbb{Z}^k \\ \operatorname{str}\beta=\lambda}} 2^{\ell(\lambda)} y(\beta) V^\beta \right\} \quad \text{by (9.7)}\ .$$

$$= \sum_{(\gamma,\beta)\in\mathbb{Z}^\ell\times\mathbb{Z}^k} 2^{\ell(\operatorname{str}\gamma)} \delta_{\operatorname{str}\gamma,\operatorname{str}\beta}\, y(\gamma)\, y(\beta)\, U^\gamma V^\beta = G, \text{ say.}$$

Using this, and substituting (7.11) for its fourth factor, (12.17) becomes

$$\sum_{\alpha,\beta} \left[Q_\beta^\perp(Q_\alpha) \right](X) U^\alpha V^\beta = G \sum_{\omega\in\mathbb{Z}^\ell} q_\omega(X) U^\omega\ .$$

Multiplying on the right-hand side and equating coefficients of $U^\alpha V^\beta$, we obtain

$$Q_\beta^\perp(Q_\alpha) = \sum_{\gamma\in\mathbb{Z}^\ell} 2^{\ell(\operatorname{str}\gamma)} \delta_{\operatorname{str}\beta,\operatorname{str}\gamma}\, y(\gamma)\, y(\beta)\, q_{\alpha-\gamma}\ . \tag{12.18}$$

Letting α and β be strict partitions λ and μ, respectively, (12.18) becomes

$$Q_{\lambda/\mu} = \sum_{\substack{\gamma\in\mathbb{Z}^\ell \\ \operatorname{str}\gamma=\mu}} y(\gamma) q_{\lambda-\gamma}\ , \tag{12.19}$$

which expresses $Q_{\lambda/\mu}$ as a polynomial in the q_i, and reduces to (9.12) when μ is empty.

As in the non-skew case, (12.19) can be used to give an inductive formula which is interpretable as a Pfaffian expansion. For this, we need a cancellation lemma involving the y-function.

Lemma 12.20. *Let* $(\lambda_1, \ldots, \lambda_\ell)$ *and* $(\mu_1, \ldots, \mu_k)$ *be strict, non-empty, partitions. Let* $\{t_i : i\in\mathbb{Z}\}$ *be a set of variables. For simplicity, set* $t_i = 0$ *for sufficiently small* i*, so that the expression (12.20) below is a finite linear combination of monomials in the* t_i*. (This is sufficient for the application, but is not essential.) For* $1\le i\le \ell$*, define*

$$C_i = \{\gamma\in\mathbb{Z}^{\ell-1} : \operatorname{str}\gamma = (\mu_1, \ldots, \mu_{k-1})\, ; \exists j<i \textit{ with } \gamma_j = \pm\mu_k\}\ .$$

Then, letting t_α mean $t_{\alpha_1}t_{\alpha_2}\ldots$, and $\hat{\lambda}_i$ mean "omit λ_i", we have

$$\sum_{i=1}^{\ell}(-1)^i t_{\lambda_i-\mu_k}\sum_{\gamma\in C_i} y(\gamma)\, t_{(\lambda_1\ldots\hat{\lambda}_i\ldots\lambda_\ell)-\gamma} = 0. \tag{12.20}$$

Remarks. (1) The set C_1 is empty, but we have included $i=1$ in the summation for simplicity later.

(2) For $i > 1$, there is, of course, a value of $j < i$ for which $\gamma_j = -\mu_k$, for γ in C_i.

(3) We shall use (12.20) with q_i substituted for t_i, but have expressed the lemma more generally to emphasize that it is merely a formal cancellation not involving the relations satisfied by the functions q_i.

Proof. Define

$$C'_i = \{\gamma\in C_i : \text{leftmost } \pm\mu_k \text{ in } (\gamma_i, \gamma_{i+1}, \ldots, \gamma_{\ell-1}) \text{ exists and is } \mu_k\}$$

$$C''_i = \{\gamma\in C_i : \text{rightmost } \pm\mu_k \text{ in } (\gamma_1, \ldots, \gamma_{i-1}) \text{ is } \mu_k\}\ .$$

Clearly C'_i and C''_i are disjoint, and $C'_i \cup C''_i = C_i$.

Define a map

$$\theta : \bigcup_p C'_p \times \{p\} \to \bigcup_q C''_q \times \{q\}$$

as follows. If $\gamma' \in C'_p$, let

$$q = 1 + \min\{j : j \ge p \text{ and } \gamma_j = \mu_k\}$$

(so $\gamma_{q-1} = \mu_k$). Define $\theta(\gamma', p) = (\gamma'', q)$, where

$$\gamma'' = (\gamma_1, \ldots, \gamma_{p-1}, \mu_k, \gamma_p, \ldots, \gamma_{q-2}, \gamma_q, \ldots, \gamma_{\ell-1})\ .$$

Then θ is bijective with inverse mapping (γ'', q) to (γ', p), where $p = \max\{j : j<q \text{ and } \gamma''_j = \mu_k\}$. Note that γ' and γ'' are obtained from each other by shuffling μ_k back and forth past the sequence $(\gamma_p, \ldots, \gamma_{q-2})$, none of whose terms is $\pm\mu_k$. Thus

$$y(\gamma'') = (-1)^{q-p-1} y(\gamma')\ .$$

It is evident that the same monomial t_α in (12.20) occurs for summation indices ($i=p$; $\gamma=\gamma' \in C'_p$) and for ($i=q$; $\gamma=\gamma'' \in C''_q$), if $\theta(\gamma', p) = (\gamma'', q)$. Its coefficients are $(-1)^p y(\gamma')$ and $(-1)^q y(\gamma'') = (-1)^{p+1} y(\gamma')$. Thus θ yields the required cancellation.

Theorem 12.21. *For any non-empty strict partitions* λ *and* μ *of lengths* ℓ *and* k *respectively,*

$$Q_{\lambda/\mu} = \sum_{i=1}^{\ell} (-1)^{i+\ell} q_{\lambda_i - \mu_k} Q_{(\lambda_1,\ldots,\hat{\lambda}_i,\ldots,\lambda_\ell)/(\mu_1,\ldots,\mu_{k-1})}\,.$$

Proof. By (12.19)

$$\begin{aligned} Q_{\lambda/\mu} &= \sum_{\substack{\beta\in\mathbb{Z}^\ell \\ \operatorname{str}\beta=\mu}} y(\beta) q_{\lambda-\beta} \\ &= \sum_{i=1}^{\ell} \sum_{A_i} y(\beta) q_{\lambda-\beta} \end{aligned}$$

where $A_i = \{\beta\in\mathbb{Z}^\ell : \operatorname{str}\beta = \mu;\ \beta_i = \mu_k;\ \forall j<i,\ \beta_j \neq \pm\mu_k\}$

$$= \sum_{i=1}^{\ell} q_{\lambda_i-\mu_k} \sum_{B_i} y(\gamma_1, \ldots, \gamma_{i-1}, \mu_k, \gamma_i, \ldots, \gamma_{\ell-1}) q_{(\lambda_1 \ldots \hat{\lambda}_i \ldots \lambda_\ell)-\gamma}\,,$$

where B_i is the set
$\{\gamma\in\mathbb{Z}^{\ell-1} : \operatorname{str}(\gamma_1, \ldots, \gamma_{i-1}, \mu_k, \gamma_i, \ldots, \gamma_{\ell-1}) = \mu;\ \forall j<i,\ \gamma_j \neq \pm\mu_k\}$

$$= \sum_{i=1}^{\ell} (-1)^{i+\ell} q_{\lambda_i-\mu_k} \sum_{D_i} y(\gamma) q_{(\lambda_1 \ldots \hat{\lambda}_i \ldots \lambda_\ell)-\gamma}\,,$$

where $D_i = \{\gamma\in\mathbb{Z}^{\ell-1} : \operatorname{str}\gamma = (\mu_1, \ldots, \mu_{k-1})\,;\ \forall j<i,\ \gamma_j \neq \pm\mu_k\}$.
Substituting q_s for all t_s in the left-hand side of (12.20) yields the same expression as the one immediately above, except that D_i is replaced by C_i. Clearly, C_i and D_i are disjoint and their union is $\{\gamma\in\mathbb{Z}^{\ell-1} : \operatorname{str}\gamma = (\mu_1, \ldots, \mu_{k-1})\}$. Thus, D_i can be replaced by $C_i \cup D_i$ in the expression immediately above. The resulting summation over $C_i \cup D_i$ yields

$$Q_{(\lambda_1,\ldots,\hat{\lambda}_i,\ldots,\lambda_\ell)/(\mu_1,\ldots,\mu_{k-1})},$$

as required, by (12.19).

Proceeding by induction on k, it follows that $Q_{\lambda/\mu}$, as given in (12.21), is an expansion of the following Pfaffian:

$$Q_{\lambda/\mu} = Pf\begin{pmatrix} A & B \\ -B^{tr} & 0 \end{pmatrix},$$

where

$$A = (Q_{\lambda_i,\lambda_j})_{\ell\times\ell}$$

and

$$B = \begin{cases} (q_{\lambda_i-\mu_{k+1-j}})_{\ell\times k} & \text{if } k+\ell \text{ is even;} \quad (12.22) \\ (q_{\lambda_i-\mu_{k+2-j}})_{\ell\times(k+1)} & \text{if } k+\ell \text{ is odd (with } \mu_{k+1}:=0). \quad (12.22)' \end{cases}$$

The special case $k = 0$ (and initial step in the induction) is given by (9.15) and (9.15)′. The inductive step is proved by interpreting (12.21) as the Laplace expansion of (12.22) down the first column, or of (12.22)′ down the second column, of $\left|\begin{matrix} B \\ 0 \end{matrix}\right)$. For expansions of Pfaffians, see DeConcini and Procesi (1976).

Notes

Corollary 12.7 is a "folk corollary", due to Richard Stanley (1984). Our proof does not seem to have occurred in the literature. Stanley points out that 12.7 can be deduced with some work by setting $t = -1$ in the analogous formula for Hall–Littlewood functions, Macdonald (1979) III 5.11. See also Stembridge (1990). Except for the use of $\mathcal{B}_n$, the material up to (12.12) largely follows a pattern set out in Macdonald's book and its forthcoming second edition. In the latter, the treatment of Macdonald's generalization (of the Jack and Hall–Littlewood functions) lays some emphasis on the overdetermination of the functions via the requirement of unitriangularity with respect to the natural partial order. Formulae (12.13) to (12.19) appear in Hoffman (1990). The Pfaffian (12.22), (12.22)′ is due to Jozefiak and Pragacz (1989). Our proof is quite different. The proof of (9.5) may be imitated to reprove (12.17). By writing the first three factors of (12.17) as a single product of $F(-w_i^{-1}w_j)$ over $i < j$, where $w_i = u_i$ for $i > 0$, and $w_i = -v_{-i}^{-1}$ for $i < 0$, the Jozefiak–Pragacz formula has a direct proof (due to I.G. Macdonald) as in the last paragraph of Chapter 9, using formula (*) there.

Appendix 12

Characters which appear without trace

Since the description of the irreducible projective representations of $\tilde{S}_n$ and $\tilde{A}_n$ in Appendix 8 was independent of character theory, we show, in this appendix, how that approach may be used to obtain information on characters. The method is reductive, reducing character calculations in $\tilde{S}_m$ to those in $\tilde{S}_n$ for $n < m$. This leads to the recovery of all the character calculations in Chapters 6 and 8 (using only the three trace calculations in Appendix 6), as well as to a variation on the proof of the Morris reduction rule (Theorem 10.1). Since these results have already been proved by other methods, this appendix is written in a style even more abbreviated than the earlier appendices. It should be regarded as a sequence of exercises, since we have given little of the computational details in the proofs. At the end we state the formula for $c_k^{\perp}(a_\lambda)$ promised in the last sentence of Appendix 8.

Recall the module T^*G, defined in Appendix 8 for objects (G, z, σ) in $\mathcal{G}$. For a negative (ungraded) representation V of G, let $\chi(V; g)$ denote the character of V at g in G. Since

$$\chi(V \oplus V'; g) = \chi(V; g) + \chi(V'; g),$$

there is a unique extension of this definition to $\chi(y; g)$ for all y in T^1G, where we require $\chi(y; g)$ to be homomorphic as a function of y.

Now let $W = (W_0, W_1)$ be a $\mathbf{Z}/2$-graded negative representation of G. Using the characters of the representations W_i of $\ker \sigma$, define

$$\Delta(W; h) := \chi(W_0; h) - \chi(W_1; h)$$

for $h \in \ker \sigma$. This also extends to a group homomorphism, denoted as $x \mapsto \Delta(x; h)$, mapping T^0G to $\mathbb{C}$.

Note that since $\chi(\kappa(W); h) = \chi(W_0; h) + \chi(W_1; h)$, knowledge of $\Delta(x; h)$ and $\chi(y; g)$, for all x, y, g and h as above, is equivalent to knowing all the characters for both $\ker \sigma$ and G.

It is easily checked that

$$\begin{aligned}
\chi(\rho y; g) &= (-1)^{\sigma(g)}\chi(y; g),\\
\Delta(\rho x; h) &= -\Delta(x; h),\\
\chi(\kappa x; g) &= 0 \qquad \text{if } \sigma(g) = 1,
\end{aligned}$$

and

$$\Delta(\kappa y; h) = 0 \qquad (\text{where } \sigma(h) = 0).$$

The behaviour of χ and Δ with respect to the operation of $\boxtimes^-$ is as follows. Notc that G' is *not* thc commutator subgroup.

Proposition A12.1. *For objects* G, G' *and elements* $x \in T^0G$, $x' \in T^0G'$, $y \in T^1G$, $y' \in T^1G'$ *we have*:

$$(i)\ \ \Delta(x \boxtimes^- x'; (g, g')) = \begin{cases} \Delta(x; g)\Delta(x'; g') & \text{if } \sigma(g) = 0 = \sigma(g'); \\ 0 & \text{if } \sigma(g) = 1 = \sigma(g'); \end{cases}$$

$$(ii)\ \ \chi(x \boxtimes^- y'; (g, g')) = \begin{cases} \chi(\kappa x; g)\chi(y'; g') & \text{if } \sigma(g) = 0 = \sigma(g'); \\ \Delta(x; g)\chi(y'; g') & \text{if } \sigma(g) = 0 \neq \sigma(g'); \\ 0 & \text{otherwise;} \end{cases}$$

$$(ii)'\ \ \chi(y \boxtimes^- x'; (g, g')) = \begin{cases} \chi(y; g)\chi(\kappa x'; g') & \text{if } \sigma(g) = 0 = \sigma(g'); \\ \chi(y; g)\Delta(x'; g') & \text{if } \sigma(g) = 1 \neq \sigma(g'); \\ 0 & \text{otherwise;} \end{cases}$$

$$(iii)\ \ \Delta(y \boxtimes^- y'; (g, g')) = \begin{cases} 0 & \text{if } \sigma(g) = 0 = \sigma(g'); \\ -2i\chi(y; g)\chi(y'; g') & \text{if } \sigma(g) = 1 = \sigma(g'). \end{cases}$$

Proof. By linearity, it suffices to prove these when x, x', y and y' are actual representations. Each one follows from the definition of $\boxtimes^-$ (after Theorem A8.4) by a brief argument similar to the calculation of characters in Theorem 5.6 (especially (ii) and (ii)$'$, which are equivalent because of Proposition A8.7). Here are the details for (iii).

Recall that $(V \boxtimes^- V')^{(t)}$ is generated by elements $\nabla^t(v, v')$. When $\sigma(g) = 0 = \sigma(g')$, the formula for the action reduces to

$$(g, g')\nabla^t(v, v') = \nabla^t(gv, g'v');$$

thus

$$\chi\big((V\boxtimes^{-}V')^{(0)};(g,g')\big)=\chi\big((V\boxtimes^{-}V')^{(1)};(g,g')\big),$$

as required. When $\sigma(g)=1=\sigma(g')$, we have

$$(g,g')\nabla^{t}(v,v')=(-1)^{t+1}i\nabla^{t}(gv,g'v').$$

Thus

$$\chi\big((V\boxtimes^{-}V')^{(1)};(g,g')\big)=i\chi(V;g)\chi(V';g')=-\chi\big((V\boxtimes^{-}V')^{(0)};(g,g')\big),$$

which yields the stated formula.

Now define elements of $\oplus\, T^{*}\tilde{S}_n$ as follows (these are not power sum functions, of course):

$$p_{2\ell} \;:=\; (1-\rho)c_{2\ell}\in T^{1}\tilde{S}_{2\ell};$$

$$p_{2\ell+1} \;:=\; (1+\ell\kappa^{2})c_{2\ell+1}+\kappa\sum_{j=1}^{\ell}(-1)^{j}(2\ell-2j+1)c_{2\ell-j+1}c_j\in T^{0}\tilde{S}_{2\ell+1}.$$

For any partition $\alpha=(\alpha_1\geq\alpha_2\geq\ldots>0)$, define p_α to be $p_{\alpha_1}p_{\alpha_2}\cdots$.

Proposition A12.2. *The set* $\{p_\alpha\otimes 1:\alpha\in\mathcal{P}\}$ *generates* $\oplus T^{*}\tilde{S}_n\otimes\mathbb{Q}$ *over* $K\otimes\mathbb{Q}$. *Equivalently, for each* u *in* $T^{*}\tilde{S}_n$, *there is a positive integer* j *such that* ju *is in the* K*-submodule spanned by* $\{p_\alpha:\alpha\in\mathcal{P}\}$.

Proof. If this holds for u_1 and for u_2, then it certainly holds both for u_1+u_2 and u_1u_2. Thus one need only verify it when u is the basic Clifford irreducible c_n defined before Proposition A8.11. This may be done by induction on n, first observing, when $n=2\ell+1$, that $1+\ell\kappa^2$ is invertible in $K\otimes\mathbb{Q}$, and when $n=2\ell$, that

$$2c_{2\ell}-p_{2\ell}=(1+\rho)c_{2\ell}$$

is in the $K\otimes\mathbb{Q}$-subalgebra generated by $\{c_1,\ldots,c_{n-1}\}$.

Proposition A12.3. *For all* $x\in T^{i}\tilde{S}_k$ *and all* y,

$$p_n^{\perp}(xy)=p_n^{\perp}(x)y+\rho^{in+kn+i}xp_n^{\perp}(y).$$

Hence $\langle p_{a+b},xy\rangle=0$ *if* $x\in T^{*}\tilde{S}_a$ *and* $y\in T^{*}\tilde{S}_b$ *with* a *and* b *both positive.*

The argument is the same as that for Proposition A8.11 (iii), using that $\phi^*_{a,b}(p_{a+b}) = 0$ for positive a and b. The latter identity is a calculation, using the definition of p_n, Proposition A8.9, and the formula for $\phi^*_{i,j}(c_{i+j})$.

Proposition A12.4. *For all ℓ such that the subscripts are positive:*

$$(i)\ \langle p_{2\ell+1}, c_{2\ell+1}\rangle = 1; \qquad (ii)\ \langle p_{2\ell+1}, p_{2\ell+1}\rangle = 1+\ell\kappa^2;$$
$$(iii)\ \langle p_{2\ell}, c_{2\ell}\rangle = 1-\rho; \qquad (iv)\ \langle p_{2\ell}, p_{2\ell}\rangle = 2-2\rho.$$

Proof. For formula (i),

$$\begin{aligned}
&\langle p_{2\ell+1}, c_{2\ell+1}\rangle \\
&\quad = c^{\perp}_{2\ell+1}(p_{2\ell+1}) \\
&\quad = (1+\ell\kappa^2)c^{\perp}_{2\ell+1}(c_{2\ell+1}) \\
&\qquad + \kappa\sum_{i=1}^{\ell}(-1)^i(2\ell-2i+1)\kappa c^{\perp}_{2\ell-i+1}(c_{2\ell-i+1})c^{\perp}_i(c_i) \\
&\qquad \text{(using Proposition A8.11 (iii))} \\
&\quad = 1 + \kappa^2\left(\ell + \sum_{i=1}^{\ell}(-1)^i(2\ell-2i+1)\right) = 1.
\end{aligned}$$

Formula (ii) is immediate from (i) and the previous proposition. The last two formulae are trivial.

In Appendix 6, s_n was the element $t_1t_2...t_{n-1}$ of $\tilde{S}_n$. In what follows, s_n can be this or any other fixed element which projects to an n-cycle in S_n. Now define numbers

$$\begin{aligned}
\varepsilon_{2k+1} &:= \chi(\kappa c_{2k+1}; s_{2k+1}), \\
\zeta_{2k+1} &:= \Delta(c_{2k+1}; s_{2k+1}), \\
\zeta_{2k} &:= \chi(c_{2k}; s_{2k}).
\end{aligned}$$

These three numbers were in fact calculated in Appendix 6, but we shall not substitute the values until later, in order to keep the formulae simple, and to emphasize the independence of these formulae from particular trace calculations.

If $j \in \mathbf{Z}$ and $\omega \in K^{(0)}$, let $\omega_{\rho=j}$ denote the image of ω under the group homomorphism $K^{(0)} \to \mathbf{Z}$ which maps 1 to 1 and ρ to j.

Proposition A12.5. *For all k,*

$$\begin{aligned}
\chi(y; s_{2k+1}) &= \varepsilon_{2k+1} < y, \kappa p_{2k+1} >_{\rho=0} && \textit{for } y \in T^1\tilde{S}_{2k+1};\\
\chi(y; s_{2k}) &= \zeta_{2k} < y, p_{2k} >_{\rho=0} && \textit{for } y \in T^1\tilde{S}_{2k};\\
\Delta(x; s_{2k+1}) &= \zeta_{2k+1} < x, p_{2k+1} >_{\rho=-1} && \textit{for } x \in T^0\tilde{S}_{2k+1}.
\end{aligned}$$

Proof. As functions of y (or x), both sides of each equation are additive, so we may assume y (or x) is in the $\mathbf{Z}$-basis $\{\mu c_\lambda : \lambda \in \mathcal{D}; \mu = 1, \rho \text{ or } \kappa\}$. Using Proposition A12.4, we see that both sides are zero when λ has length greater than 1. This is because an induced character vanishes on any element outside the conjugates of the subgroup from which one is inducing. When λ has length 1, Proposition A12.4 and the identities before Proposition A12.1 give the result.

Corollary A12.6. *For all* ℓ,

$$\begin{aligned}
\chi(\kappa p_{2\ell+1}; s_{2\ell+1}) &= (2\ell+1)\varepsilon_{2\ell+1};\\
\chi(p_{2\ell}; s_{2\ell}) &= 2\zeta_{2\ell};\\
\Delta(p_{2\ell+1}; s_{2\ell+1}) &= \zeta_{2\ell+1};
\end{aligned}$$

and all left sides above vanish if p_n *is changed to* p_α *with* α *of length greater than* 1.

Now we come to the central point, namely, formulae for reducing any character on $\tilde{S}_m$ to a character on $\tilde{S}_n$ for some $n < m$.

Theorem A12.7. *For all* $g \in \tilde{S}_n$, $x \in T^0\tilde{S}_n$ *and* $y \in T^1\tilde{S}_n$:

$$(i)\quad \chi(y;\,\phi_{2\ell+1,n}(s_{2\ell+1},\,g)) \;=\; \begin{cases} \varepsilon_{2\ell+1}\chi(p^{\perp}_{2\ell+1}(y);\,g) & \textit{if}\ \sigma(g)=0; \\ \zeta_{2\ell+1}\chi(p^{\perp}_{2\ell+1}(y);\,g) & \textit{if}\ \sigma(g)=1; \end{cases}$$

$$(ii)\quad \chi(y;\,\phi_{2\ell,n}(s_{2\ell},g)) \;=\; \begin{cases} \zeta_{2\ell}\Delta(p^{\perp}_{2\ell}(y);\,g)/2 & \textit{if}\ \sigma(g)=0; \\ 0 & \textit{if}\ \sigma(g)=1; \end{cases}$$

$$(iii)\quad \Delta(x;\,\phi_{2\ell+1,n}(s_{2\ell+1},\,g)) \;=\; \zeta_{2\ell+1}\Delta(p^{\perp}_{2\ell+1}(x);\,g) \quad (\textit{where}\ \sigma(g)=0);$$

$$(iv)\quad \Delta(x;\,\phi_{2\ell,n}(s_{2\ell},\,g)) \;=\; -i\zeta_{2\ell}\chi(p^{\perp}_{2\ell}(x);\,g) \quad (\textit{where}\ \sigma(g)=1).$$

Proof. Let $u \in T^*(\tilde{S}_{2\ell+1+n})$, and use Proposition A12.2 to find an integer k so that we can write

$$k\phi^*_{2\ell+1,n}(u) = \sum p_\alpha \boxtimes^- u_\alpha\,,$$

(summation over all partitions α of $2\ell+1$). Then

$$\begin{aligned} <kp^{\perp}_{2\ell+1}(u),\,v> &= <ku,\,(\phi_{2\ell+1,n})_*(p_{2\ell+1}\boxtimes^- v)> \\ &= \sum_\alpha <p_\alpha,\,p_{2\ell+1}> <u_\alpha,\,v> \\ &= <p_{2\ell+1},\,p_{2\ell+1}> <u_{(2\ell+1)},\,v>. \end{aligned}$$

Thus

$$kp^{\perp}_{2\ell+1}(u) = (1+\ell\kappa^2)u_{(2\ell+1)}.$$

Similarly, if $u' \in T^*\tilde{S}_{2\ell+n}$ and

$$k'\phi^*_{2\ell,n}(u') = \sum p_\beta \boxtimes^- u_\beta{}',$$

summation over partitions β of 2ℓ, we obtain

$$k'p^{\perp}_{2\ell}(u') = (2-2\rho)u'_{(2\ell)}.$$

To prove (i), take u above to be y. We have

$$\chi\big(y;\ \phi_{2\ell+1,n}(s_{2\ell+1},g)\big) = \chi\big(\phi^*_{2\ell+1,n}(y);\ (s_{2\ell+1},g)\big)$$

$$= k^{-1}\sum_{\alpha}\chi\big(p_\alpha \boxtimes \bar{\ } y_\alpha;\ (s_{2\ell+1},g)\big).$$

Now use Proposition A12.1 and Corollary A12.6. When $\sigma(g) = 0$, the above expression becomes

$$k^{-1}\left(\sum_{\alpha\ \text{even}}\chi(\kappa p_\alpha;\ s_{2\ell+1})\chi(y_\alpha;\ g) + \sum_{\alpha\ \text{odd}}\chi(p_\alpha;\ s_{2\ell+1})\chi(\kappa y_\alpha;\ g)\right)$$

$$= k^{-1}(2\ell+1)\ \varepsilon_{2\ell+1}\ \chi(\kappa y_{(2\ell+1)};\ g).$$

On the other hand, when $\sigma(g) = 1$, the same expression becomes

$$k^{-1}\left(\sum_{\alpha\ \text{even}}\Delta(p_\alpha;\ s_{2\ell+1})\chi(y_\alpha;\ g) + \sum_{\alpha\ \text{odd}}0\right) = k^{-1}\zeta_{2\ell+1}\chi(y_{(2\ell+1)};\ g).$$

But since

$$kp^{\perp}_{2\ell+1}(y) = (\ell+1+\ell\rho)y_{(2\ell+1)},$$

and

$$\chi(\rho y;\ g) = (-1)^{\sigma(g)}\chi(y;\ g),$$

we find that both

$$k^{-1}(2\ell+1)\chi(y_{(2\ell+1)};\ g)\quad (\text{when } \sigma(g) = 0)\ ,$$

and

$$k^{-1}\chi(y_{(2\ell+1)};\ g)\quad (\text{when } \sigma(g) = 1)\ ,$$

reduce to

$$\chi(p^{\perp}_{2\ell+1}(y);\ g),$$

as required.

In the proof of (ii), take u' above to be y. The proof is exactly as for (i).

To prove (iii), take u to be x; whereas for (iv), take u' to be x. In both cases, the details are closely analogous to the proof of (i).

To exploit Theorem A12.7, we need some information about $p_r^{\perp}(a_\lambda)$, where

a_λ is the special irreducible in Theorem A8.14. For this, it is useful to define the operators $\mathcal{A}_\ell$ also for $\ell \leq 0$, as follows:

$$\mathcal{A}_0(x) := x + \kappa^2 \sum_{i>0} (-1)^i c_i c_i^\perp(x),$$

$$\mathcal{A}_{-k}(x) := (-1)^k c_k^\perp(x) + \kappa \sum_{i>k} (-1)^i c_{i-k} c_i^\perp(x) \text{ for } k > 0.$$

Recall

$$\mathcal{A}_\ell(x) := c_\ell x + \kappa \sum_{i>0} (-1)^i c_{i+\ell} c_i^\perp(x) \text{ for } \ell > 0.$$

Lemma A12.8. *For all integers n:*

$$(i)\ \ p_{2\ell+1}^\perp \mathcal{A}_n - \rho^n \mathcal{A}_n p_{2\ell+1}^\perp = \begin{cases} \kappa \mathcal{A}_{n-2\ell-1} & \text{if } n \neq 0 \text{ or } 2\ell+1; \\ \mathcal{A}_0 & \text{if } n = 2\ell+1; \\ \kappa^2 \mathcal{A}_{-2\ell-1} & \text{if } n = 0; \end{cases}$$

$$(ii)\ \ p_{2\ell}^\perp \mathcal{A}_n - \rho^{n+1} \mathcal{A}_n p_{2\ell}^\perp = \begin{cases} 0 & \text{if } n \neq 2\ell; \\ (1-\rho)\mathrm{Id} & \text{if } n = 2\ell. \end{cases}$$

Proof. Using Proposition A12.3 and the formula

$$p_r^\perp(c_n) = \begin{cases} \kappa c_{n-r} & \text{if } r < n \text{ and } r \text{ is odd;} \\ 0 & \text{if } r < n \text{ and } r \text{ is even;} \\ 1 & \text{if } r = n \text{ is odd;} \\ 1-\rho & \text{if } r = n \text{ is even,} \end{cases}$$

these are straightforward calculations similar to Proposition 9.3.

Now define elements a_β, for any sequence $\beta = (\beta_1,\ldots,\beta_s)$ of integers, just as a_λ was defined for strict partitions λ:

$$a_\beta := \mathcal{A}_{\beta_1} \ldots \mathcal{A}_{\beta_s}(1).$$

Then the following corollary is proved by induction on s.

Corollary A12.9. *Let λ be a strict partition.*

(*i*) *Let $\beta(i)$ be the sequence obtained from λ by subtracting $2\ell+1$ from λ_i. Then*

$$p^{\perp}_{2\ell+1}(a_\lambda) = \sum_{i:\lambda_i \neq 2\ell+1} \kappa a_{\beta(i)} + \sum_{i:\lambda_i = 2\ell+1} \rho^{\lambda_1+\ldots+\lambda_{i-1}} a_{\beta(i)}.$$

(*ii*) *Letting $\hat{\lambda}_i$ mean "omit λ_i"*

$$p^{\perp}_{2\ell}(a_\lambda) = \sum_{i:\lambda_i = 2\ell} \rho^{\lambda_1+\ldots+\lambda_{i-1}+i-1}(1-\rho)a_{\lambda_1,\ldots,\hat{\lambda}_i,\ldots,\lambda_s}.$$

Note that two of the summations are over a singleton or empty set. In particular, $p^{\perp}_{2\ell}(a_\lambda) = 0$ unless 2ℓ is a part of λ.

If g is an element of $\tilde{S}_k$ which projects to a permutation of cycle type $(k_1, \ldots, k_\ell)$ in S_k, then either g or zg (or both) are conjugate to $\phi_{k_1,\ldots,k_\ell}(s_{k_1}, \ldots, s_{k_\ell})$ in $\tilde{S}_k$. The formulae in A12.7 and A12.9 now have several corollaries.

(I) *If λ is even, and g projects to a cycle type with at least one even part, then $\chi(\kappa a_\lambda; g) = 0$.*

This is immediate from Theorem A12.7 (ii), and Corollary A12.9 (ii).

(II) *Suppose that g projects to a cycle type which is different from λ, and which has at least one even part. Then $\chi(a_\lambda; g) = 0$ if λ is odd, and $\Delta(a_\lambda; g) = 0$ if both λ and g are even.*

This follows by induction on the length of λ, using (ii) and (iv) of Theorem A12.7, and the fact that $p^{\perp}_{2\ell}(a_\lambda)$ is zero if 2ℓ is not a part of λ.

(III) *Suppose λ is even, and g projects to a cycle type which is different from λ, and which has entirely odd parts. Then $\Delta(a_\lambda; g) = 0$. (For odd λ, obviously $\Delta(\kappa a_\lambda; g) = 0$.)*

Using (iii) of Theorem A12.7, this follows by induction on the length of λ, making use of the fact that $p^{\perp}_{2\ell+1}(a_\lambda)$ has the form κb if $2\ell+1$ is not a part of λ.

(IV) *Let $\lambda = (\lambda_1, \ldots, \lambda_\ell)$ be a strict partition. Let $g = \phi_{\lambda_1,\ldots,\lambda_\ell}(s_{\lambda_1}, \ldots, s_{\lambda_\ell})$, so that g projects to cycle type λ. Define*

$$f(\lambda) = (-1)^A(-2i)^B\zeta_{\lambda_1}\ldots\zeta_{\lambda_\ell},$$

where

a = *the number of odd parts in* λ; b = *the number of even parts in* λ;
$A = a(a-1)/2$; B = *integer part of* $(b/2)$.

Then

$$f(\lambda) = \begin{cases} \chi(a_\lambda;\, g) & \text{if } \lambda \text{ is odd;} \\ \Delta(a_\lambda;\, g) & \text{if } \lambda \text{ is even.} \end{cases}$$

This again proceeds inductively. Defining $F(\lambda)$ to be the right side of the last formula, one obtains, using Theorem A12.7, that $F(\lambda)$ is

$$\begin{cases} (-1)^{\lambda_1+\ldots+\lambda_{\ell-1}}\zeta_{\lambda_\ell} F(\lambda_1, \ldots, \lambda_{\ell-1}) & \begin{cases} \lambda_\ell \text{ odd, } \lambda \text{ even, using (iii);} \\ \lambda_\ell \text{ odd, } \lambda \text{ odd, using (i);} \end{cases} \\ \zeta_{\lambda_\ell} F(\lambda_1, \ldots, \lambda_{\ell-1}) & \lambda_\ell \text{ even, } \lambda \text{ odd, using (ii);} \\ -2i\zeta_{\lambda_\ell} F(\lambda_1, \ldots, \lambda_{\ell-1}) & \lambda_\ell \text{ even, } \lambda \text{ even, using (iv).} \end{cases}$$

These recursions are easily solved to give $F = f$.

(V) *Let* $g = \phi_{k_1,\ldots,k_\ell}(s_{k_1}, \ldots, s_{k_\ell})$ *where all* k_i *are odd, and* $k = \sum k_i$. *Then*

$$\chi(c_k;\, g) = 2^{(\ell-2)/2}\, \varepsilon_{k_1}\ldots\varepsilon_{k_\ell} \quad \text{if } k \text{ is even;}$$

$$\chi(\kappa c_k;\, g) = 2^{(\ell-1)/2}\, \varepsilon_{k_1}\ldots\varepsilon_{k_\ell} \quad \text{if } k \text{ is odd.}$$

This is immediate, using Theorem A12.7 (i), by induction on ℓ.

As mentioned earlier, the numbers ζ and ε have already been calculated. Proposition A6.5 yields

$$\varepsilon_{2\ell+1} = 1;\ \zeta_{2\ell+1} = \pm i^\ell\sqrt{2\ell+1};\ \zeta_{2\ell} = \pm i^\ell\sqrt{\ell}.$$

Substituting these into **(IV)** and **(V)** above, we see that **(I)** to **(V)** amount to reproofs of all the character calculations in Theorems 6.6, 6.7, 6.8, 6.9, 6.10, 6.11, 6.12 and 8.7.

To show how parts (i) of Theorem A12.7 and of Corollary A12.9 contain essentially Theorem 10.1, we need to express the right side of Corollary A12.9 (i) in terms of irreducibles. (Note that **(V)** is a consequence of Theorem 10.1, though this would be circular in the earlier treatment.)

Lemma A12.10. *The following composition formulae hold*:

(*i*) $\mathcal{A}_0\mathcal{A}_0 = \mathrm{Id}$.

(*ii*) $\mathcal{A}_n\mathcal{A}_0 + \rho\mathcal{A}_0\mathcal{A}_n = 0$ *if* $n \neq 0$.

(*iii*) $\mathcal{A}_{-n}\mathcal{A}_n + \mathcal{A}_n\mathcal{A}_{-n} = (-1)^n\mathrm{Id}$ *if* $n \neq 0$.

(*iv*) $\mathcal{A}_k\mathcal{A}_\ell + \rho^{k+\ell}\mathcal{A}_\ell\mathcal{A}_k = 0$ *if none of* k, ℓ, *or* $k+\ell$ *is zero.*

(*v*) $\mathcal{A}_k\mathcal{A}_k = 0$ *if* $k \neq 0$.

Each of these can be proved directly from definitions in a similar manner to the proof of Theorem 9.1. However, it is easier to deduce most of them from the case of (iv) in which $k > 0$, using the following inductive principle: If V is a K-linear operator on $\oplus T^*\tilde{S}_n$ for which $V(1) = 0$, and if $V(x) = 0$ implies $V\mathcal{A}_k(x) = 0$ for all $k > 0$, then V is zero. This is clear because the K-basis $\{a_\lambda\}$ is obtained by iterating the operators $\mathcal{A}_k$ on 1. To deduce (iii), for example, take

$$V = \mathcal{A}_{-n}\mathcal{A}_n + \mathcal{A}_n\mathcal{A}_{-n} + (-1)^{n+1}\mathrm{Id}\,.$$

Note also that (v) follows immediately from (iv). Though similar to Theorem 9.1, the proof of (iv) when $k > 0$ is somewhat more tedious. It separates into two cases: $0 > \ell \neq -k$ and $\ell > 0$.

Recall that all the terms in Corollary A12.9 had a factor $a_{k_1,\ldots,k_i-r,\ldots,k_s}$. The following is a slight sharpening of Theorem 9.2, in that the powers of ρ could not be recovered directly from the latter.

Proposition A12.11. *If* r *is an odd positive integer, and* $k_1 > \ldots > k_s > 0$, *then* $a_{k_1,\ldots,k_i-r,\ldots,k_s} = a'$ *(say) may be re-expressed in terms of irreducibles according to the following five cases* ($i \in \mathcal{I}_j$, $0 \leq j \leq 4$):

$\mathcal{I}_0 = \{i : k_i > r$ *and* $k_i - r = k_u$ *for some* $u\}$ *gives* $a' = 0$;

$\mathcal{I}_1 = \{i : k_i > r$ *and* $k_j < k_i - r < k_{j+1}$ *for some* $j\}$

(*where* $k_{s+1} = 0$)

gives $a' = (-1)^{j-i}\rho^{N_1}a_{k_1,\ldots,\hat{k}_i,\ldots,k_j,k_i-r,k_{j+1},\ldots,k_s}$,

where $\hat{k}_i$ *means* "*omit* k_i," , *and*

$N_1 = (j-i)(k_i-r) + k_{i+1} + \ldots + k_j$;

$\mathcal{I}_2 = \{i : k_i = r\}$ *gives* $a' = (-1)^{s-i}a_{k_1,\ldots,\hat{k}_i,\ldots,k_s}$;

$\mathcal{I}_3 = \{i : k_i < r \text{ and } k_i + k_j = r \text{ for some } j > i\}$

$$\text{gives } a' = (-1)^{j-i} \rho^{N_3} a_{k_1,\ldots,\hat{k}_i,\ldots,\hat{k}_j,\ldots,k_s},$$

$$\text{where } N_3 = (j-i-1)(k_i - r) + k_{i+1} \ldots + k_{j-1};$$

$\mathcal{I}_4 = \{i : k_i < r \text{ and } k_i + k_j \neq r \text{ for all } j > i\}$ gives $a' = 0$.

Proof. In each case, move $k_i - r$ to the right, past other subscripts, using (iv) or (ii) of Lemma A12.10. If $k_i - r > 0$, use (iv). Either move it until reaching an equal entry, giving 0 by (v) of Lemma A12.10; or until reaching a smaller entry, giving the formula in case $\mathcal{I}_1$. If $k_i - r = 0$, use (ii). Move it to the right-hand end and delete it, giving case $\mathcal{I}_2$. If $k_i - r < 0$, use (iv). Either move it till reaching an entry $r - k_i$, then apply A12.10 (iii), giving the formula in case $\mathcal{I}_3$ after applying the next argument to the other term; or move it to the right hand end, giving the zero of case $\mathcal{I}_4$.

Applying the results A12.9 and A12.11 to the formula A12.7 (i) now reproves Theorem 10.1.

We shall conclude this appendix by displaying the formula for $c_k^{\perp}(a_\lambda)$ which produces the expressions for the a_λ as linear combinations of c_μ. This is done by substituting the formula into the definition of a_λ and by induction on the length of λ.

Firstly, in analogy with Proposition 9.3, a direct calculation with the definitions of the operators $\mathcal{A}_\ell$ yields the following for all $n > 0$ and $k > 0$:

$$c_k^{\perp} \mathcal{A}_n - \rho^{n+k+1} \mathcal{A}_n c_k^{\perp}$$
$$= \kappa^2 \sum_{\substack{0<j<k \\ j \neq n}} \mathcal{A}_{n-j} c_{k-j}^{\perp} + \begin{cases} \kappa \mathcal{A}_{n-k} & \text{if } n > k; \\ \mathcal{A}_0 & \text{if } n = k; \\ \kappa \mathcal{A}_0 c_{k-n}^{\perp} + \kappa \mathcal{A}_{n-k} & \text{if } 0 < n < k. \end{cases}$$

From this, in analogy to Proposition 12.1, we have the following formula. Assume $k > 0$ and $n_1 > n_2 > 0$. Interpret $c_0^{\perp}$ as 1, and $c_\ell^{\perp}$ as 0 for $\ell < 0$. Then

$$c_k^{\perp}\mathcal{A}_{n_1}\mathcal{A}_{n_2}$$

$$= \rho^{n_1+k+1}\mathcal{A}_{n_1}c_k^{\perp}\mathcal{A}_{n_2} + \kappa^2 \sum_{0<k_1<\min(k,n_1-n_2)} \mathcal{A}_{n_1-k_1}c_{k-k_1}^{\perp}\mathcal{A}_{n_2}$$

$$+ \mathcal{A}_{n_2}c_{k-n_1+n_2}^{\perp}\mathcal{A}_{n_2} + \kappa\mathcal{A}_{n_1-k}\mathcal{A}_{n_2}\delta_{0,\lceil k/(n_1-n_2)\rceil}.$$

Now an induction on s proves the desired formula, which we describe using the combinatorial definitions in Chapter 12. Let X be the set of all zig-zag strips Z of size k in $S(\lambda)$ for which $S(\lambda) \setminus Z = S(\mu(Z))$ for some strict partition $\mu(Z)$. For each Z in X, define

$$N(\lambda, Z) = 2fr(\lambda, Z) - \delta_{\alpha_\ell,\lambda_\ell} - 1,$$

where $\lambda = (\lambda_1, ..., \lambda_\ell)$ and Z has α_i nodes in the ith row. It follows that $N(\lambda, Z) > 0$, except when $k = \lambda_j$ for some j and

$$(\alpha_1, ..., \alpha_\ell) = (0, ..., 0, \lambda_j-\lambda_{j+1}, ..., \lambda_{\ell-1}-\lambda_\ell, \lambda_\ell).$$

(Equivalently, $\mu(Z)$ is obtained by deleting λ_j from λ.) In the exceptional case, $N(\lambda, Z) = 0$ and we define

$$M(\lambda, Z) = (j-1)(\lambda_j+1) + \lambda_1 + \lambda_2 + ... + \lambda_{j-1}.$$

Then

$$c_k^{\perp}(a_\lambda) = \sum_Z \rho^{M(\lambda, Z)}\kappa^{N(\lambda,Z)}a_{\mu(Z)},$$

summation over all Z in the set X above.

Note that the value of $M(\lambda, Z)$ is irrelevant when $N(\lambda, Z) > 0$.

Note. The proof above of Theorem 10.1 appears in Hoffman (1988). This approach to characters is similar to the proof of the Murnaghan–Nakayama formula in Zelevinsky (1979), pp. 64 and 91.

13

THE SHIFTED KNUTH ALGORITHM

This chapter will be devoted entirely to giving a second, purely combinatorial, proof of Theorem 12.11. We describe a number of algorithms and their inverses, and combine them to provide an explicit bijection. This material is the backbone of the combinatorial approach to Q-functions. We shall make one further foray into this side of the subject, in the second half of the next chapter. The reader should consult the papers referred to in the notes to this and the next chapter for further details, as well as for alternative methods.

First let us recall the definition of a marked shifted tableau and introduce some notation for manipulating such an object. Adjoin ∞ to P, giving a totally ordered set

$$P \cup \{\infty\} = \{1' < 1 < 2' < 2 < \ldots < \infty\}.$$

Given a strict partition λ, a *marked shifted tableau of shape* λ is an indented doubly infinite array

$$\begin{array}{cccccc} T[1,1] & T[1,2] & T[1,3] & \cdot & \cdot & \cdot \\ & T[2,2] & T[2,3] & \cdot & \cdot & \cdot \\ & & T[3,3] & \cdot & \cdot & \cdot \\ & & & \cdot & & \\ & & & & \cdot & \\ & & & & & \cdot \end{array}$$

where each $T[i,j]$ is in $P \cup \{\infty\}$, and such that the following three conditions hold:

(T1) $T[i,j] = \infty$ if either $j \geq i+\lambda_i$ with $1 \leq i \leq \ell$, or $j \geq i > \ell$. Otherwise $T[i,j]$ is finite.

(T2) The entries show weak increase along each row and down each column, so that, when defined,

$$T[i,j] \leq T[i,j+1] \quad \text{and} \quad T[i,j] \leq T[i+1,j].$$

(T3) Any given unmarked letter occurs at most once in each column whereas marked letters occur at most once in each row.

When writing a marked shifted tableau of shape λ, it is customary to omit the ∞'s and only write the finite part of the tableau, as in the examples in Chapters 10 and 12.

Notation. Given a letter $p \in P$, denote by $|p|$ the unmarked number corresponding to p. For example, $|3'| = |3| = 3$.

Remark. For each integer k, the set of positions at which the entries of a marked shifted tableau are either k or k' is, by definition, a zig-zag strip. Since any north-west to south-east diagonal of T intersects a zig-zag strip in at most one node,

$$|T[i,j]| < |T[i+1,j+1]| \; .$$

Thus even after removing marks, the entries of a marked shifted tableau are strictly increasing along each diagonal.

We can now state the main result of this chapter. It was proved independently by Sagan (1987) and Worley (1984), and is a restatement of Theorem 12.11 with effectiveness added.

Theorem 13.1. *There is an effective bijective correspondence between matrices $A = (a_{ij})$ over $P \cup \{0\}$ and ordered pairs (S, T) of marked shifted tableaux of the same shape, such that T has no marked letters on its main diagonal. The correspondence has the property that $\sum_i a_{ij}$ is the number of entries s of S for which $|s| = j$, whereas $\sum_j a_{ij}$ is the number of entries t of T for which $|t| = i$.*

The proof will be divided into steps which we shall now summarize to give an overview of the strategy involved. In Step 1, two algorithms BUMP and INSERT are described and examples of their use given. Step 2 is concerned with a further algorithm SHIFTED_KNUTH. It is then possible to show how to produce (S, T) from a given A. Most of Step 3 is devoted to a proof that the arrays so produced are indeed marked shifted tableaux. These three steps constitute half the proof. In Steps 4 and 5, algorithms BUMPOUT, DELETE and SHIFTED_KNUTH_INV are discussed. The proof is completed by checking that these are inverses of BUMP, INSERT and SHIFTED_KNUTH respectively.

Proof of Theorem 13.1: Step 1. The first algorithm is the process of bumping. This is a procedure which takes a weakly increasing vector v over $P \cup \{\infty\}$ and bumps a given element x of P into v as follows. Scan the entries of v, from left to right, to find the first which is larger than x. This entry of v is then replaced by x and we remove (or "bump") this entry from v. If x is greater than or equal to each entry of v, we extend v by placing x to the right of all finite entries and remove ∞. The output of BUMP is thus a vector over $P \cup \{\infty\}$ together with an element of $P \cup \{\infty\}$.

Note that the vector w obtained from bumping x into v is also weakly increasing. (In fact, the algorithm BUMP works perfectly well on any vector over $P \cup \{\infty\}$. The only reason for restricting to weakly increasing vectors is to be able to construct an inverse for the algorithm.)

Examples. Take v to be the vector

$$1'\ 1\ 2'\ 2\ 2\ 3\ 6.$$

Bumping 3 into this gives

$$1'\ 1\ 2'\ 2\ 2\ 3\ 3.$$

removing 6. Bumping 8 into v gives

$$1'\ 1\ 2'\ 2\ 2\ 3\ 6\ 8.$$

which extends v (that is, removes ∞).

An algorithm closely related to BUMP is EQBUMP. When x is inserted into v using EQBUMP, v is extended by adding x if it is greater than each entry of v. Otherwise x removes the leftmost entry of v which is greater than or equal to x.

We now come to the algorithm INSERT, which is the central one studied in this chapter and the next. Given an element x of P and a marked shifted tableau T, we INSERT x into T by a sequence of bumps applied to rows and (possibly) columns of T. When bumping y into v:

use BUMP, if y is unmarked and v is a row;
use BUMP, if y is marked and v is a column;
use EQBUMP, if y is unmarked and v is a column; and
use EQBUMP, if y is marked and v is a row.

In summary, one uses that replacement which removes the first possible entry consistent with maintaining the row (or column) as a possible row (or column) of a marked shifted tableau. Thus the row (or column) continues to weakly increase and to have no repeated marked integers (or unmarked integers in the case of columns). Insertion proceeds by bumping first $x = x_1$ into the top row to remove x_2. Then bump x_2 into the second row to remove x_3. Continue in this way until one of the following two possibilities occurs.

The first possibility, known as an *unmixed* insertion, arises when a bump simply adds the element to the right hand end of a row. One ends the insertion at this point.

The second possibility, a *mixed* insertion, arises when a bump replaces an entry which is on the diagonal. There is a special situation to consider in this case when x_i bumps x_{i+1} from the diagonal, where $x_i = k'$ and $|x_{i+1}| = k$. We then make the following alterations:

(a) If $x_i = x_{i+1} = k'$, let the new element on the diagonal be k' (as expected), but use k (rather than k') for the next (column) bump.

(b) If $x_i = k'$ and $x_{i+1} = k$, lct thc new element on the diagonal be k (rather than k'), and use k (as expected) for the next (column) bump.

The procedure in (a) and (b) will be called UNMARK.

The insertion is then continued, in all mixed cases, by a sequence of column bumps. Suppose that the last row bumped was the ith, so that the diagonal element in position $[i, i]$ was replaced. The sequence starts by bumping from the $(i+1)$st column, then the $(i+2)$nd, and so on. The insertion is completed as soon as some bump simply adds the element to the bottom of a column.

An insertion generates a CELL, that is, a pair $[j, \ell]$ indicating the position at which the insertion of x into T terminates by removing ∞. It also generates a boolean function *unmix*, which is true or false according as the insertion is unmixed or mixed.

We now present some examples of this algorithm.

Example 1. Let us consider inserting 3 into

$$\begin{array}{ccccccc}
1' & 1 & 2' & 2 & 2 & 5' & 6 \\
 & 2' & 2 & 3' & 4 & 5 & \\
 & & 3' & 4 & 5 & & \\
 & & & 6 & 7' & & \\
 & & & & 7' & &
\end{array}$$

For this insertion

3 bumps $5'$ from row 1;

$5'$ bumps 5 from row 2;

5 bumps ∞,to become the fourth entry in row 4.

This gives the marked shifted tableau as shown:

$$\begin{array}{ccccccc}
1' & 1 & 2' & 2 & 2 & 3 & 6 \\
 & 2' & 2 & 3' & 4 & 5' & \\
 & & 3' & 4 & 5 & 5 & \\
 & & & 6 & 7' & & \\
 & & & & 7' & &
\end{array}$$

The CELL associated with this insertion is $[3, 6]$ and *unmix* is true.

Examples 2 and 3. We now give two simple examples to illustrate the use of UNMARK. First, consider the insertion of 1 into

$$\begin{array}{ccc} 1 & 1 & 2' \\ & 2 & \end{array}$$

1 bumps $2'$ from row 1;
$2'$ bumps 2 from row 2, but the new [2, 2] entry is 2, by (b);
2 bumps ∞, to become the second entry in column 2.

This gives

$$\begin{array}{ccc} 1 & 1 & 1 \\ & 2 & 2 \end{array}$$

with associated CELL being [2, 3] and *unmix* being false.

The second situation is illustrated by inserting 1 into

$$\begin{array}{ccc} 1 & 1 & 2' \\ & 2' & \end{array}$$

1 bumps $2'$ from row 1;
$2'$ bumps $2'$ from row 2, but the new bumping element is 2, by (a);
2 bumps ∞, to become the second entry in column 2.

This gives

$$\begin{array}{ccc} 1 & 1 & 1 \\ & 2' & 2 \end{array}$$

with CELL and *unmix* as above.

Note that, without UNMARK, this last example would not have produced a shifted tableau.

Example 4. As a final example, we insert $2'$ into the tableau of example 1. In this case,

$2'$ bumps $2'$ from row 1;
$2'$ bumps $2'$ from row 2, but continue with bumping element 2;
2 bumps 2 from column 3;
2 bumps 2 from column 4;
2 bumps 2 from column 5;
2 bumps $5'$ from column 6;
$5'$ bumps 6 from column 7;
6 bumps ∞ from column 8.

This gives

$$\begin{array}{cccccccc}
1' & 1 & 2' & 2 & 2 & 2 & 5' & 6 \\
 & 2' & 2 & 3' & 4 & 5 & & \\
 & & 3' & 4 & 5 & & & \\
 & & & 6 & 7' & & & \\
 & & & & 7' & & &
\end{array}$$

with CELL being [1,8] and *unmix* being false.

Proof of Theorem 13.1: Step 2. Suppose that we are given an $n \times m$ matrix A with entries in $P \cup \{0\}$. The matrix A can be encoded into a two-line notation E as follows: for $1 \leq i \leq n$ and $1 \leq j \leq m$, the pair $\begin{smallmatrix} i \\ j \end{smallmatrix}$ is repeated $|a_{ij}|$ times (with the top line weakly increasing). If a_{ij} was marked, then we mark the leftmost j in the pairs $\begin{smallmatrix} i \\ j \end{smallmatrix}$. Thus the matrix

$$A = \begin{pmatrix} 1' & 0 & 2 \\ 2 & 1 & 2' \\ 1' & 1' & 0 \end{pmatrix}$$

would be encoded as

$$E = \begin{matrix} 1 & 1 & 1 & 2 & 2 & 2 & 2 & 2 & 3 & 3 \\ 1' & 3 & 3 & 1 & 1 & 2 & 3' & 3 & 1' & 2' \end{matrix}.$$

It is clear that the matrix A can be recovered from E.

The algorithm to produce S and T is SHIFTED_KNUTH, described as follows. The array S is obtained by successively applying the algorithm INSERT to the elements in the lower line of E, in their natural (left-to-right) order, beginning with the empty tableau. The array T is obtained by first writing the upper line of E into the shifted diagram which has the shape of S, in the order in which the cells of S were created. Then mark exactly those entries whose cells were created by mixed insertions.

Example. We apply the procedure to the example

$$E = \begin{matrix} 1 & 1 & 1 & 2 & 2 & 2 & 2 & 2 & 3 & 3 \\ 1' & 3 & 3 & 1 & 1 & 2 & 3' & 3 & 1' & 2' \end{matrix}.$$

considered earlier. First S is produced in steps from the lower row of E as shown:

$$1' \qquad \begin{array}{cc} 1' & 3 \end{array} \qquad \begin{array}{ccc} 1' & 3 & 3 \end{array} \qquad \begin{array}{ccc} 1' & 1 & 3 \\ & 3 & \end{array} \qquad \begin{array}{ccc} 1' & 1 & 1 \\ & 3 & 3 \end{array}$$

$$\begin{array}{lll} 1'\ 1\ 1\ 2 & 1'\ 1\ 1\ 2\ 3' & 1'\ 1\ 1\ 2\ 3'\ 3 \\ \quad 3\ 3 & \quad 3\ 3 & \quad 3\ 3 \end{array}$$

$$\begin{array}{ll} 1'\ 1\ 1\ 1\ 2\ 3'\ 3 & 1'\ 1\ 1\ 1\ 2'\ 3'\ 3 \\ \quad 3\ 3 & \quad 2\ 3\ 3 \end{array}$$

Since the only mixed insertions are the last two, the tableau T is

$$\begin{array}{l} 1\ 1\ 1\ 2\ \ 2\ 2\ 3' \\ \quad 2\ 2\ 3' \end{array}$$

Proof of Theorem 13.1: Step 3. Notice that the construction of E ensures that the number of occurrences of s in S with $|s| = j$ is the number of p in the lower line of E with $|p| = j$. The latter is the sum of the entries in the jth column of A. Similarly, the number of entries t in T with $|t| = i$ is the ith row sum of A. Also, no entry on the main diagonal of T is marked, since a mixed insertion never creates a diagonal cell. It remains to show that S and T are indeed marked shifted tableaux.

Proposition 13.2. *The array S obtained by applying the algorithm SHIFTED_KNUTH to the lower line of the encoding of a matrix is always a marked shifted tableau.*

Proof. We shall prove more: after each bumping operation in an insertion, the result is a marked shifted tableau. The choices between BUMP and EQBUMP in the definition of INSERT were made in such a way that each row bump preserves the required properties of rows, and each column bump preserves the properties of columns. We therefore need to show both that a row bump cannot "ruin" any column, and that column bumps do not violate the row properties. The proofs are similar, so we shall give the first one. Suppose, for a contradiction, that the row bump operation bumps x into the $[i, j]$ position of a marked shifted tableau to produce an array R, and that the jth column of R violates (T2) or (T3). The violation clearly involves x. Let y be the entry $R[i+1, j]$. We cannot have that $x > y$ nor that x and y are equal and unmarked, because the same would have been true with x replaced by the element which it removed. Thus we must have that $i > 1$ and the violation arises because the entry $z = R[i-1, j]$ is either greater than x or is unmarked and equals x. The $(i-1)$st bumping operation removed x from, say, position $[i-1, k]$. We obtain a contradiction to the fact that the ith row of R is correctly ordered by observing that: (i) $R[i, k]$ is either larger than $R[i, j] = x$ or is marked and is equal to x; and (ii) $k < j$. The reason for (i) is that $R[i, k]$ is directly below x in the tableau to which the present bumping operation is applied to produce R. The reason for (ii) is the following: because $R[i-1, k]$ bumps x in the previous operation, we see that all entries

$R[i-1, \ell]$ with $\ell < k$ are strictly less than x and that $R[i-1, k] < x$ if x is unmarked. Thus $z \neq R[i-1, \ell]$ for any $\ell < k$. But $z = R[i-1, j]$. Thus $j > \ell$, as required. Note that case (b) of UNMARK is needed in checking that the first column bump of a mixed insertion does not "ruin" the row involved.

The last step in establishing the map from matrices to pairs of tableaux is to show that T is also a marked shifted tableau. The following result is needed in that proof.

Lemma 13.3. *Let R be a marked shifted tableau, and let x, y be in P. Let R' be $I_x(R)$, the tableau obtained by inserting x into R, and let R'' be $I_y(R')$. Suppose that $[i, j]$ and $[k, \ell]$ are the cells added to the boundary in passing from R to R' and from R' to R'' respectively. Assume $x < y$.*

(i) If the insertion of x is unmixed , then so is the insertion of y, and $j < \ell$.

(ii) If the insertion of y is mixed, then so is the insertion of x, and $k > i$.

Proof. (i) Suppose that in the insertion of x, the element x_1 is removed from row 1 by x, then x_2 removed from row 2 by x_1, ending with x_{i-1} being removed from row $(i-1)$ and inserted at the end of row i. It is easily seen that as p increases, the cell from which x_p is removed travels (weakly) to the left and downwards. Let $y_1, y_2, \ldots$ be the analogous entries for the insertion of y. Since $x < y$, we have $x_1 < y_1$ and the cell from which x_1 is removed is to the left of that from which y_1 is removed. Since $x_1 < y_1$, the same conclusion applies to the second row, and so on.

If the insertion of y terminates before the ith row is involved, then it is certainly unmixed and creates a cell in the kth row to the right of the element x_k, where $k < i$. Since x_k is weakly to the right of (and weakly above) x_{i-1}, the new cell created by the insertion of y is to the right of that created by the insertion of x; that is, $\ell > j$ as required.

On the other hand, suppose that the insertion of y reaches the stage of removing an element y from row $(i-1)$. Since $x_{i-1} < y_{i-1}$, the next bump simply places y_{i-1} at the end of row i, next to x_{i-1}. So $k = i$, and we again have an ummixed insertion with $\ell = j+1$. Thus $\ell > j$ as required.

(ii) The first part is just the contrapositive of the first part of (i). A version of the argument used in (i), which takes into account the column bumps in the insertion of x, shows that the cell of the main diagonal element removed in the last row bump of the insertion of x is above the corresponding main diagonal cell for y. In a mixed insertion, it is easily seen that the successive elements in the column bump portion of the insertion are from cells which move weakly upwards (and to the right). Now an argument very similar to that in (i) shows that the path of such cells in the insertion of x remains above the corresponding path for y. Inspecting the ends of these paths then yields the required inequality, $k > i$.

We can now prove that T is a marked shifted tableau.

Proposition 13.4. *Let T be the second array of the pair (S, T) obtained by applying the algorithm SHIFTED_KNUTH to an encoding E of a matrix. Then T is a marked shifted tableau with the same shape as that of S.*

Proof. It is part of the definition of SHIFTED_KNUTH that S and T have the same shape. We show by induction on n, the number of columns of E, that T is a marked shifted tableau. Suppose that E^- consists of the first $n-1$ columns of E, and let the shifted tableaux corresponding to E^- be (S^-, T^-). Let the last two columns of E be

$$\begin{pmatrix} y^- & y \\ x^- & x \end{pmatrix},$$

so that $S = I_x(S^-)$. If $y > y^-$, then y is greater than every entry in T^-. Since T^- is a marked shifted tableau by induction, placing y on the boundary of T^- to form T gives another marked shifted tableau. If $y = y^-$, then $x^- \leq x$. We can therefore apply Lemma 13.3 (ii) to see that, if I_x is mixed, then all previous insertions obtained from columns with y in the upper line must be mixed. Thus only marked y's occur in T^- and the y added by the last use of INSERT is in a cell in a lower row of T^- than that containing y^-. Thus the tableau properties (T2) and (T3) are preserved. If, however, I_x and I_{x^-} are both unmixed, then, by Lemma 13.3, an unmarked y is added to T^- in a position to the right of all other y's so that the tableau properties hold. Finally, if I_x is unmixed, but I_{x^-} is mixed, then the largest element in T^- is y^-, so that y can be added anywhere on the border without violating (T2) or (T3).

We have shown how to produce a pair of marked shifted tableaux from a matrix A. It remains to discuss how to recover A from S and T. This is done using "reversing" algorithms, which are discussed next, namely BUMPOUT, DELETE and SHIFTED_KNUTH_INV.

Proof of Theorem 13.1: Step 4. We now give algorithms which are inverses to those given in Step 1. Firstly, the algorithm BUMPOUT has as input a weakly increasing vector v and an element x of $P \cup \{\infty\}$ which is greater than some entry of v. It operates as follows.

Scan the entries of v from right to left to find the first which is less than x. If x is finite, this entry is replaced by x. If x is ∞, we contract v by removing its rightmost entry. The output of BUMPOUT is the new vector, together with the element removed.

Examples. Take v to be

$$1' \ 1 \ 2' \ 2 \ 2 \ 3 \ 3.$$

Apply BUMPOUT with $x = 6$. Since $6 > 3$, we obtain

$$1' \ 1 \ 2' \ 2 \ 2 \ 3 \ 6$$

and remove 3. Taking v to be

$$1' \ 1 \ 2' \ 2 \ 2 \ 3 \ 6 \ 8$$

and x to be ∞, we remove 8 and obtain

$$1' \ 1 \ 2' \ 2 \ 2 \ 3 \ 6.$$

Notice that the vectors produced by these two applications of BUMPOUT are the starting points for the two examples of BUMP we gave in Step 1 of the proof. This illustrates the fact that these two algorithms are mutually inverse.

Proposition 13.5. *The algorithms BUMP and BUMPOUT are mutual inverses.*
The proof of this is obvious.

There is an algorithm EQBUMPOUT (which is the inverse of EQBUMP), as follows. The algorithm EQBUMPOUT has as input a weakly increasing vector v over P and any element x of $P \cup \{\infty\}$ which is greater than or equal to some entry of v. It scans the entries of v from right to left to find the first which is less than or equal to x. If x is finite, this entry is replaced by x. If x is ∞, we contract v by removing its rightmost entry. The proof that EQBUMP is the inverse of EQBUMPOUT is also trivial.

We now come to the algorithm DELETE, which is the inverse of INSERT. This takes a marked shifted tableau T over P, a CELL on the boundary of T (so that the cell contains a finite entry, but the cells below and to the right of it contain ∞) and a boolean function *unmix*. The output of DELETE is a tableau $D(T)$, and an element y of P, determined as follows:

Within DELETE, when an element z alters a row or a column v:
use BUMPOUT, if z is unmarked and v is a row;
use EQBUMPOUT, if z is marked and v is a row;
use EQBUMPOUT, if z is unmarked and v is a column;
use BUMPOUT, if z is marked and v is a column.
In applying DELETE, first check the value of *unmix*.

(1) If *unmix* is true, apply BUMPOUT to ∞ and the row containing CELL, removing y_1. Since the diagonals are increasing, y_1 is larger than some element of the row directly above. Now apply BUMPOUT or EQBUMPOUT to that row, adding y_1 and removing y_2. Continue this (upwards) until reaching the first row. The output is then the element y removed from the first row, together with the tableau produced.

(2) If unmix is false, apply EQBUMPOUT to ∞ and the column containing CELL. We proceed, similarly to the previous paragraph, column by column, moving to the left. The process continues until we remove, from the $(j+1)$st column, say, an entry larger than all entries in the jth column. This must occur because of the increasing diagonals. The next step removes an element from the main diagonal at position (j, j).

At this point, the following modification, called MARK, is applied in the special case when z replaces w on the main diagonal, where z is unmarked and $|z| = |w|$.

(a) If $w = z$ is unmarked, the new main diagonal entry is z (as expected), but the process goes on to the next step using w' (rather than w).

(b) If $w = z'$ is marked, the new main diagonal entry is z' (rather than z), and the process continues to the next step using w (as expected).

This final removal from columns takes place on the main diagonal entry of the jth column. Then DELETE continues with a removal from the $(j-1)$st row, and is completed with a sequence of removals from rows, exactly as was done for the other value of *unmix*.

Remark. It may easily be seen that MARK is the inverse of the corresponding function UNMARK, which is part of the algorithm INSERT.

Example 1. We apply DELETE to the tableau

$$\begin{array}{ccccccc} 1' & 1 & 1 & 1 & 2' & 3' & 3 \\ & 2 & 3 & 3 & & & \end{array}$$

with *unmix* false and CELL being [2, 4]:
the 3 is removed from column 4;
then this 3 displaces 3 from the third column;

this 3 displaces 2 on the diagonal; and finally
2 displaces $2'$ from the first row.
Thus we have deleted $2'$ from T and $D(T)$ is

$$\begin{array}{ccccccc} 1' & 1 & 1 & 1 & 2 & 3' & 3 \\ & 3 & 3 & & & & \end{array}$$

Example 2. Take the marked shifted tableau

$$\begin{array}{cccccccc} 1' & 1 & 2' & 2 & 2 & 2 & 5' & 6 \\ & 2' & 2 & 3' & 4 & 5 & & \\ & & 3' & 4 & 5 & & & \\ & & & 6 & 7' & & & \\ & & & & 7' & & & \end{array}$$

with *unmix* false and CELL as [1, 8]. In this case,
6 is removed from the eighth column;
6 displaces $5'$ from the seventh column;
$5'$ displaces 2 from the sixth column;
2 displaces 2 from the fifth column;
2 displaces 2 from the fourth column;
2 displaces 2 from the third column;
2 displaces $2'$ from the diagonal, and the BUMPOUTing element is $2'$, but, by (b), the new diagonal element is $2'$;
$2'$ displaces $2'$ from the first row.
We obtain $2'$, and $D(T)$ is

$$\begin{array}{ccccccc} 1' & 1 & 2' & 2 & 2 & 5' & 6 \\ & 2' & 2 & 3' & 4 & 5 & \\ & & 3' & 4 & 5 & & \\ & & & 6 & 7' & & \\ & & & & 7' & & \end{array}$$

Notice that this is the tableau we began with in Example 3 of Step 1, illustrating the fact that the algorithm DELETE is the inverse of INSERT.

Proposition 13.6. *The array obtained by applying DELETE to a marked shifted tableau is another marked shifted tableau. Also the algorithms INSERT and DELETE are inverse to each other.*

Proof. The proof of the first claim is similar to that of Proposition 13.2. The second follows since BUMP, EQBUMP and UNMARK are the inverses of BUMPOUT, EQBUMPOUT and MARK respectively.

Proof of Theorem 13.1: Step 5. We now describe the algorithm SHIFTED_KNUTH_INV, which takes a pair (S, T) of marked shifted tableaux of the same shape, such that T has no marked diagonal entries, and produces a matrix over $P \cup \{0\}$. In fact, the algorithm produces an encoding E of such a matrix. This is achieved by successive applications of DELETE. The proof of Proposition 13.4 enables us to decide what value of CELL to use. To do this, inspect the positions in T at which the greatest entry, p, occurs. If p is unmarked, take CELL to be the location of the rightmost entry p and take *unmix* to be true. If p is marked, take CELL to be the lowest location of p and take *unmix* to be false. Then apply DELETE to S to produce y, say. The construction of E is begun by having a rightmost column with $|p|$ in the first line and y in the second. The procedure is then repeated on the pair (S^-, T^-), where S is the tableau obtained by this application of DELETE, and T^- is the tableau obtained by omitting from T the entry p in CELL. This produces the next-to-last column of E, and so on.

Example. Consider the case where

$$S = \begin{array}{ccccc} 1 & 1 & 1 & 3' & 3 \\ & 2' & 2 & & \end{array} \quad \text{and } T = \begin{array}{ccccc} 1 & 1 & 1 & 2' & 3' \\ & 2 & 3' & & \end{array}.$$

Since the greatest element in T is $3'$, the first location of CELL is the lower of these, [2, 3], and *unmix* is false. Applying DELETE to S therefore gives 1. Thus S and T are

$$\begin{array}{ccccc} 1 & 1 & 2' & 3' & 3 \\ & 2' & & & \end{array} \quad \text{and} \quad \begin{array}{ccccc} 1 & 1 & 1 & 2' & 3' \\ & 2 & & & \end{array}.$$

This give the rightmost column of E as $\begin{array}{c} 3 \\ 1 \end{array}$.

To continue, we take CELL to be [1, 5] and *unmix* to be false, giving tableaux

$$\begin{array}{cccc} 1 & 2' & 3' & 3 \\ & 2' & & \end{array} \quad \text{and} \quad \begin{array}{cccc} 1 & 1 & 1 & 2' \\ & 2 & & \end{array}$$

with partial encoding $\begin{array}{cc} 3 & 3 \\ 1' & 1 \end{array}$.

Notice that this application of DELETE has bumped out $1'$ rather than 1, since MARK was applied. Then taking CELL to be [2, 2] and *unmix* to be true gives tableaux

$$\begin{array}{cccc} 1 & 2' & 3' & 3 \end{array} \quad \text{and} \quad \begin{array}{cccc} 1 & 1 & 1 & 2' \end{array}$$

with partial encoding $\begin{matrix} 2 & 3 & 3 \\ 2' & 1' & 1 \end{matrix}$.

The next is the last mixed deletion, with CELL being [1, 4] giving tableaux

$$2'\ 3'\ 3 \quad \text{and} \quad 1\ 1\ 1$$

and partial encoding $\begin{matrix} 2 & 2 & 3 & 3 \\ 1 & 2' & 1' & 1 \end{matrix}$.

The final three applications of DELETE give the encoding

$$E = \begin{matrix} 1 & 1 & 1 & 2 & 2 & 3 & 3 \\ 2' & 3' & 3 & 1 & 2' & 1' & 1 \end{matrix}.$$

Thus

$$A = \begin{pmatrix} 0 & 1' & 2' \\ 1 & 1' & 0 \\ 2' & 0 & 0 \end{pmatrix}.$$

The proof of Theorem 13.1 is completed by the following result.

Proposition 13.7. *The algorithm SHIFTED_KNUTH_INV is the inverse of SHIFTED_KNUTH.*

Proof. It can be seen that SHIFTED_KNUTH followed by SHIFTED_KNUTH_INV is well-defined and is the identity. The only point to be checked with the reverse claim is that the array E produced by SHIFTED_KNUTH_INV is an encoding of a matrix A over $P \cup \{0\}$. Thus we need to show that for the repetitions of a fixed integer in the upper line of E, the corresponding interval in the second line of E (i) is non-increasing and (ii) has at most one marked version of any number. The first of these claims follows from Lemma 13.3 and the fact that DELETE is the inverse of INSERT. For the second claim, note that each row of S has at most one occurrence of any marked letter. Thus, when the output letter, p say, is removed from the first row, this row can still only have at most one marked value of p. Any subsequent deletions in SHIFTED_KNUTH_INV are to the left of this, so no further letters whose value is a marking of p can be deleted from the first row. This completes the proof.

Remark. The algorithms of Theorem 13.1 enable us to give another proof of Corollary 10.8. This result says that

$$\sum_{\lambda \in \mathcal{D}(n)} 2^{n-\ell(\lambda)} g_\lambda^2 = n!\,,$$

where g_λ denotes the number of standard shifted tableaux of shape λ. We obtain this second proof by restricting the inverse of the shifted Knuth correspondence to the set of those pairs (S, T) for which (i) S is a standard shifted tableau, and (ii) T becomes one after erasing all its marks. Thus the entries of S are all the distinct (unmarked) integers from 1 to n, as are the |entries of T|, where S and T have shape λ, a strict partition of n. A standard shifted tableau can be converted to a marked shifted tableau by choosing any subset of its entries to be marked. However, T is not allowed to have any of its main diagonal entries marked. Thus the number of such pairs (S, T) for fixed n is the left hand side of the formula to be proved. The bijection of Theorem 13.1 puts such pairs into bijective correspondence with the two line arrays of the form

$$\begin{pmatrix} 1 & 2 & \dots & n \\ \sigma(1) & \sigma(2) & & \sigma(n) \end{pmatrix}$$

for elements σ in S_n. Thus we have a so-called "bijective proof" of Corollary 10.8.

Notes

The algorithms in this chapter were discovered independently by Sagan (1987) and Worley (1984). The original invention of insertion, for ordinary tableaux, was by Robinson (1938), and an improved description, generalization and analysis was given by Schensted (1961). This led to the Knuth correspondence for ordinary tableaux, Knuth (1970). Compared to its marked shifted counterpart, each aspect of the theory for ordinary tableaux is an order of magnitude simpler. The first bijective proof of Corollary 10.8 was given by Sagan (1979). The algorithm used for that proof is less satisfactory than the Sagan–Worley algorithm discussed in this chapter. To the extent that they differ, our treatment is closer to Worley (1984) than to Sagan (1987) in the details of the algorithms. See also Haiman (1989). For a third proof of 10.8, see Stanley (1988A).

14

DEEPER INSERTION, EVACUATION AND THE PRODUCT THEOREM

Since the algebra $\Delta_{\mathbb{Q}}$ is spanned by the Q-functions, the product $Q_\mu Q_\nu$ is a linear combination of the functions Q_λ. Using the map och and Theorem 8.6, it can be proved that the coefficients involved are non-negative integers. The purpose of this chapter is to sketch a direct proof of this fact. But much more is done: the enumerative descriptions of these integers from Stembridge (1987, 1989) and Worley (1987) are given. This is the analogue of what is known as the Littlewood–Richardson rule for Schur functions, and for linear representations of S_n or (certain) representations of $GL(n, \mathbb{C})$.

There is basically only one approach known to us at present for proving this theorem. It involves some deeper properties of the algorithm INSERT from Chapter 13. In particular, an "evacuation" operator $\mathcal{J}$ is needed to prove a certain fibring property of insertion. In general, the object needing combinatorial study is the set of sequences which, when INSERT is applied successively to terms, will all produce some fixed given tableau. The theory includes a certain equivalence relation and invariant, analogous to what are known as Knuth relations and the Greene invariant for unshifted insertion and tableaux. Basic results concerning these will be quoted with references to the literature. We have chosen to only sketch the proof in this manner, since the result will not be used further within this book, and since the excursion into combinatorics needed to give complete details would add considerably to the length of the book. A shorter, algebraic proof would be desirable. On the other hand, the overall strategy of the present proof is attractive. It seems likely that further simplifications of shifted insertion theory will be forthcoming. See Haiman (1989), for example.

Since there are so many definitions in this chapter, we shall label them more elaborately, for ease of reference.

Definition of f_x and k^*. For any finite array $\mathcal{A}$ of elements from $P = \{1', 1, 2', 2, \ldots\}$, (such as a sequence or a tableau) and any $x \in P$, let $f_x(\mathcal{A})$ be the number of occurrences of x in $\mathcal{A}$. For $k \in \mathbb{N}$, let

$$f_{k^*}(\mathscr{A}) = f_k(\mathscr{A})+f_{k'}(\mathscr{A}) ,$$

so that the *content* of $\mathscr{A}$ is the sequence

$$[f_{1^*}(\mathscr{A}), f_{2^*}(\mathscr{A}), \ldots] .$$

In general, k^* will mean "either k' or k".

Definition of the lattice conditions. Let β be a finite sequence of elements from P. We say that β satisfies the (marked shifted) *lattice conditions* if and only if the following two conditions hold for all positive integers k:

MLT (k): For all $j \geq 0$, if, among the leftmost j terms of β, the number of occurrences of $k+1$ equals that of k, then the $(j+1)$st term from the left in β is neither $(k+1)'$ nor $(k+1)$.

MLT′ (k): For all $j \geq 0$, if $[f_k(\beta)$ plus the number of occurrences of k' among the rightmost j terms of β] is equal to $[f_{k+1}(\beta)$ plus the number of occurrences of $(k+1)'$ among the rightmost j terms of β], then the $(j+1)$st term from the right is neither k nor $(k+1)'$.

We have given a verbose, non-analytic, formulation of the lattice conditions which coincides with how one would check particular examples: one scans the sequence from left to right, then from right to left, once for each k, accumulating the frequency difference between k and $k+1$ in the first phase, and between k' and $(k+1)'$ in the second phase. For example, any sequence beginning as $1'\,2\ldots$ fails the conditions MLT (1) very early in the scanning. Indeed, a sequence which satisfies condition MLT (1) must have an (unmarked) 1 to the left of all terms larger than 1. On the other hand, $1\,1\,2'\,2'$ and $1\,1'\,2'\,1\,2'\,2\,2\,1'$ both satisfy MLT (1), but not MLT′ (1), the failures occurring when one reads the 1 in the interior of each example during the second, right to left, phase.

Definition of word. Let $T \in \mathrm{MST}(\lambda/\mu;\, \omega)$ for some strict partitions λ and μ, and some sequence ω of positive integers. Define word (T) to be the sequence from P obtained by reading the entries of T from right to left along each row, starting from the top row and proceeding downwards. Thus word (T) has the same content, ω, as T.

Definition of MST_L **and** MST_L^u. If λ, μ and ν are strict partitions, define $\mathrm{MST}_L(\lambda/\mu;\, \nu)$ to be the set of $T \in \mathrm{MST}(\lambda/\mu;\, \nu)$ such that for all $k \geq 1$, word (T) satisfies MLT (k) and MLT′ (k). Define

$$\mathrm{MST}_L^u(\lambda/\mu;\nu) = \mathrm{MST}_L(\lambda/\mu;\nu) \cap \mathrm{MST}^u(\lambda/\mu;\nu)\,,$$

which is the set of those T satisfying the lattice conditions all of whose main diagonal entries are unmarked.

Remark. An easy exercise shows that if ν were not a strict partition, then $\mathrm{MST}_L(\lambda/\mu;\nu)$ would be empty.

The following is the main result of this chapter.

Theorem 14.1. *For all strict partitions μ and ν,*

$$Q_\mu Q_\nu = \sum_{\lambda\in\mathcal{D}} |\mathrm{MST}_L^u(\lambda/\mu;\nu)|\, Q_\lambda\,.$$

We shall first give some corollaries and discussion of this, and then state another theorem, 14.6, from which 14.1 is easily deduced. The bulk of the chapter will then be taken up with the sketch of combinatorial arguments to prove 14.6.

Corollary 14.2. *For all strict partitions λ, μ and ν,*

$$|\mathrm{MST}_L^u(\lambda/\mu;\nu)| = |\mathrm{MST}_L^u(\lambda/\nu;\mu)|\,.$$

This is immediate, since $Q_\mu Q_\nu = Q_\nu Q_\mu$; it is not at all obvious directly. An explicit bijection will appear later in this chapter (see the remark after Theorem 14.15).

Proposition 14.3. *Let $T\in \mathrm{MST}_L(\lambda/\mu;\nu)$ for some strict partitions λ, μ and ν. Suppose $f_{k^*}(T) > 0$, and obtain S from T by changing the marking of k^* in its rightmost appearance in word (T) (that is, in its lower leftmost appearance in T). Then $S\in \mathrm{MST}_L(\lambda/\mu;\nu)$.*

Proof. By considering the entries immediately above, below, and to either side of the k^* at issue, it is trivially checked that $S\in \mathrm{MST}(\lambda/\mu;\nu)$. Let $\beta = \mathrm{word}(T)$ and $\gamma = \mathrm{word}(S)$. It remains to verify the lattice conditions for γ. This is done in eight cases, combining the two possibilities for the change of marking with the four possibilities of failure of MLT (k), MLT$'$ (k), MLT $(k-1)$ and MLT$'$ $(k-1)$ for γ, in each case deducing a failure of the lattice conditions for β. For example, suppose a k is changed to k' to get from β to γ and suppose MLT $(k-1)$ holds but MLT$'$ $(k-1)$ fails for γ. It cannot fail after reading this k', which is the rightmost k^*, since the counts with respect to both k and $k-1$ in β and γ are identical at these points. But it cannot fail earlier, because at such points,

$$f_k(\gamma)+\#k' = f_k(\gamma) = f_k(\beta)-1 \leq f_{k-1}(\beta)-1$$
$$= f_{k-1}(\gamma)-1 < f_{k-1}(\gamma)\,.$$

Similar arguments deal with all the other cases.

Remarks. For T as in Proposition 14.3, it is obvious that every main diagonal entry is such a rightmost k^* for some k. This has two consequences:

(i)

$$|\mathrm{MST}^u_L(\lambda/\mu;\,\nu)| = 2^{\ell(\mu)-\ell(\lambda)}|\mathrm{MST}_L(\lambda/\mu;\,\nu)|$$

and so, by Corollary 14.2

$$2^{\ell(\mu)}|\mathrm{MST}_L(\lambda/\mu;\,\nu)| = 2^{\ell(\nu)}|\mathrm{MST}_L(\lambda/\nu;\,\mu)|\,.$$

(ii) If $\ell(\nu) < \ell(\lambda)-\ell(\mu)$, then MST $(\lambda/\mu;\,\nu)$ is empty, since the number of available k^* to make a rightmost appearance on the main diagonal is too small to fill all the places there in skew shape λ/μ.

Definition of MST^w_L. Define $\mathrm{MST}^w_L(\lambda/\mu;\,\nu)$ to be the subset of those $T \in \mathrm{MST}_L(\lambda/\mu;\,\nu)$ in which the rightmost occurrence of k^* in word (T) is unmarked for all k (not just those where the occurrence is on the main diagonal).

Since exponent u denoted "unmarked", it seemed appropriate to use w for "doubly unmarked".

It is clear from 14.3 that

$$|\mathrm{MST}^w_L(\lambda/\mu;\,\nu)| = 2^{-\ell(\nu)}|\mathrm{MST}_L(\lambda/\mu;\,\nu)|\,,$$

and so, from 14.2, that

$$|\mathrm{MST}^w_L(\lambda/\mu;\,\nu)| = |\mathrm{MST}^w_L(\lambda/\nu;\,\mu)|\,.$$

Corollary 14.4. *The coefficient of* Q_λ *in* $Q_\mu Q_\nu$ *is zero if* $\ell(\mu)+\ell(\nu) < \ell(\lambda)$ *(or if* $|\mu|+|\nu| \neq |\lambda|$, *of course), and is an integer divisible by* $2^{\ell(\mu)+\ell(\nu)-\ell(\lambda)}$ *when* $\ell(\mu)+\ell(\nu) \geq \ell(\lambda)$.

This is immediate from Remark (ii) and since

$$|\mathrm{MST}^u_L(\lambda/\mu;\,\nu)| = 2^{\ell(\mu)+\ell(\nu)-\ell(\lambda)}|\mathrm{MST}^w_L(\lambda/\mu;\,\nu)|\,.$$

Remark. The fact that the coefficient is an integer seems to have been taken for granted before Worley (1984) (7.2.3). His proof is based on (ii) in Theorem 14.6 below, and was discovered before Theorem 14.1 was known. The proof suggested in the first sentence of this chapter would have been known to Schur, had he thought about the question, but seems not to occur in the literature.

Corollary 14.5. *For all* λ, μ *and* ν *in* $\mathcal{D}$:

(*i*) $<Q_\mu Q_\nu, Q_\lambda> = 2^{\ell(\mu)}|\mathrm{MST}_L(\lambda/\mu;\nu)|$;

(*ii*) $<Q_{\lambda/\mu}, Q_\nu> = |\mathrm{MST}_L(\lambda/\mu;\nu)|$.

For all λ *and* μ *in* $\mathcal{D}$:

(*iii*) $Q_{\lambda/\mu} = \sum_{\nu\in\mathcal{D}} |\mathrm{MST}_L^w(\lambda/\mu;\nu)|\, Q_\nu$; *thus*, $Q_{\lambda/\mu}\in\Delta$;

(*iv*) $Q_\mu^\perp(Q_\lambda) = \sum_{\nu\in\mathcal{D}} |\mathrm{MST}_L(\lambda/\nu;\mu)| Q_\nu$;

(*v*) *If* $P_\lambda := 2^{-\ell(\lambda)}Q_\lambda$, *a function with integer coefficients, then*

$$P_\mu P_\nu = \sum_{\lambda\in\mathcal{D}} |\mathrm{MST}_L^w(\lambda/\mu;\nu)|P_\lambda\,.$$

Remarks. Each of these is easily seen to be equivalent to Theorem 14.1. The set $\{P_\lambda\}$ is the dual basis to $\{Q_\lambda\}$, and also comes from certain Hall–Littlewood functions. The coefficient of X^α in $P_\lambda(X)$ is the integer $|\mathrm{MST}^u(\lambda;\alpha)|$. The expert might protest that the coefficient in (iv) ought to be written as $2^{\ell(\mu)-\ell(\nu)}|\mathrm{MST}_L(\lambda/\mu;\nu)|$; but possibly (iv) in its present form is saying something about how an algebraic proof should proceed. When μ is empty, (iii) reduces to the fact that $\mathrm{MST}_L^w(\lambda;\nu)$ has one element when $\lambda=\nu$, and none otherwise. See T_ν in (iii)′ of the next theorem.

Theorem 14.6. *Let* λ *and* μ *be strict partitions. Define*

$$\mathrm{MST}(\lambda/\mu;*) = \bigcup_\omega \mathrm{MST}(\lambda/\mu;\omega)\,,$$

the set of all marked shifted tableaux of shape λ/μ. *Denote by* $\mathrm{MST}(*;*)$ *the set of all non-skew marked shifted tableaux. Then there is a function*

$$\mathcal{G}_{\lambda/\mu}:\mathrm{MST}(\lambda/\mu;*)\to\mathrm{MST}(*;*)$$

which has the following properties:

(*i*) $\mathcal{G}_{\lambda/\mu}$ *preserves content; that is, for all* ω,

$$\mathcal{G}_{\lambda/\mu}\big(\mathrm{MST}(\lambda/\mu;\omega)\big)\subset \mathrm{MST}(*;\omega) := \bigcup_{\nu\in\mathcal{D}}\mathrm{MST}(\nu;\omega)\,.$$

(ii) For each strict ν, *the map* $\mathcal{G}_{\lambda/\mu}$ *is a "fibring" over* MST(ν; $*$); *that is, if* $T_1 \in$ MST(ν; ω_1) *and* $T_2 \in$ MST(ν; ω_2) *then*

$$|\mathcal{G}^{-1}_{\lambda/\mu}(T_1)| = |\mathcal{G}^{-1}_{\lambda/\mu}(T_2)| \,.$$

(iii) For all strict ν, *there exists* $T_\nu \in$ MST(ν; ν) *such that*

$$|\mathcal{G}^{-1}_{\lambda/\mu}(T_\nu)| = |\mathrm{MST}^w_L(\lambda/\mu;\nu)| \,.$$

More precisely:

(iii)′ Let T_ν *be the unique tableau of shifted shape* ν *all of whose entries on the i-th row are equal to i (none is marked). Then*

$$\mathcal{G}^{-1}_{\lambda/\mu}(T_\nu) = \mathrm{MST}^w_L(\lambda/\mu;\nu) \,.$$

Below we shall define $\mathcal{G}_{\lambda/\mu}$ using insertion, so that (i) is obvious. The sketch proofs of (ii) and (iii)′ will occupy most of the chapter, with some appeals to results in the literature. First we show how to deduce (14.1) from (i), (ii) and (iii):

Proof of Theorem 14.1. For λ and μ in $\mathcal{D}$ and ω in $\mathbb{N}^k$,

$$\langle Q_{\lambda/\mu}, q_\omega\rangle = |\mathrm{MST}(\lambda/\mu;\omega)| \quad \text{by 12.12}$$

$$= \sum_{\nu\in\mathcal{D}} \sum_{T\in \mathrm{MST}(\nu;\omega)} |\mathcal{G}^{-1}_{\lambda/\mu}(T)| \quad \text{by 14.6 (i)}$$

$$= \sum_{\nu\in\mathcal{D}} |\mathcal{G}^{-1}_{\lambda/\mu}(T_{\nu,\omega})|\ |\mathrm{MST}(\nu,\omega)|$$

for any fixed $T_{\nu,\omega} \in$ MST(ν; ω), by (14.6) (ii) with $\omega_1 = \omega_2 = \omega$

$$= \sum_{\nu\in\mathcal{D}} |\mathcal{G}^{-1}_{\lambda/\mu}(T_\nu)|\ |\mathrm{MST}(\nu;\omega)|$$

by 14.6 (ii), taking $\omega_1 = \omega$; $T_1 = T_{\nu,\omega}$; $\omega_2 = \nu$; $T_2 = T_\nu$

$$= \sum_{\nu\in\mathcal{D}} |\mathrm{MST}^w_L(\lambda/\mu;\nu)|\ |\mathrm{MST}(\nu;\omega)|$$

by 14.6 (iii)

$$= \langle \sum_{\nu\in\mathcal{D}} |\mathrm{MST}^w_L(\lambda/\mu;\nu)| Q_\nu, q_\omega\rangle$$

by 12.12 again. Since $\{q_\omega\}$ spans Δ, we obtain (14.5) (iii) which is equivalent to (14.1).

Definition of ins. Let T be a (non-skew) tableau. (Henceforth we shall usually drop the adjectives "marked shifted" since these are all that will be considered.) Let α be a finite sequence from P. Define $\text{ins}(\alpha, T)$ to be the tableau obtained by successively inserting terms of α, starting from T; that is,

$$\text{ins}[(\alpha_1, \alpha_2, \ldots), T] = \ldots\text{ins}[\alpha_2, \text{ins}(\alpha_1, T)],$$

where $\text{ins}(a, T)$ is the result of applying INSERT to $a \in P$ and T. When T is empty, write $\text{ins}(\alpha)$ in place of $\text{ins}(\alpha, T)$.

Definition of rev. The *reverse* of a sequence $\alpha = (\alpha_1, \ldots, \alpha_\ell)$ is

$$\text{rev}(\alpha) = (\alpha_\ell, \alpha_{\ell-1}, \ldots, \alpha_1).$$

Definition of $\mathcal{G}_{\lambda/\mu}$.

$$\mathcal{G}_{\lambda/\mu}(S) = \text{ins rev word}(S).$$

Since word, rev and ins preserve content (but not the marking in the case of ins), part (i) of Theorem 14.6 is immediate. The conjunction "rev word" will appear frequently below. It might have been more efficient to have defined the word of a tableau as the reverse of what we are calling word (T). Our choice was motivated by the wish to be helpful to readers who are already familiar with unshifted insertion and the Littlewood–Richardson rule. The letter $\mathcal{G}$ was chosen for "game" as in "jeu de taquin"; see Sagan (1987, Chapter 11), and Worley (1984, Ch. 3 and 6.4). For a comprehensive treatment, the definition of $\mathcal{G}$ would be given using "jeu" *ab initio*. But a thorough explanation of "jeu" would require many extra pages.

Definition of $<_r$ **and** $<_c$. Let $<_r$ be the relation on P which must be satisfied by entries along each row of a tableau; that is,

$$a <_r b \text{ if and only if either } a<b \text{ or } a=b \text{ is unmarked}.$$

Define $<_c$ similarly, using "column" in place of "row":

$$a <_c b \text{ if and only if either } a<b \text{ or } a=b \text{ is marked}.$$

Definition of $\sqcup$. For $\psi \in P^m$ and $\phi \in P^n$, let $\psi \sqcup \phi \in P^{m+n}$ denote their concatenation:

$$\psi \sqcup \phi := (\psi_1, \ldots, \psi_m, \phi_1, \ldots, \phi_n) .$$

Definition of ascent pairs. Given a sequence α from P, an ascent pair (ψ,ϕ) for α is a pair of subsequences ψ of rev α and ϕ of α such that, if $\psi \in P^m$ and $\phi \in P^n$, then

$$\psi_1 <_c \psi_2 <_c \ldots <_c \psi_m \leq \phi_1 <_r \phi_2 <_r \ldots <_r \phi_n .$$

The last condition says

(i) $\psi \sqcup \phi$ is weakly increasing in the usual order on P;

(ii) for all $k \in \mathbb{N}$, at most one (unmarked) k appears in ψ;

(iii) for all $k \in \mathbb{N}$, at most one (marked) k' appears in ϕ.

The *length* of the above ascent pair is $m+n$. By a slight abuse of notation, we shall usually write an ascent pair as $\psi \sqcup \phi$, rather than (ψ, ϕ).

For example, the only ascent pairs whose terms are all equal to k^* are of the forms

$$(k', k', \ldots, k') \sqcup (k', k, k, \ldots, k) ,$$
$$(k', \ldots, k') \sqcup (k, \ldots, k) ,$$

or

$$(k', \ldots, k', k) \sqcup (k, \ldots, k) .$$

Such ascent pairs for α can have any length up to $f_{k^*}(\alpha)+1$ (as long as k^* actually appears in α). The unique pair of maximal length takes the first or third form above, depending respectively on whether the leftmost k^* in α is marked or unmarked.

A good way to remember the definition of ascent pairs is to think of the pair of sequences obtained by descending some early column of a tableau and by proceeding rightwards across the row from the main diagonal entry at the bottom of that column.

Definition of A_k. Let $\alpha \in P^n$ and $k \in \mathbb{N}$. Define $A_k(\alpha)$ to be the length of the longest subsequence of (rev α) $\sqcup$ (α) which can be decomposed into a union of k sequences of the form $\psi \sqcup \phi$, using ascent pairs (ψ, ϕ) for α.

Thus

$$0 \leq A_1(\alpha) \leq A_2(\alpha) \leq \ldots \leq A_n(\alpha) = A_{n+1}(\alpha) = \ldots = 2n .$$

Definition of shape. The function shape from MST($*$; $*$) to $\mathcal{D}$ is the obvious one: it maps all of MST(λ; $*$) to $\{\lambda\}$.

Theorem 14.7. *If* α *is a sequence from* P *and*

$$\text{shape ins}(\alpha) = (\lambda_1, \ldots, \lambda_\ell),$$

then, for $1 \leq k \leq \ell$,

$$A_k(\alpha) = \lambda_1 + \ldots + \lambda_k + \binom{k+1}{2}.$$

In particular, if α *and* β *are in* P^n *and* $A_k(\alpha) = A_k(\beta)$ *for all* k, *then*

$$\text{shape ins}(\alpha) = \text{shape ins}(\beta)\,.$$

Remarks and references. The definitions differ slightly, but this is equivalent to part of Theorem 10.2 in Sagan (1987). Our definition agrees with that in Worley (1984) (6.3.1). Theorem 14.7 is implicit in Worley's thesis. For example, combining (6.4.2) and (6.4.16) yields that any α is Knuth equivalent to rev word ins(α), and so by (6.3.6),

$$A_k(\alpha) = A_k(\text{rev word ins } \alpha)\,.$$

But now (6.3.9) yields that if

$$A_k(\text{rev word ins } \alpha) = A_k(\text{rev word ins } \beta)$$

then shape ins α = shape ins β.

Definition of labelling. For each $\alpha \in P^n$ define a permutation $\sigma_\alpha \in S_n$ by induction on n, as follows. Let $k = \max\{\ell \in \mathbb{N} : \ell^* \text{ appears in } \alpha\}$. If k appears in α, let $\sigma_\alpha(n) = i$ where α_i is the rightmost occurrence of k in α. If k does not appear in α, let $\sigma_\alpha(n) = i$ where α_i is the leftmost occurrence of k' in α. Let $\alpha[1, n-1] \in P^{n-1}$ be the subsequence of α obtained by removing $\alpha_{\sigma_\alpha(n)}$ from α. Now inductively, for $j<n$, define

$$\sigma_\alpha(j) = \begin{cases} \sigma_{\alpha[1,n-1]}(j) & \text{if } \sigma_{\alpha[1,n-1]}(j) < \sigma_\alpha(n); \\ \sigma_{\alpha[1,n-1]}(j)+1 & \text{if } \sigma_{\alpha[1,n-1]}(j) \geq \sigma_\alpha(n). \end{cases}$$

Let $\alpha[i, j] \in P^{j-i+1}$ be the subsequence of α consisting of all α_k for which $k = \sigma_\alpha(\ell)$ for some ℓ with $i \leq \ell \leq j$.

In effect, we are "labelling" the term $\alpha_{\sigma_\alpha(j)}$ with label j, and taking $\alpha[i, j]$ to be the subset of terms with labels between i and j. Labelling is done by first labelling all occurrences of $1'$ from the right to left, then all occurrences of 1 from left to right, then $2'$ from right to left, etc. ..., using labels $1, 2, \ldots, n$ in increasing order.

Theorem 14.8. *Let* α *and* β *be sequences in* P^n. *Then* $\text{ins}\,\alpha = \text{ins}\,\beta$ *if and only if* (*i*), (*ii*) *and* (*iii*) *hold, where*:

(*i*) *for all* k, i *and* j *with* $k > 0$ *and* $1 \leq i \leq j \leq n$,

$$A_k(\alpha[i,j]) = A_k(\beta[i,j])\ ;$$

(*ii*) *the contents of* α *and* β *are the same*;

(*iii*) *for* k *such that* $f_{k^*}(\alpha)$ $[= f_{k^*}(\beta)]$ *is positive, the leftmost* k^* *in* α *is marked if and only if the leftmost* k^* *in* β *is marked.*

Remarks and references. The invariant of a sequence given in (i), (ii) and (iii) is known as the Greene invariant, being analogous to the somewhat simpler invariant, Greene (1974), for ordinary (unshifted) insertion. Modulo slightly varying definitions, this theorem is equivalent to Theorem 10.3 of Sagan (1987). It is also implicit in Worley (1984), whose definitions we have used. More explicitly, by the remarks after Theorem 14.7 and by (6.3.8) of Worley (1984), the Greene invariants $G(\alpha)$ and $G[\text{rev word ins}(\alpha)]$ agree. Thus

$$\begin{aligned} G(\alpha) = G(\beta) &\Leftrightarrow G(\text{rev word ins}\,\alpha) = G(\text{rev word ins}\,\beta) \\ &\Leftrightarrow \text{ins}\,\alpha = \text{ins}\,\beta \end{aligned}$$

as required, where the second $\Rightarrow$ is (6.3.10) of Worley (1984).

Definition of the alphabet ${}_{-}P$. The alphabet P is produced from $\mathbb{N}$. Evidently $\mathbb{N}$ may be replaced by any other totally ordered infinite set and all results concerning marked shifted tableaux will continue to hold when content is taken in the corresponding marked alphabet. A notion of isomorphism could be introduced to formalize this, producing verbosity without new insight. For the next few pages it will be convenient (though not strictly necessary) to replace $\mathbb{N}$ by $\mathbb{Z}$ as the "superset of content". The corresponding marked alphabet ${}_{-}P$ is the set

$$\{\ldots < (-2)' < (-2) < (-1)' < (-1) < 0' < 0 < 1' < 1 < 2' < \ldots\}\,.$$

Definition of opp. For $k \in \mathbb{Z}$, define $\text{op}(k) = (-k)'$ and $\text{op}(k') = -k$. For $\alpha \in {}_{-}P^n$, define

$$\text{opp}(\alpha_1, \ldots, \alpha_n) = \big(\text{op}(\alpha_1), \text{op}(\alpha_2), \ldots, \text{op}(\alpha_n)\big)\,.$$

Evidently opp is an involution on ${}_{-}P^n$.

Proposition 14.9. *For all* $\alpha \in {}_{-}P^n$ *and* $k \in \mathbb{N}$, $A_k(\text{opp}\,\alpha) = A_k(\alpha)$. *Thus*

$$\text{shape ins opp}(\alpha) = \text{shape ins}(\alpha)\,.$$

Proof. By Theorem 14.7, the second assertion follows from the first. To prove the first assertion, it clearly suffices to define a length preserving bijection between the sets of ascent pairs for the two sequences α and $\mathrm{opp}(\alpha)$. Such a bijection is given, for subsequences ψ of $\mathrm{rev}(\alpha)$ and ϕ of α, by

$$\psi \sqcup \phi \mapsto (\mathrm{rev\ opp}\,\phi) \sqcup (\mathrm{rev\ opp}\,\psi)\,,$$

as may bc casily checked.

Proposition 14.10. *If* α *and* β *are sequences in* $\_P^n$ *and* $\mathrm{ins}(\alpha) = \mathrm{ins}(\beta)$, *then*

$$\mathrm{ins\ opp}(\alpha) = \mathrm{ins\ opp}(\beta)\,.$$

Proof. Conditions (i), (ii) and (iii) in Theorem 14.8 are satisfied by (α, β), so we shall use that to deduce the same conditions for $[\mathrm{opp}(\alpha), \mathrm{opp}(\beta)]$, as required.

(i) Obviously, opp reverses labelling, in that

$$(\mathrm{opp}\,\alpha)[i,j] = \mathrm{opp}(\alpha[n-j+1, n-i+1])\,.$$

It is understood, of course, that labelling begins with the smallest α_k, not necessarily $1'$, now that negative terms are allowed. Thus, using Proposition 14.9,

$$\begin{aligned} A_k\big((\mathrm{opp}\,\alpha)[i,j]\big) &= A_k\big(\mathrm{opp}(\alpha[n-j+1, n-i+1])\big) \\ &= A_k(\alpha[n-j+1, n-i+1])\,. \end{aligned}$$

The identical equalities hold with β in place of α. But the two right-hand sides agree by (14.8) (i) for α and β, and so the left-hand sides also agree, as required.

(ii) For all $k \in \mathbb{Z}$, since α and β have the same content,

$$f_{k^*}(\mathrm{opp}\,\alpha) = f_{(-k)^*}(\alpha) = f_{(-k)^*}(\beta) = f_{k^*}(\mathrm{opp}\,\beta)\,,$$

and so $\mathrm{opp}\,\alpha$ and $\mathrm{opp}\,\beta$ also have the same content.

(iii) Since $\mathrm{left}(\alpha) = \mathrm{left}(\beta)$, where

$$\mathrm{left}(\alpha) := \{k \in \mathbb{Z} : f_{k^*}(\alpha) > 0 \text{ and the leftmost } k^* \text{ in } \alpha \text{ is } k'\}\,,$$

we have

$$\begin{aligned} \mathrm{left\ opp}(\alpha) &= \{-k : k \in \mathrm{left}(\alpha); f_{k^*}(\alpha) > 0\} \\ &= \{-k : k \in \mathrm{left}(\beta); f_{k^*}(\beta) > 0\} = \mathrm{left\ opp}(\beta)\,, \end{aligned}$$

as required.

Theorem 14.11. *If S is a non-skew marked shifted tableau, then*

$$\text{ins rev word}(S) = S.$$

Remarks and references. This follows easily from Theorem 11.2 of Sagan (1987), and immediately from 6.4.1 and 6.4.2 of Worley (1984).

Now we can give the evacuation operator $\mathcal{J}$.

Theorem 14.12. *There is a unique function $\mathcal{J}$ for which the square on the left below commutes. Furthermore, $\mathcal{J}$ is an involution and the triangle commutes.*

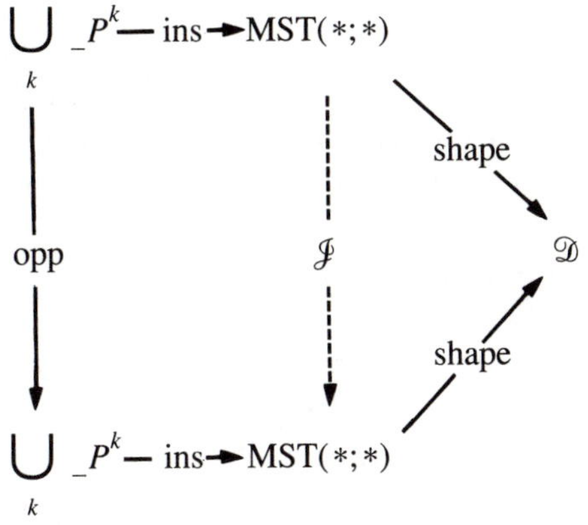

Proof. By Theorem 14.11, ins is surjective. This shows that $\mathcal{J}$, if it exists, is unique. It also is used in proving the second assertion as follows. To show that $\mathcal{J} \circ \mathcal{J}$ is the identity, note that opp $\circ$ opp is the identity and that the square, obtained by replacing $\mathcal{J}$ by $\mathcal{J} \circ \mathcal{J}$ and opp by opp $\circ$ opp in the diagram above, is commutative. To show that the triangle commutes, use the fact that both the square and the whole pentagon do, the latter by Proposition 14.9.

To prove that $\mathcal{J}$ exists, we give a formula:

$$\mathcal{J}(T) := \text{ins opp rev word}(T)\,.$$

By Theorem 14.11,

$$\text{ins}(\text{rev word}(\text{ins}\,\alpha)) = \text{ins}(\alpha)\,.$$

That the square commutes then follows using Proposition 14.10:

$$\mathcal{J}\,\text{ins}\,\alpha = \text{ins opp}(\text{rev word ins}\,\alpha) = \text{ins opp}(\alpha)\,.$$

Definition of $\alpha_{\pm}$. Given a sequence α from $_{-}P$, let α_{-} be the subsequence of all terms equal to $(-k)'$ or $(-k)$ for positive integers k. Let α_{+} be the complementary subsequence of α.

We shall illustrate and argue the next few results using "sectors", such as

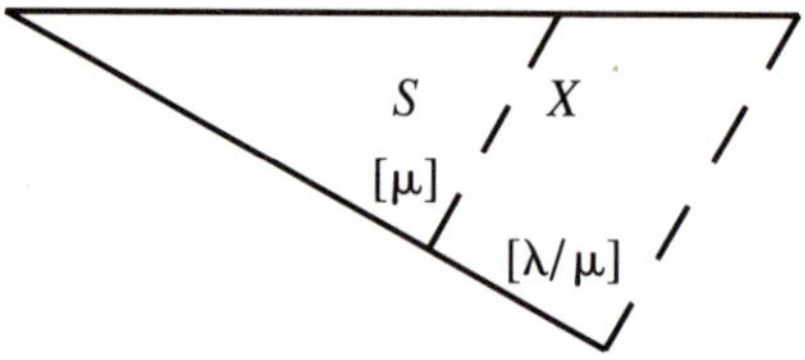

This denotes a non-skew tableau of shape λ, for which S is that part of the tableau with shape μ (where $S(\mu) \subset S(\lambda)$), and X is the complementary part, a skew tableau of shape λ/μ. Here the notations S, X, μ and λ may be names of previously discussed objects; if not, the sector defines the notation.

Proposition 14.13. *For any* $\alpha \in {}_{-}P^{n}$,

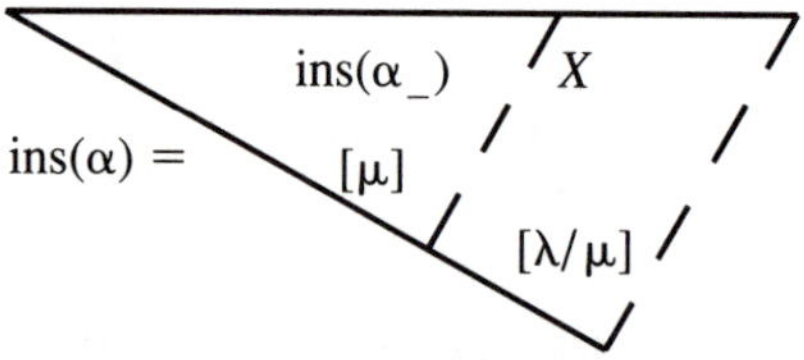

for some X, μ *and* λ, *where* X *has the same content as* α_{+}.

Proof. The assertion is that the array consisting of the negative entries in ins(α) is ins(α_{-}). This follows by induction on n. The inductive step is immediate if the rightmost term in α is non-negative, since all bumps in its insertion will take place at cells with non-negative entries in ins($\tilde{\alpha}$), where $\tilde{\alpha} = (\alpha_1,\ldots,\alpha_{n-1})$. If $\alpha_n < 0'$, by induction, inserting α_n into ins($\tilde{\alpha}$) begins with the same sequence of bumps as does inserting it into ins($\tilde{\alpha}_{-}$), followed by possibly by some bumps that involve only the non-negative entries of ins($\tilde{\alpha}$), as required.

Proposition 14.14. *Let*

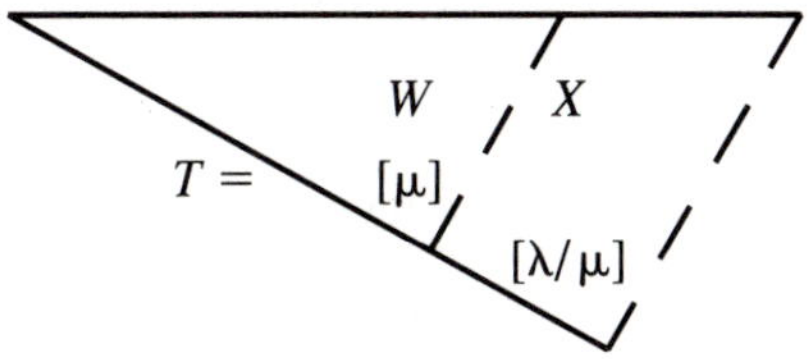

where $f_{k^*}(W) = 0$ *for* $k \geq 0$ *and* $f_{k^*}(X) = 0$ *for* $k \leq 0$ (*so that* 0^* *does not appear in* T). *Then*

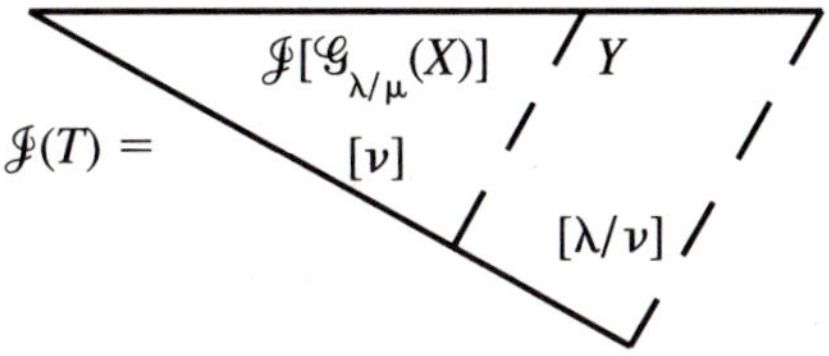

for some Y and ν, *where again the two halves have strictly negative and strictly positive content.*

(*Recall that* $\mathcal{G}_{\lambda/\mu}(X) = \text{ins rev word}(X)$.)

Proof. By the definition of opp, if β is a sequence which has no terms equal to 0^*, then

$$\big(\text{opp rev}(\beta)\big)_{\pm} = \text{opp rev}(\beta_{\mp})\,.$$

Letting $\beta = \text{word}(T)$, this yields

$$\big(\text{opp rev word}(T)\big)_{-} = \text{opp rev}[(\text{word}\,T)_{+}]\,.$$

The equalities denoted (*) below use the definition of $\mathcal{J}$. By (14.13), there exist ν and Y for which

$$\mathcal{J}(T) \underset{(*)}{=} \text{ins opp rev word}(T) =$$

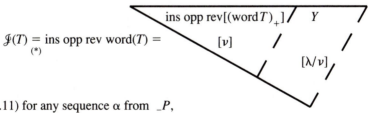

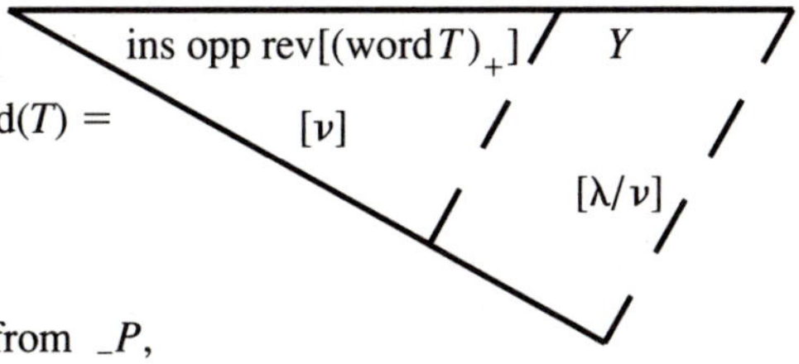

Now by (14.11) for any sequence α from _P,

$$\text{ins}\{\text{rev word}(\text{ins rev}\,\alpha)\} = \text{ins}\{\text{rev}\,\alpha\}\,.$$

Thus, by (14.10),

$$\text{ins opp}\{\text{rev word ins rev}\,\alpha\} = \text{ins opp}\{\text{rev}\,\alpha\}\,.$$

Letting $\alpha = (\text{word}\,T)_+ = \text{word}(X)$ in the last equality, we obtain

$$\begin{aligned}\mathcal{J}\mathcal{G}_{\lambda/\mu}(X) &= \mathcal{J}\big(\text{ins rev word}(X)\big)\\ &= \text{ins opp rev word ins rev word}(X) \quad \text{by } (*)\\ &= \text{ins opp rev}\big((\text{word}\,T)_+\big),\end{aligned}$$

as required.

Remark. When working with the alphabet $\_P$, content sequences ω are indexed by $\mathbb{Z}$, i.e.

$$\omega = (\ldots, \omega_{-2}, \omega_{-1}, \omega_0, \omega_1, \ldots) \in \mathbb{N}^{\mathbb{Z}}\,,$$

with $\omega_i = 0$ for almost all i, so that $T \in \text{MST}(\lambda/\mu;\, \omega)$ has ω_k entries equal to k^*.

Theorem 14.15. *Let λ, μ and ν be strict partitions with $|\lambda| = |\mu| + |\nu|$. Let ω and ω' be content sequences with $|\omega| = |\nu|$ and $|\omega'| = |\mu|$. Let $S \in \text{MST}(\nu;\, \omega)$ and $S' \in \text{MST}(\mu;\, \omega')$. Then*

$$|\mathcal{G}^{-1}_{\lambda/\mu}(S)| = |\mathcal{G}^{-1}_{\lambda/\nu}(S')|\,.$$

Proof. By adding some fixed positive integer to all entries, we may assume that $f_{k^*}(S) = 0 = f_{k^*}(S')$ for all $k \leq 0$. Define a map

$$\mathcal{J}_{S,S'} : \mathcal{G}^{-1}_{\lambda/\mu}(S) \to \mathcal{G}^{-1}_{\lambda/\nu}(S')$$

as follows. Let $X \in \mathcal{G}^{-1}_{\lambda/\mu}(S)$. Define

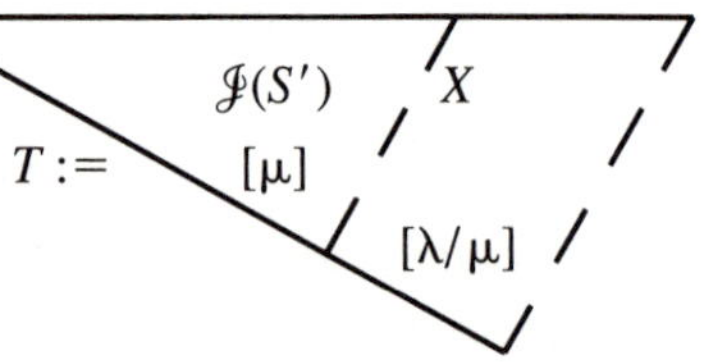

so that the two parts of the sector contain strictly negative and strictly positive entries of T, respectively. Since $\mathcal{G}_{\lambda/\mu}(X) = S$, from 14.14 we find that

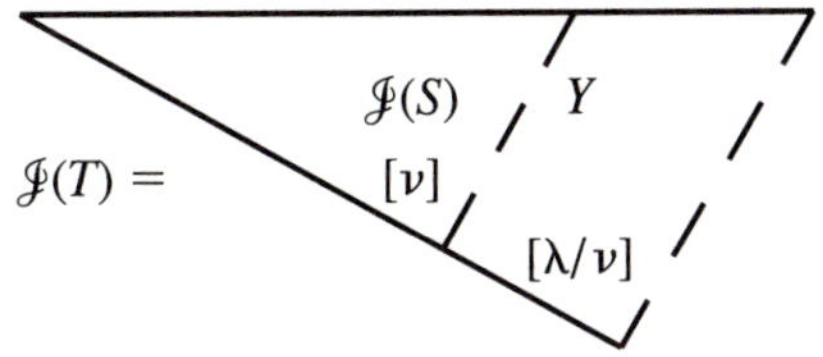

for some Y with strictly positive entries. By 14.14 again,

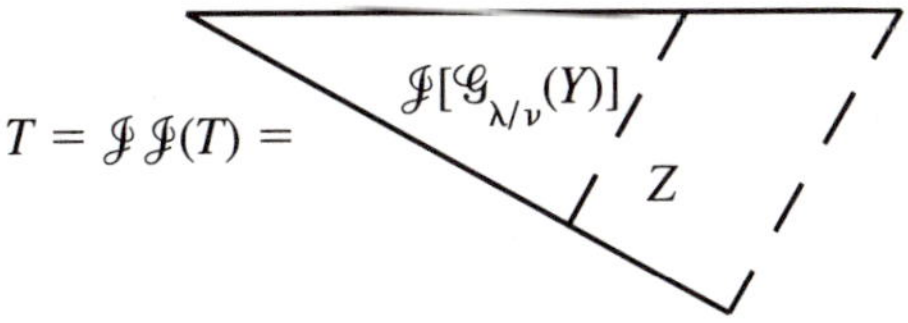

for some Z, where the two halves consist of the negative and positive entries of T. Comparing this to the first sector above, it is clear that $\mathcal{G}_{\lambda/\nu}(Y) = S'$, that is, $Y \in \mathcal{G}^{-1}_{\lambda/\nu}(S')$. Define $\mathcal{J}_{S,S'}(X) = Y$. By symmetry, $\mathcal{J}_{S',S}(Y) = X$. Thus $\mathcal{J}_{S,S'}$ is bijective, completing the proof.

Deduction of (ii) in Theorem 14.6. There is nothing to prove when $|\nu| \neq |\lambda| - |\mu|$. Otherwise, choose some ω' with $\mathrm{MST}(\mu; \omega')$ non-empty (for example, $\omega' = \mu$). Choose any $S' \in \mathrm{MST}(\mu; \omega')$. In the statement of 14.6, $|\omega_1| = |\omega_2| = |\nu|$. Applying 14.15 both with $S = T_1$ and with $S = T_2$, and with $\omega = \omega_1$ and $\omega = \omega_2$ respectively, we obtain

$$|\mathcal{G}^{-1}_{\lambda/\mu}(T_1)| = |\mathcal{G}^{-1}_{\lambda/\nu}(S')| = |\mathcal{G}^{-1}_{\lambda/\mu}(T_2)| \,,$$

as required.

Remark. Once 14.6 (iii)′ is known, Theorem 14.15 and its proof give an explicit bijection

$$\mathrm{MST}^{w}_{L}(\lambda/\mu; \nu) \to \mathrm{MST}^{w}_{L}(\lambda/\nu; \mu) \,,$$

by taking $\omega = \nu$; $S = T_\nu$; $\omega' = \mu$; $S' = T_\mu$. This is readily converted to an explicit bijection with w replaced by u; that is, one which gives an effective proof of Corollary 14.2.

The remainder of the chapter is directed towards sketching the proof of (iii)′ in Theorem 14.6. First we shall state the lengthy Theorem 14.16, which sets this material in its appropriate context, and deduce 14.6 (iii)′ from it. Then the various equivalences in 14.16 are proved in a number of propositions.

Definition of descent pairs. If α is a sequence from P, a descent pair for α is any pair (ϕ, ψ), where $(\text{rev}\psi, \text{rev}\phi)$ is an ascent pair for $\text{rev}(\alpha)$.

Theorem 14.16. *Let α be a sequence from P whose content is a strict partition ν of length ℓ. Then the following five conditions are equivalent:*

(*i*) *The lattice conditions are satisfied by α; that is, for all $k \in \mathbb{N}$, the sequence α satisfies* MLT(k) *and* MLT$'$(k).

(*ii*) *For all $k \in \mathbb{N}$, any descent pair for α, each of whose terms is either k^* or $(k+1)^*$, has length at most ν_k+1.*

(*iii*) *For all $k \in \mathbb{N}$, any descent pair for α, none of whose terms is less than k', has length at most ν_k+1.*

(*iv*) *For $1 \le k < \ell$, we have*

$$A_k(\text{rev}\alpha) = \nu_1 + \ldots + \nu_k + \binom{k+1}{2}.$$

(*v*) *The tableau* ins rev(α) *has shape equal to ν.*

Remarks. Because α has content equal to ν, the assertion in (v) is equivalent to asserting that

$$\text{ins rev}\,\alpha = \begin{array}{cccccc} 1^* & 1 & 1 & \cdots & & 1 \\ & 2^* & 2 & \cdots & 2 & \\ & & 3^* & \cdots & 3 & \\ & & & \cdot & & \\ & & & & \cdot & \\ & & & & & \cdot \end{array}$$

By the example after the definition of ascent pairs, there is a descent pair of maximal length ν_k+1 in (ii) and (iii), in fact one whose terms are all k^*.

Proposition 14.17. *Let $\alpha \in P^n$ and let $k \in \mathbb{N}$ be such that k^* occurs in α. Then the leftmost occurrence of k^* in α and the rightmost occurrence of k^* in* word ins(α) *have the same marking.*

Proof. This is immediate by induction on n, since an insertion, into a tableau T which has the property to be proved for ins(α), will not destroy that property. If the k^* at issue is marked, a row bump cannot lower a k which is on the same row as the rightmost k' in word(T), and a column bump cannot raise this k' above the lowest row containing a k. If the k^* at issue is unmarked, a similar argument works.

Deduction of (iii)′ in Theorem 14.6 from 14.16 and 14.17. Suppose that $S \in \mathrm{MST}_L^w(\lambda/\mu;\, \nu)$. Using the definitions, word(S) has content ν and satisfies the lattice conditions, and

$$\mathcal{G}_{\lambda/\mu}(S) = \text{ins rev word}(S)\,.$$

Since condition (i) implies (v) in 14.16, $\mathcal{G}_{\lambda/\mu}(S)$ has shape ν, and thus its kth row for each k has all entries equal to k^*. When it exists, the leftmost k^* in rev word(S) is unmarked, since S is "doubly unmarked". Thus the rightmost k^* in word$\mathcal{G}_{\lambda/\mu}(S)$ (i.e. the main diagonal k^* in $\mathcal{G}_{\lambda/\mu}(S)$ if $k \leq \ell(\nu)$) is also unmarked, by 14.17. Therefore $\mathcal{G}_{\lambda/\mu}(S) = T_\nu$, as required. The proof of the converse is exactly the reverse: If $\mathcal{G}_{\lambda/\mu}(S) = T_\nu$ where $S \in \mathrm{MST}(\lambda/\mu;\, \nu)$, a subscript L may be added to MST since (v) implies (i) in 14.16, and a superscript w may be added by 14.17.

Equivalence of (i) and (ii) in Theorem 14.16. In fact, we shall prove that, for each *fixed k*, the two conditions are equivalent.

By removing all terms less than k' and then subtracting $(k-1)$, it becomes clear that we can assume $k=1$ to simplify notation.

To prove that (ii) implies (i), we prove the contrapositive in the following four cases (a) to (d).

(a) When MLT(1) fails and the rightmost 1^* in α is unmarked, the following choices for ϕ and ψ give a descent pair $\phi \sqcup \psi$ which makes (ii) false:

Take ϕ to be the sequence consisting of the offending 2 or $2'$ at some point where MLT(1) fails, preceded by each occurrence of 2 in α to the left of the offending 2^*, together with all copies of 1 to its right. Take ψ to consist of the rightmost 1 in α, followed by all occurrences in α of $1'$. Then $\phi \sqcup \psi$ is a descent pair of length ν_1+2, as required.

(b) When MLT(1) fails and the rightmost 1^* in α is marked, change the counterexample in (a) by adding the rightmost $1'$ in α to the right end of ϕ and letting ψ be $1'$ repeated $f_{1'}(\alpha)$ times.

(c) When MLT′(1) fails and the rightmost 2^* in α is unmarked, let ϕ consist of all copies of 2 in α. Choose some term equal to 1 or $2'$ in rev(α), where MLT′(1) fails for α. Let ψ begin with the leftmost 2 in rev(α) followed by each $2'$ occurring before the term chosen above; then the offending 1 or $2'$ where MLT′(1) fails; and completed by all occurrences of $1'$ in rev(α) to the right of the offending term chosen.

(d) When MLT$'$(1) fails and the rightmost 2^* in α is marked, alter the counterexample in (c) by including into ϕ the rightmost $2'$ in α and removing the leftmost element of ψ.

To prove that (i) implies (ii), suppose that (ii) is false. Then there is a descent sequence $\phi \sqcup \psi$ of length $(2+\nu_1)$ in α which has the form

$$2, 2, ..., 2 \downarrow 2 \downarrow 2' \downarrow 2', \ldots, 2', 1, 1, 1, \ldots, 1 \downarrow 1 \downarrow 1' \downarrow 1', 1', \ldots, 1' .$$

The downward-pointing arrows in this sequence replace commas at the only possible points where ϕ could end and ψ begin. There are certain further restrictions on frequences for any choice of position for the arrow.

Suppose that the leftmost arrow applies. Then the frequency of "1" in this sequence is at most one (since ψ has no repeated unmarked numbers). Consider any point, in the right-to-left phase in checking MLT$'$(1) for α, after counting the $2'$ which is rightmost, and before counting the $1'$ which is leftmost, in ψ above. The count for 2^* at that point is at least

$$f_{2^*}(\phi \sqcup \psi)-1 = 2+\nu_1-f_{1^*}(\phi \sqcup \psi)-1 ,$$

which is at least

$$\nu_1-f_{1'}(\psi)-f_1(\psi)+1 .$$

Since none of the copies of $1'$ in ψ will have been counted at that point, the count for 1^* is at most $\mu_1-f_{1'}(\alpha)$. So if $f_1(\psi) = 0$, the count of 2^* is larger than that of 1^*. Thus MLT$'$(1) will have failed at least by the time of counting the rightmost $2'$ in ψ. If $f_1(\psi) = 1$, the counts of 1^* and 2^* are at best equal, so MLT$'$(1) will fail at least when the "1" in ψ is encountered. Note that this case is the "same" one as in the third of the four cases in the proof of the converse.

For each of the other five positions for the arrow, there is an analogous, often simpler, argument to show that MLT(1) or MLT$'$(1) fails. Each of these correlates with one or two of the four cases in proving the converse.

Equivalence of (ii) and (iii) in Theorem 14.16. Obviously (iii) implies (ii) for fixed k. We need (ii) for all k to deduce (iii). However, it suffices, by subtracting $k-1$ from all entries, to deduce (iii) for $k=1$ (from (ii) for all k). This reduces to the following. Given a descent pair $\delta = \phi \sqcup \psi$ for α, which has all its terms from $\{1^*, 2^*, ..., (m+1)^*\}$, we use only terms from $\{1^*, ..., m^*\}$ to construct a descent pair of at least equal length to that of δ. By successively applying this, one eventually produces a descent pair with all its terms 1^*, and so of length at most ν_1+1, which is therefore an upper bound for the length of δ.

In the case where the rightmost m^* in α is unmarked, proceed as follows, defining subsequences of α:

$$\gamma = \begin{cases} (\text{the leftmost } m^* \text{ in } \delta), & \text{if that } m^* \text{ is unmarked and is in } \phi; \\ \text{empty subsequence} & \text{otherwise.} \end{cases}$$

δ_L = union of γ with all terms in δ which are equal to $(m+1)^*$.

$\delta_R = \delta \backslash \delta_L$ = the complement of δ_L in δ.

κ = the longest descent pair for α of the form

$$(m, m, ..., m) \sqcup (m, m', ..., m')$$

(The copies of m on either side of $\sqcup$ are both the rightmost m^* in α; the length of the pair is $1+\nu_{m^*}$.)

κ_L = all terms of κ which are strictly leftward in α of the leftmost term of δ_R.

$\kappa_R = \kappa \backslash \kappa_L$.

It is easily verified that both $\delta_L \cup \kappa_R$ and $\kappa_L \cup \delta_R$ are descent pairs for α. Since $\delta_L \cup \kappa_R$ has all its terms from $\{m^*, (m+1)^*\}$, condition (ii) gives

$$|\delta_L \cup \kappa_R| \leq 1+\nu_m .$$

Thus

$$\begin{aligned} |\kappa_L \cup \delta_R| &= |\kappa_L| + |\delta_R| \\ &= |\kappa \backslash \kappa_R| + |\delta \backslash \delta_L| \\ &= |\delta| + |\kappa| - |\delta_L| - |\kappa_R| \\ &= |\delta| + (1+\nu_m) - |\delta_L \cup \kappa_R| \\ &\geq |\delta| . \end{aligned}$$

Since $\kappa_L \cup \delta_R$ has all its terms from $\{1^*, 2^*, ..., m^*)$, it is the required descent pair for α.

In the case where the rightmost m^* in α is m', the proof is similar, changing only the following two definitions. The form of κ is

$$(m, m, ..., m, m') \sqcup (m', ..., m') .$$

The sequence γ is non-empty only when the leftmost m^* in δ is m' and is in ψ, and γ then has that m' as its only term.

This completes the proof.

Equivalence of (iv) and (v) in Theorem 14.16. That (v) implies (iv) is immediate from the first assertion in Theorem 14.7. So is the converse, using the fact that ins rev(α) must have $|\nu|$ entries in total.

Equivalence of (iii) with either (iv) or (v) in Theorem 14.16. A tedious, but not difficult, combinatorial argument exists to show that (iii) and (iv) are equivalent. There is however, a more satisfying argument showing the equivalence of (iii) and (v); it depends on another basic piece of general theory about insertion which we shall quote from the literature.

Theorem 14.18. *If $u <_r w <_c v$, then*

$$\operatorname{ins}\big((u, v, w); T\big) = \operatorname{ins}\big((v, u, w); T\big)$$

for any non-skew tableau T.

Remarks and references. This is one from a small set of relations which are analogous to the Knuth relations for unshifted insertion. This set generates the equivalence relation $\sim$ on sequences from P defined by: $\alpha \sim \beta$ if and only if $\operatorname{ins}(\alpha) = \operatorname{ins}(\beta)$. It is part of Theorem 12.2 in Sagan (1987) and of Theorem 6.2.2 in Worley (1984).

Lemma 14.19. *If β is a sequence from P, $x \in P$, $x \neq 1^*$ and $\ell \in \mathbb{N}$, then*

$$\operatorname{ins}[\beta \sqcup (1^*, 1^{\ell-1}, x, 1)] = \operatorname{ins}[\beta \sqcup (x, 1^*, 1^{\ell})]\,,$$

where $1^\ell = (1, \ldots, 1) \in \mathbb{Z}^\ell$, and the 1^* on both sides of the equality is the same element (*either* 1 *or* $1'$).

Proof. Since $1^* <_r 1 <_c x$, this is immediate from Theorem 14.18 when $\ell = 1$. Proceed by induction on ℓ:

$$\begin{aligned}\operatorname{ins}[\beta \sqcup (1^*, 1^{\ell}, x, 1)] &= \operatorname{ins}[(\beta \sqcup 1^*) \sqcup (1, 1^{\ell-1}, x, 1)]\\ &= \operatorname{ins}[(\beta \sqcup 1^*) \sqcup (x, 1^{\ell+1})]\end{aligned}$$

by the inductive hypothesis

$$\begin{aligned}&= \operatorname{ins}\{1^{\ell}; \operatorname{ins}[\beta \sqcup (1^*, x, 1)]\}\\ &= \operatorname{ins}\{1^{\ell}; \operatorname{ins}[\beta \sqcup (x, 1^*, 1)]\}\end{aligned}$$

by the case $\ell = 1$

$$= \text{ins}[\beta \sqcup (x, 1^*, 1^{\ell+1})],$$

as required.

Definitions. Let T be a non-skew tableau. We shall say that T *splits* if and only if the first row of T has no entries other than 1^*. Let lower (T) be the tableau obtained by deleting the first row of T. If α is a sequence from P, let higher (α) be the subsequence consisting of all terms not equal to 1^*.

Theorem 14.20. *If* $\alpha \in P^n$ *and* ins(α) *splits, then* lower ins(α) = ins higher(α).

Proof. Proceed by induction on n. If all terms of α are 1^*, the result is trivial. This starts the induction, and allows us to assume that α has terms x with $x > 1$ in the inductive step. We have $\alpha_n = 1^*$ since ins(α) splits. Therefore $\alpha = \beta \sqcup (x, 1^{*\ell})$ for $x \in P$, $x > 1$, $\ell \geq 1$, where $1^{*\ell} \in P^\ell$ may have terms equal to both 1 and $1'$. Now we divide into two cases according to whether any term in $1^{*\ell}$, other than possibly the leftmost term, is marked.

If not, then $\alpha = \beta \sqcup (x, 1^*, 1^{\ell-1})$. Now

$$\text{higher}(\alpha) = (\text{higher } \beta) \sqcup (x),$$

and $A_1(\alpha) = f_{1^*}(\alpha) + 1$ because α splits and using Theorem 14.7. Thus

$$\begin{aligned} \text{ins higher}(\alpha) &= \text{ins}[x; \text{ins higher}(\beta)] \\ &= \text{ins}\{x; \text{ins higher}[\beta \sqcup (1^*, 1^{\ell-1})]\} \\ &= \text{ins}\{x; \text{lower ins } [\beta \sqcup (1^*, 1^{\ell-1})]\} \end{aligned}$$

by the inductive hypothesis applied to $\gamma = \beta \sqcup (1^*, 1^{\ell-1}) \in P^{n-1}$, for which ins γ splits because

$$A_1\big(\beta \sqcup (1^*, 1^{\ell-1})\big) \leq A_1(\alpha) = f_{1^*}(\alpha) + 1 = f_{1^*}\big(\beta \sqcup (1^*, 1^{\ell-1})\big) + 1 .$$

Thus

$$\text{ins higher}(\alpha) = \text{lower ins}[\beta \sqcup (1^*, 1^{\ell-1}, x, 1)] ,$$

since inserting x and then 1 into the split tableau ins(γ) certainly produces a lower part which agrees with inserting x into lower ins(γ). Now applying Lemma 14.19,

$$\begin{aligned} \text{ins higher}(\alpha) &= \text{lower ins}[\beta \sqcup (x, 1^*, 1^\ell)] \\ &= \text{lower ins}[\alpha \sqcup (1)] \\ &= \text{lower ins}(\alpha) , \end{aligned}$$

as required, again using that ins α splits.

In the other case,

$$\alpha = \beta \sqcup (1^{*a}, 1', 1^{b})$$

where $a+b+1 = \ell$ and $a > 0$. Let $\gamma = \beta \sqcup (1^{*a}, 1^{b})$. First we prove by contradiction that ins(γ) splits. If not, then

$$A_1(\gamma) > f_{1^*}(\gamma)+1 = f_{1^*}(\alpha)\,,$$

and γ has an ascent pair of length $f_{1^*}(\alpha) + 1$, say ω. A contradiction to the fact that ins (α) splits arises from constructing an ascent pair for α of length $f_{1^*}(\alpha)+2$ by replacing the leftmost term of ω by two terms, as follows:

If ω begins with $1'$, replace it by $1'$, $1'$.

If ω begins with 1, replace it by $1'$, 1^* for some 1^* in 1^{*a}.

If ω begins with $y > 1$ replace it by $1'$, y.

Thus ins(γ) splits and the inductive hypothesis applies to it, yielding

$$\begin{aligned} \text{ins higher}(\alpha) &= \text{ins higher}(\gamma) \\ &= \text{lower ins}(\gamma) \\ &= \text{lower ins}(\alpha)\,. \end{aligned}$$

This completes the proof.

At this point, we can finally prove that (iii) and (v) in Theorem 14.16 are equivalent. Proceed by induction on $\ell(\nu)$. The result is trivial when $\ell(\nu) = 1$. For the inductive step, let β be the sequence in P obtained by subtracting 1 from all terms in higher(α). Let $\tilde{\nu} = (\nu_2, \nu_3, \ldots)$; that is, $\tilde{\nu}_i = \nu_{i+1}$. Then β has content $\tilde{\nu}$. Furthermore, ins(β) is obtained by subtracting 1 from all entries in ins higher(α). The following four statements are equivalent:

(1) The restriction to $k = 1$ of 14.16 (iii) for α.

(2) $A_1(\alpha) = \nu_1+1$.

(3) The first row of ins(α) has length ν_1.

(4) ins(α) splits.

The equivalence of (2) and (3) is immediate from Theorem 14.7. Given that ins(α) splits, the following five statements are equivalent:

(5) The restriction to all $k \geq 2$ of 14.16 (iii) for α.

(6) The entire quantification ($k \geq 1$) of 14.16 (iii) for β.

(7) shape ins (β) = $\tilde{\nu}$.

(8) shape ins higher (α) = $\tilde{\nu}$.

(9) shape lower ins (α) = $\tilde{\nu}$.

The inductive hypothesis is used for the equivalence of (6) and (7); the definition of β for both pairs (5), (6) and (7), (8); and Theorem 14.20 for the pair (8), (9). The induction is now complete, since 14.16 (iii) is (1) plus (5), whereas

14.16 (v) is (3) plus (9).

Notes

The lattice conditions are due to Stembridge (1987, 1989), where Theorem 14.1 appears. With the lattice conditions replaced by 14.16 (ii), it also appears in Worley (1987). The deduction of 14.1 from 14.6 is analogous to a standard proof of the Littlewood– Richardson rule. The proof of 14.6 (ii) is due to Worley (1984); except for not using "jeu", we have followed his proof. As noted at certain points in the chapter, the other deeper properties of insertion are due independently to Worley (1984) and Sagan (1987). As for Theorem 14.16, the equivalence of (i) with (v) is in Stembridge (1987, 1989), and of (ii) with (v) in Worley (1987). Our proof follows the latter, except for not using "jeu".

Appendix 14

The product of irreducible projective representations of S_n and A_n

By Theorem 8.6, the K-algebra $\oplus T^*(\tilde{S}_n)$ described in Appendix 8 is closely related (at least as an abelian group) to the $\mathbb{R}$-algebra $\Delta \otimes \mathbb{R}$ spanned by the Q-functions. One therefore might expect a product theorem for the irreducibles in $\oplus T^*(\tilde{S}_n)$ which is similar to the product theorem for Q-functions.

For strict partitions λ and μ, define $h(\lambda, \mu)$ in $\mathbb{Z}/2$ to be 0 if μ is not a subsequence of λ. Otherwise, it is defined inductively on the length of μ as follows:

$$h\big(\lambda, (\lambda_p)\big) = \lambda_1 + \lambda_2 + \ldots + \lambda_{p-1} + (p+1)(\lambda_p+1)\,;$$

and if $\mu = \mu' \sqcup (\lambda_p)$,

$$h(\lambda, \mu) = h(\lambda, \mu') + h\big(\lambda \backslash \mu', (\lambda_p)\big)\,.$$

It is straightforward to obtain that $h(\lambda, \mu)$ is given by

$$\sum_{j=1}^{t} (i_j+j)(\lambda_{i_j}+1) + (t+j+1)(\lambda_{i_{j-1}+1} + \lambda_{i_{j-1}+2} + \ldots + \lambda_{i_j-1})\,,$$

with $i_0 = 0$, if μ is the subsequence $(\lambda_{i_1}, \ldots., \lambda_{i_t})$ of λ.

To state the result, recall that $\{a_\lambda\}$ is the special irreducible basis given in Theorem A8.14.

Theorem A14.1. *In* $\oplus T^*(\tilde{S}_n)$,

$$a_\mu^{\perp}(a_\lambda) = \sum_\nu F(\lambda;\, \mu, \nu) a_\nu, \textit{ and}$$

$$a_\mu a_\nu = \sum_\lambda F(\lambda;\, \mu, \nu) a_\lambda\,,$$

where

$$F(\lambda;\,\mu,\,\nu) = \rho^{h(\lambda,\mu)}\kappa^{\ell(\mu)+\ell(\nu)-\ell(\lambda)}|\,\mathrm{MST}_L^w(\lambda/\mu;\,\nu)|\,.$$

These two formulae are immediately seen to be equivalent, by the definition of $x^{\perp}$ and the orthonormality of the set $\{a_\lambda\}$. This theorem implies the product theorem for Q-functions, using Theorem 8.6. Its extra information consists in explicitly involving the projective representations of the alternating group and in keeping track of associates and conjugates of representations. By doing these using characters, one could conversely deduce Theorem A14.1 from Theorem 14.1.

Note. The above theorem is stated as formula (F) at the end of Hoffman (1989).

REFERENCES

Adams, J. F. (1969). Lectures on Lie Groups. W.A. Benjamin, New York.

Andrews, G. E. (1976). The theory of partitions. Encyclopedia of Math. Appl., Vol. 2. Addison-Wesley, Reading, Mass.

Asano, K. (1933). Über die Darstellungen einer endlichen Gruppe durch reelle Kollineationen. Proc. Imp. Acad. Japan 9, 574–576.

Asano, K. and Shoda, K. (1935). Zur Theorie die darstellungen einer endlichen Gruppe durch Kollineationen. Comp. Math. 2, 230–240.

Asano, K., Osima, M. and Takahashi, M. (1937). Über die Darstellung von Gruppen durch Kollineationen in Körpern der Characteristic p. Proc. Phys. Math. Soc. Japan 19, 199–209.

Atiyah, M. F., Bott, R. and Shapiro, A. (1964). Clifford modules. Topology 3 (Supplement 1), 3–38.

Bean, M. and Hoffman, P. N. (1988). Zelevinsky algebras related to projective representations. Trans. Amer. Math. Soc. 309, 99–111.

Benson, D. J. (1987). Some remarks on the decomposition numbers for the symmetric groups. In Part I of: The Arcata Conference on Representations of Finite Groups (editor P. Fong), Proceedings of Symposia in Pure Mathematics, Vol. 47, 381–394. American Mathematical Society, Providence, Rhode Island.

Benson, D. J. (1988). Spin modules for symmetric groups. J. London Math. Soc. (2) 38, 250–262.

Beyl, F. R. and Tappe, J. (1980). Group Extensions, Representations and the Schur Multiplier. Lecture Notes in Mathematics, Vol. 958. Springer-Verlag, Berlin.

Brauer, R. (1941). On the connection between ordinary and modular characters of groups of finite order. Ann. Math. 42, 936–958.

Brauer, R. and Robinson, G. de B. (1947). On a conjecture by Nakayama. Trans. Roy. Soc. Can. ser. III, 41, 11–19, 20–25.

Brauer, R. and Weyl, H. (1935). Spinors in n-dimensions. Amer. J. Math. 57, 425.

Cabanes, M. (1988). Local structure of the p-blocks of $\hat{S}_n$. Math. Zeit. 198, 519–543.

Clifford, A. H. (1937). Representations induced in an invariant subgroup. Ann. Math. 38, 533–550.

Conway, J. H., *et al.* (1985). Atlas of Finite Groups: Maximal Subgroups and Ordinary Characters for Simple Groups. Clarendon Press, Oxford.

DeConcini, C. and Procesi, C. (1976). A characteristic-free approach to invariant theory. Adv. Math. 21, 330–354.

Dehuai, Luan and Wybourne, B. G. (1981). The symmetric group: branching rules, products and plethysms for spin representations. J. Phys. A 14, 327–348.

Dehuai, Luan and Wybourne, B. G. (1981). The alternating group: branching rules, products and plethysms for ordinary and spin representations. J. Phys. A 14, 1835–1848.

El Sharabasy, M. A. (1985). Hall–Littlewood functions and representations of some Weyl groups. Ph. D. Thesis, Univ. College of Wales, Aberystwyth.

Frenkel, I. B. (1986). Lectures on infinite-dimensional Lie algebras and groups. Yale University (lecture notes).

Gagola, S. M. and Garrison, S. C. (1982). Real characters, double covers and the multiplier. J. Algebra 74, 20–51.

Geissinger, L. (1977). Hopf algebras of symmetric functions and class functions. In: Combinatoire et Represéntation du Groupe Symétrique, Lecture Notes in Mathematics, Vol. 579, p. 168. Springer-Verlag, Berlin.

Greub, W. (1978). Multilinear Algebra. Springer-Verlag, New York.

Goulden, I. P. and Jackson, D. M. (1983). Combinatorial Enumeration. John Wiley and Sons, New York.

Haiman, Mark D. (1989). On mixed insertion, symmetry, and shifted Young tableaux. J. Combin. Theory, Ser. A 50, 196–225.

Hoffman, K. and Kunze, R. (1971). Linear Algebra. Prentice Hall, Englewood Cliffs, New Jersey.

Hoffman, P. N. (1979). τ-rings and Wreath Product Representations. Lecture Notes in Mathematics, Vol. 746. Springer-Verlag, Berlin.

Hoffman, P. (1983). An algebra of S_n and A_n projective representations. Lecture, British Mathematical Colloquium, April 1983, Aberdeen University.

Hoffman, P. N. (1988). Projective representations of S_n: ring structure and inductive formulae. Topics in Algebra, Banach Centre Publications, Vol. 26, Part 2. PWN Polish Scientific Publishers, Warsaw.

Hoffman, P. N. (1989). A Bernstein-type formula for projective representations of A_n and S_n. Adv. Math. 74, 135–143.

Hoffman, P. N. (1990). Remarks on Schur, Q-, and $H-L$ functions (preprint).

Hoffman, P. N. (1990A). Projective and multigraded representations of monomial and multisigned groups. I. Graded representations of a twisted product. Can. J. Math. (to appear).

Hoffman, P. N. and Humphreys, J. F. (1985). Twisted products and projective representations of monomial groups. Expositiones Math. 3, 91–95.

Hoffman, P. N. and Humphreys, J. F. (1986). Hopf algebras and projective representations of $G \wr S_n$ and $G \wr A_n$. Can. J. Math. 38, 1380–1458.

Hoffman, P. N. and Humphreys, J. F. (1987). Primitives in the Hopf algebra of projective S_n-representations. J. Pure Appl. Algebra 47, 155–164.

Hoffman, P. N. and Humphreys, J. F. (1989). Projective representations of generalized symmetric groups using *PSH*-algebras. Proc. London Math. Soc. (3)59, 483–506.

Hoffman, P. N. and Humphreys, J. F. (1990). Real projective representations of finite groups. Math. Proc. Camb. Phil. Soc. 107, 27–32.

Humphreys, J. F. (1977). Projective modular representations of finite groups. J. London Math. Soc. (2) 16, 51–66.

Humphreys, J. F. (1983). Lecture notes on Q-functions and a characteristic map for projective representations of S_n (unpublished).

Humphreys, J. F. (1984). Conjugacy classes of double covers of monomial groups. Math. Proc. Camb. Phil. Soc. 96, 195–201.

Humphreys, J. F. (1985). Certain projective representations of direct products. J. London Math. Soc. (2) 32, 449–460.

Humphreys, J. F. (1986). Blocks of projective representations of the symmetric groups. J. London Math. Soc. (2) 33, 441–452.

Humphreys, J. F. (1988). Blocks of certain classes of finite groups. Arch. Math. 50, 481–491.

Huppert, B. (1967). Endliche Gruppen I. Springer-Verlag, Berlin.

Isaacs, I. M. (1976). Character Theory of Finite Groups. Academic Press, New York.

James, G. D. (1978). The Representation Theory of the Symmetric Groups. Lecture Notes in Mathematics, Vol. 682. Springer-Verlag, Berlin.

James, G. D. and Kerber, A. (1981). The Representation Theory of the Symmetric Group. Addison-Wesley, Reading, Mass.

Jing, Naihuan (1989A). Vertex operators, symmetric functions and the spin group Γ_n. J. Algebra (to appear).

Jing, Naihuan (1989B). Vertex operators and the Hall–Littlewood symmetric functions (preprint).

Jozefiak, T. (1988). Semisimple superalgebras. Proceedings of the Varna Conference 1986, Lecture Notes in Mathematics, Vol. 1352. Springer-Verlag, Berlin.

Jozefiak, T. (1989). Characters of projective representations of symmetric groups. Expositiones Math. 7, 193–247.

Jozefiak, T. and Pragacz, P. (1989). A determinantal formula for skew Schur Q-functions (preprint).

Karpilovsky, G. (1985). Projective Representations of Finite Groups. Marcel Dekker, New York.

Karpilovsky, G. (1987). The Schur Multiplier. London Math. Soc. Monographs, New Series 2. Oxford Science Publications, Oxford.

Knuth, D. E. (1970). Permutations, matrices, and generalized Young tableaux. Pacific J. Math. 34, 709–727.

Lang, S. (1965). Algebra. Addison-Wesley, Reading, Mass.

Littlewood, D. E. (1961). On certain symmetric functions. Proc. London Math. Soc. (3) 11, 485–498.

Liulevicius, A. (1980). Arrows, symmetries and representation rings. J. Pure Appl. Algebra 19, 259–273.

Macdonald, I. G. (1979). Symmetric Functions and Hall Polynomials. Oxford University Press, Oxford.

Macdonald, I. G. (1991+). Symmetric Functions and Hall Polynomials (expanded 2nd edn.). Oxford University Press, Oxford.

MacLane, S. (1967). Homology. Springer-Verlag, New York.

Mackey, G. W. (1980). Harmonic analysis as the exploitation of symmetry — a historical survey. Bull. Amer. Math. Soc. 3, 543–699.

Mazet, P. (1979). Sur le multiplicateur de Schur du groupe de Mathieu M_{22}. C. R. Acad. Paris Ser. A-B, 289, No. 14, 659–661.

Meier, N. and Tappe, J. (1976). Ein neuer beweis der Nakayama-Vermutung uber die Blockstruktur symmetrischer Gruppen. Bull. London Math. Soc. 8, 34–37.

Michler, G. O. and Olsson, J. B. (1990). The Alperin–McKay conjecture holds in the covering groups of symmetric and alternating groups, $p \neq 2$. J. reine angew. Math. 405, 78–111.

Michler, G. O. and Olsson, J. B. (1990A). Weights for covering groups of symmetric and alternating groups, $p \neq 2$ (to appear).

Morris, A. O. (1962). The spin representation of the symmetric group. Proc. London Math. Soc. (3) 12, 55–76.

Morris, A. O. (1962A). On Q-functions. J. London Math. Soc. 37, 445–455.

Morris, A. O. (1965). The spin representation of the symmetric group. Can. J. Math. 17, 543–549.

Morris, A. O. (1974). Projective characters of exceptional Weyl groups. J. Algbra 29, 567–586.

Morris, A. O. (1976). Projective representations of reflection groups I. Proc. London Math. Soc. (3)32, 403–420.

Morris, A. O. (1977). A survey of Hall–Littlewood functions and their applications to representation theory. In: Combinatoire et Représéntation du Groupe Symétrique. (D. Foata editor) Lecture Notes in Mathematics Vol. 579. Springer-Verlag, Berlin.

Morris, A. O. (1979). The projective characters of the symmetric group — an alternative proof. J. London Math. Soc. (2) 19, 57–58.

Morris, A. O. (1980). Projective representations of reflection groups II. Proc. London Math. Soc. (3) 40, 553–576.

Morris, A. O. and Olsson, J. B. (1988). On p-quotients for spin characters. J. Algebra 119, 51–82.

Morris, A. O. and Yaseen, A. K. (1986). Some combinatorial results involving shifted Young diagrams. Math. Soc. Camb. Phil. Soc. 99, 23–31.

Morris, A. O. and Yaseen, A. K. (1988). Decomposition matrices for spin characters of symmetric groups. Proc. Roy. Soc. Edinb. 108A, 145–164.

Nakayama, T. (1940). On some modular properties of irreducible representations of symmetric groups I, II. Jap. J. Math. 17, 165–184 , 411–423.

Nazarov, M. L. (1988). An orthogonal basis of irreducible projective representations of the symmetric groups. Functional Anal. Appl. 22, 77–78.

Nazarov, M. L. (1989). Young's orthogonal form of irreducible projective representations of the symmetric group. Proc. London Math. Soc. (to appear).

Olsson, J. B. (1987). Frobenius symbols for partitions and degrees of spin characters. Math. Scand. 61, 223–247.

Olsson, J. B. (1989). On the p-blocks of the symmetric and alternating groups and their covering groups. J. Algebra 128, 188–213.

Porteous, I. R. (1969). Topological Geometry. Van Nostrand Reinhold, London.

Pragacz, P. (1987). Enumerative geometry of degeneracy loci. Ann. ENS (to appear)

Read, E. W. (1974). Projective characters of the Weyl groups of type F_4. J. London Math. Soc. (2) 8, 83–93.

Read, E. W. (1974). Linear and projective characters of the finite reflection group of type H_4. Quart. J. Math. 25, 73–79.

Read, E. W. (1975). On projective representations of the finite reflection groups of type B_ℓ and D_ℓ. J. London Math. Soc. (2) 10, 129–142.

Read, E. W. (1976). On the Schur multipliers of the finite imprimitive unitary reflection groups $G(m,p,n)$. J. London Math. Soc. (2) 13, 150–154.

Read, E. W. (1977). On the projective characters of the symmetric groups. J. London Math. Soc. (2) 15, 456–464.

Robinson, G. de B. (1938). On the representations of the symmetric group. Amer. J. Math. 60, 745–760.

Sagan, B. E. (1979). An analog of Schensted's algorithm for shifted Young tableaux. J. Combin. Theory, Ser. A, 27, 10–18.

Sagan, B. E. (1987). Shifted tableaux, Schur Q-functions and a conjecture of Stanley. J. Combin. Theory Ser. A 45, 62–103.

Salam, M. and Wybourne, B. G. (1989). Q-functions and $O_n \to S_n$ branching rules for ordinary and spin irreps. J. Phys. A. Math. Gen. 22, 3771–3778.

Schensted, C. (1961). Longest increasing and decreasing subsequences. Can. J. Math. 13, 179–191.

Schur, I. (1904). Über die Darstellung der endlichen Gruppen durch gebrochene lineare Substitutionen. J. Reine Angew. Math. 127, 20–50.

Schur, I. (1907). Untersuchungen über die Darstellungen der endlichen Gruppen durch gebrochenen lineare Substitutionen. J. Reine Angew. Math. 132, 85–137.

Schur, I. (1911). Über die Darstellung der symmetrischen und der alternierenden Gruppen durch gebrochene lineare Substitutionen. J. Reine Angew. Math. 139, 155–250.

Schur, I. (1927). Über die reelen Kollineationsgruppen die der symmetrischen oder der alternierenden Gruppe isomorph sind. J. Reine Angew. Math. 158, 63–79.

Sergeev, A. N. (1984). The tensor algebra of the identity representation as a module over the Lie superalgebras $GL(n,m)$ and $Q(n)$. Mat. Sbornik 123, 422–430. (A.M.S. Translations 51 (1985), 419–427.)

Serre, J. P. (1977). Linear Representations of Finite Groups. GTM Vol. 42. Springer-Verlag, New York.

Springer, T. A. and Zelevinsky, A. V. (1984). Characters of $GL(n, \mathbb{F}_q)$ and Hopf algebras. J. London Math. Soc. (2) 30, 27–43.

Stanley, R. P. (1984). Problem 4. In: Combinatorics and algebra. Contemporary Mathematics, Vol. 34, ed. C. Greene, p. 304. American Mathematical Society, Providence, R. I.

Stanley, R. P. (1988). Some combinatorial properties of Jack symmetric functions (preprint).

Stanley, R. P. (1988A). Variations on differential posets (preprint).

Stembridge, J. R. (1987). Shifted tableaux and the projective representations of symmetric groups (preprint).

Stembridge, J. R. (1988). On symmetric functions and the spin characters of S_n. Topics in Algebra, Banach Centre Publications, Vol. 26, Part 2. PWN Polish Scientific Publishers, Warsaw.

Stembridge, J. R. (1989). Shifted tableaux and the projective representations of symmetric groups. Adv. Math. 74, 87–134.

Stembridge, J. R. (1990). Nonintersecting paths, pfaffians, and plane partitions. Adv. Math. 83, 96–131.

Thrall, R. M. (1952). A combinatorial problem. Mich. Math. J. 1, 81–88.

Turull, A. (1990). The Schur index of projective characters of symmetric and alternating groups (preprint).

Wales, D. B. (1979). Some projective representations of S_n. J. Algebra 61, 63–79.

Wall, C. T. C. (1964). Graded Brauer groups. J. Reine Angew. Math. 213, 187–199.

Wiegold, J. (1981). The Schur multiplier: An elementary approach. In: Groups, St. Andrews 1981 (edit. C.M. Campbell and E.F. Robertson), London Math. Soc. Lecture Notes, Vol. 71. Cambridge University Press, Cambridge.

Worley, D. R. (1984). A theory of shifted Young tableaux. Ph. D. thesis, M. I. T.

Worley, D. (1987). A shifted analog of the Littlewood–Richardson rule (preprint).

Zelevinsky, A. V. (1981). Representations of Finite Classical Groups — a Hopf Algebra Approach. Lecture Notes in Mathematics, Vol. 869. Springer-Verlag, Berlin.

CHARACTER TABLES

Degree 4

Class	(1^4)	(13)	
Order	1	8	
$\langle 4\rangle$	2	1	(4) 6 $\sqrt{2}$
$\langle 31\rangle^*$	4	-1	

Degree 5

Class	(1^5)	$(1^2 3)$	(5)	
Order	1	20	24	
$\langle 5\rangle^*$	4	2	1	
$\langle 41\rangle$	6	0	-1	(14) 30 $\sqrt{2}$
$\langle 32\rangle$	4	-1	1	(23) 20 $\sqrt{3}$

Degree 6

Class	(1^6)	$(1^3 3)$	(15)	(3^2)	
Order	1	40	144	40	
$\langle 6\rangle$	4	2	1	1	(6) 120 $\sqrt{(2)}i$
$\langle 51\rangle^*$	16	2	-1	-2	
$\langle 42\rangle^*$	20	-2	0	2	
$\langle 321\rangle$	4	-1	1	-2	(123) 120 $\sqrt{3}$

Degree 7

Class	(1^7)	$(1^4 3)$	$(1^2 5)$	(13^2)	(7)	
Order	1	70	504	280	720	
⟨7⟩*	8	4	2	2	1	
⟨61⟩	20	4	0	−1	−1	(16) 840 $\sqrt{(3)}i$
⟨52⟩	36	0	−1	0	1	(25) 504 $\sqrt{(5)}i$
⟨43⟩	20	−2	0	2	−1	(34) 420 $\sqrt{(6)}i$
⟨421⟩*	28	−4	2	−2	0	

Degree 8

Class	(1^8)	$(1^5 3)$	$(1^3 5)$	$(1^2 3^2)$	(17)	(35)	
Order	1	112	1344	1120	5760	2688	
⟨8⟩	8	4	2	2	1	1	(8) 5040 2
⟨71⟩*	48	12	2	0	−1	−2	
⟨62⟩*	112	8	−2	−2	0	2	
⟨53⟩*	112	−4	−2	4	0	−1	
⟨521⟩	64	−4	1	−2	1	−1	(125) 4032 $\sqrt{(5)}i$
⟨431⟩	48	−6	2	0	−1	1	(134) 3360 $\sqrt{(6)}i$

Degree 9

Class	(1^9)	$(1^6 3)$	$(1^4 5)$	$(1^3 3^2)$	$(1^2 7)$	(135)	(3^3)	(9)	
Order	1	168	3,024	3,360	25,920	24,192	2,240	40,320	
⟨9⟩*	16	8	4	4	2	2	2	1	
⟨81⟩	56	16	4	2	0	−1	−2	−1	(18) 45,360 2
⟨72⟩	160	20	0	−2	−1	0	2	1	(27) 25,920 $\sqrt{7}$
⟨63⟩	224	4	−4	2	0	1	1	−1	(36) 20,160 3
⟨54⟩	112	−4	−2	4	0	−1	−4	1	(45) 18,144 $\sqrt{10}$
⟨621⟩*	240	0	0	−6	2	0	−6	0	
⟨531⟩*	336	−24	4	0	0	−1	6	0	
⟨432⟩*	96	−12	4	0	−2	2	−6	0	

Degree 10

Class	(1^{10})	$(1^7 3)$	$(1^5 5)$	$(1^4 3^2)$	$(1^3 7)$	$(1^2 35)$	(13^3)	(19)	(37)	(5^2)	
Order	1	240	6,048	8,400	86,400	120,960	22,400	403,200	172,800	72,576	
⟨10⟩	16	8	4	4	2	2	2	1	1	1	(10) 362,880 $\sqrt{(5)}i$
⟨91⟩*	128	40	12	8	2	0	−2	−1	−2	−2	
⟨82⟩*	432	72	8	0	−2	−2	0	0	2	2	
⟨73⟩*	768	48	−8	0	−2	2	6	0	−1	−2	
⟨64⟩*	672	0	−12	12	0	0	−6	0	0	2	
⟨721⟩	400	20	0	−8	1	0	−4	1	−1	0	(127) 259,200 $\sqrt{7}$
⟨631⟩	800	−20	0	−4	2	0	1	−1	1	0	(136) 201,600 3
⟨541⟩	448	−28	2	4	0	−2	2	1	0	−2	(145) 181,440 $\sqrt{10}$
⟨532⟩	432	−36	8	0	−2	1	0	0	−1	2	(235) 20,960 $\sqrt{15}$
⟨4321⟩*	96	−12	4	0	−2	2	−6	0	2	−4	

Degree 11

Class	(1^{11})	$(1^8 3)$	$(1^6 5)$	$(1^5 3^2)$	$(1^4 7)$	$(1^3 35)$	$(1^2 3^3)$	$(1^2 9)$	(137)	(15^2)	$(3^2 5)$	(11)	
Order	1	330	11,088	18,480	237,600	443,520	123,200	2,217,600	1,900,800	798,336	443,520	3,628,800	
⟨11⟩*	32	16	8	8	4	4	4	2	2	2	2	1	
⟨10,1⟩	144	48	16	12	4	2	0	0	−1	−1	−2	−1	(1,10) 3,991,680 $\sqrt{(5)}i$
⟨92⟩	560	112	20	8	0	−2	−2	−1	0	0	2	1	(29) 2,217,600 $3i$
⟨83⟩	1,200	120	0	0	−4	0	6	0	1	0	0	−1	(38) 1,603,200 $2\sqrt{(3)}i$
⟨74⟩	1,440	48	−20	12	−2	2	0	0	−1	0	−2	1	(47) 1,425,600 $\sqrt{(14)}i$
⟨65⟩	672	0	−12	12	0	0	−6	0	0	2	3	−1	(56) 1,330,560 $\sqrt{(15)}i$
⟨821⟩*	1,232	112	8	−16	0	−2	−8	2	0	2	−4	0	
⟨731⟩*	3,168	48	−8	−24	4	2	0	0	−1	−2	4	0	
⟨641⟩*	3,168	−96	−8	12	4	−4	0	0	2	−2	−2	0	
⟨632⟩*	2,464	−112	16	−8	0	2	2	−2	0	4	−2	0	
⟨542⟩*	1,760	−128	20	8	−4	−2	4	2	−2	0	2	0	
⟨5321⟩	528	−48	12	0	−4	3	−6	0	1	−2	0	0	(1235) 1,330,560 $\sqrt{15}$

Degree 12

Class	(1^{12})	(1^93)	(1^75)	(1^63^2)	(1^57)	(1^435)	(1^33^3)	(1^39)	(1^237)	(1^25^2)	(13^25)	$(1,11)$	(39)	(3^4)	(57)	
Order	1	440	19,008	36,960	570,240	1,330,560	492,800	8,870,400	11,404,800	4,790,016	5,322,240	43,545,600	17,740,800	246,400	13,685,760	
⟨12⟩	32	16	8	8	4	4	4	2	2	2	2	1	1	2	1	(12) 39,916,800 $\sqrt{6}$
⟨11,1⟩*	320	112	40	32	12	8	4	2	0	0	−2	−1	−2	−4	−2	
⟨10,2⟩*	1,408	320	72	40	8	0	−4	−2	−2	−2	0	0	2	4	2	
⟨93⟩*	3,520	464	40	16	−8	−4	8	−2	2	0	4	0	−1	4	−2	
⟨84⟩*	5,280	336	−40	24	−12	4	12	0	0	0	−4	0	0	−12	2	
⟨75⟩*	4,224	96	−64	48	−4	4	−12	0	−2	4	2	0	0	12	−1	
⟨921⟩	1,792	224	28	−8	0	−4	−10	1	0	2	−2	1	−1	−8	0	(129) 26,611,200 $3i$
⟨831⟩	5,600	280	0	−40	0	0	−2	2	0	0	0	−1	1	8	0	(138) 19,958,400 $2\sqrt{(3)}i$
⟨741⟩	7,776	0	−36	0	6	0	0	0	0	−4	0	1	0	0	−1	(147) 17,107,200 $\sqrt{(14)}i$
⟨651⟩	3,840	−96	−20	24	4	−4	−6	0	2	0	1	−1	0	−12	1	(156) 15,966,720 $\sqrt{(15)}i$
⟨732⟩	5,632	−64	8	−32	4	4	2	−2	−1	2	2	0	−1	−8	1	(237) 11,404,800 $\sqrt{(21)}i$
⟨642⟩	7,392	−336	28	12	0	−4	6	0	0	2	−2	0	0	12	0	(246) 9,979,200 $2\sqrt{(6)}i$
⟨543⟩	1,760	−128	20	8	−4	−2	4	2	−2	0	2	0	1	−16	−1	(345) 7,983,360 $\sqrt{(30)}i$
⟨6321⟩*	3,520	−208	40	−8	−8	8	−10	−2	2	0	−2	0	2	−8	−2	
⟨5421⟩*	2,816	−224	44	8	−12	4	−8	2	0	−4	2	0	−2	8	2	

Degree 13

Class	(1^{13})	$(1^{10}3)$	(1^85)	(1^73^2)	(1^67)	(1^535)	(1^43^3)	(1^49)	(1^337)	(1^35^2)	(1^23^25)	(1^211)	(139)	(13^4)	(157)	(3^27)	(35^2)	(13)	
Order	1	572	30,888	68,640	1,235,520	3,459,456	1,601,600	28,828,800	49,420,800	20,756,736	34,594,560	283,046,400	230,630,400	3,203,200	177,914,880	49,420,800	41,513,472	479,001,600	
⟨13⟩*	64	32	16	16	8	8	8	4	4	4	4	2	2	4	2	2	2	1	
⟨12,1⟩	352	128	48	40	16	12	8	4	2	2	0	0	−1	−2	−1	−2	−2	−1	(1,12) 518,918,400 $\sqrt{5}$
⟨11,2⟩	1,728	432	112	72	20	8	0	0	−2	−2	−2	−1	0	0	0	2	2	1	(2,11) 283,046,400 $\sqrt{11}$
⟨10,3⟩	4,928	784	112	56	0	−4	4	−4	0	−2	4	0	1	8	0	0	−1	−1	(3,10) 207,567,360 $\sqrt{15}$
⟨94⟩	8,800	800	0	40	−20	0	20	−2	2	0	0	0	−1	−8	0	−2	0	1	(49) 172,972,800 $3\sqrt{2}$
⟨85⟩	9,504	432	−104	72	−16	8	0	0	−2	4	−2	0	0	0	1	2	2	−1	(58) 155,675,520 $2\sqrt{5}$
⟨76⟩	4,224	96	−64	48	−4	4	−12	0	−2	4	2	0	0	12	−1	−1	−4	1	(67) 148,262,400 $\sqrt{21}$
⟨10,21⟩*	4,992	768	128	24	8	−8	−24	0	−2	2	−4	2	0	−12	2	−4	−2	0	
⟨931⟩*	18,304	1,472	96	−80	−8	−12	−16	4	2	4	0	0	−1	4	−2	4	2	0	
⟨841⟩*	32,032	896	−112	−56	0	4	8	4	0	−8	−4	0	2	4	0	0	−4	0	
⟨751⟩*	27,456	−96	−176	96	16	−4	−24	0	2	−4	4	0	0	−12	−1	−2	4	0	
⟨832⟩*	22,464	432	16	−144	8	8	0	0	−2	4	4	−2	0	0	2	−4	2	0	
⟨742⟩*	41,600	−800	0	−40	20	0	16	−4	−2	0	0	2	−2	8	0	2	0	0	
⟨652⟩*	22,464	−864	16	72	8	−16	0	0	4	4	−2	−2	0	0	2	2	−4	0	
⟨643⟩*	18,304	−928	96	40	−8	−12	20	4	−4	4	0	0	2	−8	−2	−2	2	0	
⟨7321⟩	9,152	−272	48	−40	−4	12	−8	−4	1	2	0	0	1	−16	−1	2	−2	0	(1237) 148,262,400 $\sqrt{(21)}i$
⟨6421⟩	13,728	−768	112	12	−20	8	−12	0	2	−2	−2	0	0	12	0	−2	2	0	(1246) 129,729,600 $2\sqrt{(6)}i$
⟨5431⟩	4,576	−352	64	16	−16	2	−4	4	−2	−4	4	0	−1	−8	1	2	−2	0	(1345) 103,783,680 $\sqrt{(30)}i$

Degree 14

Class	(1^{14})	$(1^{11}3)$	(1^95)	(1^83^2)	(1^77)	(1^635)	(1^53^3)	(1^59)	(1^473)	(1^45^2)	(1^33^25)	(1^311)	(1^239)	(1^23^4)	(1^257)	(13^27)	(135^2)	(113)	(3^35)	$(3,11)$	(59)	(7^2)	
Order	1	728	48,048	120,120	2,471,040	8,072,064	4,484,480	80,720,640	172,972,800	72,648,576	161,441,280	1,320,883,200	1,614,412,800	22,422,400	1,245,404,160	691,891,200	581,188,608	6,706,022,400	107,627,520	2,641,766,400	1,937,295,360	889,574,400	
⟨14⟩	64	32	16	16	8	8	8	4	4	4	4	2	2	4	2	2	2	1	2	1	1	1	(14) 6,227,020,800 $i\sqrt{7}$
⟨13,1⟩*	768	288	112	96	40	32	24	12	8	8	4	2	0	0	0	−2	−2	−1	−4	−2	−2	−2	
⟨12,2⟩*	4,160	1,120	320	224	72	40	16	8	0	0	−4	−2	−2	−4	−2	0	0	0	4	2	2	2	
⟨11,3⟩*	13,312	2,432	448	256	40	8	8	−8	−4	−8	4	−2	2	16	0	4	2	0	2	−1	−2	−2	
⟨11,2,1⟩	6,720	1,200	240	96	28	0	−24	0	−4	0	−6	1	0	−12	2	−2	0	1	−6	−1	0	0	(1,2,11) 3,962,649,600 $\sqrt{11}$
⟨10,4⟩*	27,456	3,168	224	192	−40	−8	48	−12	4	−4	8	0	0	0	0	−4	−2	0	−8	0	2	2	
⟨10,3,1⟩	28,224	3,024	336	0	0	−24	−36	0	0	4	0	2	0	0	0	0	−1	−1	6	1	0	0	(1,3,10) 2,905,943,040 $\sqrt{15}$
⟨9,5⟩*	36,608	2,464	−208	224	−72	16	40	−4	0	8	−4	0	−2	−16	2	0	4	0	10	0	−1	−2	
⟨9,4,1⟩	59,136	3,168	−16	−96	−28	−8	12	6	4	−4	−4	0	0	0	−2	2	−2	1	−2	0	−1	0	(1,4,9) 2,421,619,200 $3\sqrt{2}$
⟨9,3,2⟩	40,768	1,904	112	−224	0	−4	−16	4	0	8	4	−2	−1	4	0	0	4	0	−4	−1	1	0	(2,3,9) 1,614,412,800 $3\sqrt{3}$
⟨8,6⟩*	27,456	1,056	−336	240	−40	24	−24	0	−8	16	0	0	0	24	0	2	−4	0	−6	0	0	2	
⟨8,5,1⟩	68,992	1,232	−392	112	0	8	−16	4	0	−8	−2	0	2	−8	0	0	2	−1	−4	0	1	0	(1,5,8) 2,179,457,280 $2\sqrt{5}$
⟨8,4,2⟩	96,096	528	−96	−240	28	12	24	0	−4	−4	0	0	0	12	2	−2	−2	0	6	0	0	0	(2,4,8) 1,362,160,800 $4\sqrt{2}$
⟨8,3,2,1⟩*	40,768	−112	112	−224	0	32	−16	−8	0	8	4	−2	2	−32	0	0	−2	0	−4	2	−2	0	
⟨7,6,1⟩	31,680	0	−240	144	12	0	−36	0	0	0	6	0	0	0	−2	−3	0	1	6	0	0	−2	(1,6,7) 2,075,673,600 $\sqrt{21}$
⟨7,5,2⟩	91,520	−1,760	−160	128	44	−20	−8	−4	4	0	2	0	−2	−4	1	2	0	0	−2	0	−1	2	(2,5,7) 1,245,404,160 $\sqrt{35}$
⟨7,4,3⟩	59,904	−1,728	96	0	12	−12	36	0	−6	4	0	2	0	0	−2	0	2	0	−6	1	0	−2	(3,4,7) 1,037,836,800 $\sqrt{42}$
⟨7,4,2,1⟩*	87,360	−2,880	320	−96	−28	40	−24	−12	4	0	−4	2	0	0	−2	2	0	0	4	−2	2	0	
⟨6,5,3⟩	40,768	−1,792	112	112	0	−28	20	4	0	8	−2	−2	2	−8	0	0	−2	0	5	−1	1	0	(3,5,6) 968,647,680 $3\sqrt{5}$
⟨6,5,2,1⟩*	49,920	−2,400	240	96	−32	0	−24	0	8	0	−6	−2	0	24	2	−2	0	0	−6	2	0	−4	
⟨6,4,3,1⟩*	54,912	−3,168	448	96	−80	8	−12	12	−4	−8	4	0	0	0	0	−2	2	0	2	0	−2	4	
⟨5,4,3,2⟩*	9,152	−704	128	32	−32	4	−8	8	−4	−8	8	0	−2	−16	2	4	−4	0	−2	0	2	−4	

INDEX OF NOTATION

Roman letters

Greek letters

Other symbols

INDEX